ENVIRONMENTAL ISSUES AND SUSTAINABLE FUTURES

A Critical Guide to Recent Books, Reports, and Periodicals

by Michael Marien

Editor, *Future Survey*

A *Future Survey* Guidebook

Published in association with

International Center for the Environment and Public Policy
A Program of the California Institute of Public Affairs

Commission on Environmental Strategy and Planning
IUCN—The World Conservation Union

World Future Society
Bethesda, Maryland U.S.A.
1996

Published by:
World Future Society
7910 Woodmont Avenue, Suite 450
Bethesda, MD 20814 U.S.A.
(phone: 301/656-8274; fax: 301/951-0394)

The World Future Society is a non-profit educational and scientific association founded in 1966 for the study of alternative futures. It acts as an impartial clearinghouse for a variety of different views and does not take positions on what will or should happen in the future.

In association with:

International Center for the Environment and Public Policy
California Institute of Public Affairs
Thaddeus C. Trzyna, President
P.O. Box 189040
Sacramento, CA 95818 USA (phone: 916/442-2472; fax: 916/442-2478)

Commission on Environmental Strategy and Planning
IUCN—The World Conservation Union
Rue Mauverney 28
CH-1196 Gland, Switzerland (phone: 41-22/999-0001)
Commission Chairman's Office: PO Box 189040, Sacramento, CA 95818

Library of Congress Cataloging-in-Publication Data:

Marien, Michael.
Environmental issues and sustainable futures : a critical guide to recent books, reports, and periodicals / by Michael Marien.
p. cm. --- (A Future survey guidebook ; 2)
Includes bibliographical references and index
ISBN 0-930242-51-3

1. Environmental sciences--Handbooks, manuals, etc.
2. Environmental sciences--Bibliography.
3. Sustainable development--Handbooks, manuals, etc.
4. Sustainable development--Bibliography.

I. Title. II. Series.

GE123.M37 1996 96-10658
363.7--dc20 CIP

World Future Society ISBN 0-930242-51-3

California Institute of Public Affairs ISBN 1-880028-06-9

ENVIRONMENTAL ISSUES AND SUSTAINABLE FUTURES
A Critical Guide to Recent Books, Reports, and Periodicals

CONTENTS

PREFACE

by Lester W. Milbrath

When I finished writing *Envisioning a Sustainable Society: Learning Our Way Out*, I recall feeling both satisfied and dissatisfied. I was satisfied that I had presented a coherent vision of how a society could live sustainably. I felt dissatisfied because I realized that the learning required to achieve a sustainable society was so immensely complex that no one mind could comprehend it. I said to myself, "We need a hundred books like this."

I did not anticipate how quickly my wish would be granted. In the years 1989-1992, leading up to the 20th anniversary of Earth Day in 1990 and to the Earth Summit in Rio in 1992, at least three hundred books relevant to environmental policy were published. Michael Marien has prepared an annotated guide to 450 of the most important books and reports on this topic, most of them published between 1991 and 1995. That still does not include all the books published on environmental policy in English, and excludes the many relevant books published in other languages. In addition, Marien has identified 170 journals and magazines publishing articles relevant to environmental policy (doubtless even more exist).

Marien has invented the word "infoglut" to characterize this flood of words and ideas. Even full time practitioners cannot keep up and do not know what key facts they may be missing. The knowledge needed to transform unsustainable modern society into a sustainable society probably lies hidden within that huge collection of words and ideas, but how are we to draw out the nuggets of keen insight and apply them to contemporary policy making? The average citizen probably is not aware of this rich ferment of thinking. Those who do confront the topic in all of its richness and complexity can soon feel overwhelmed.

As the editor of *Future Survey*, Mike Marien is a master abstracter/annotator. He has corralled much important material on recently published works in the monthly issues of his journal. Many people who think about futures depend on *Future Survey* to keep them current. Mike hasn't solved the infoglut problem, but he has made it possible for busy people to use their time a little better.

He has now performed the additional service of pulling out the most important abstracts dealing with environment and sustainability appearing in the last five years and reassembling them in one guide. He has conveniently grouped them to facilitate searching and has provided indexes by author, title, and publisher. The publisher index can guide authors to houses that might publish a new work. Similarly, the list of journals can help authors place an article for publication. Persons wondering what journals to subscribe to, or search in, can also find guidance in that list.

But I caution readers not to overly depend on the category labels. Many works could fit into several categories yet, since each work is listed only once in the guide; placements are somewhat arbitrary. Listings in the subject index are inevitably arbitrary as well. Highly relevant items are easily missed. Browsing is thus required to spot materials that might otherwise be overlooked. I also recommend browsing as it leads to a deeper appreciation of the full richness of this collection.

Here are some brief impressions from several hours spent scanning, dipping in, reflecting back, and projecting forward.

- The cumulative weight of all this thought, engaged in by so many people, boggles my mind. Environmentalism is indeed a powerful and successful movement that is changing mankind's place in the world.
- The buildup of materials in preparation for the UNCED conference in Rio (Earth Summit) in 1992 is especially impressive for its richness. Despite this rich buildup, most of this material was ignored by top policy makers, the media, and the public. After Rio, the pace slackened, but the flow is still rather impressive.
- There certainly is no shortage of ideas. The wisdom needed to move our society to sustainability is present and is expressed on many fronts in many formats.
- In such a mass of material and cacophony of voices, any one voice, no matter how wise, is smothered.
- We are often not reading on the same page, not even in the same book, so we cannot carry on a focussed public discussion. The cacophony and the complexity may lead many to simply give up and leave environmental policy to others.
- Are we making progress? That depends on our expectations. The productivity of the effort certainly exceeds what past developments in intellectual history would suggest. Yet, the cumulative result falls far short of what is needed for helping our society to learn how to become sustainable. We should not confuse progress in avant garde thinking with progress in general thinking and action.
- Toe-holds for environmental/sustainable ways of thinking have been gained in many sectors (see the table of contents). It is especially gratifying to see that the bastion of classical economic thinking has been breached. Ecological economics is swiftly gaining adherents, is producing impressive thinking, and is beginning to affect the mainstream discipline.
- During the past 15 years there have been impressive gains in global environmental action, policies, and laws. [See Section III on Policies and Planning.] Non-governmental organizations (NGOs) are now important players in this sector. Environmentalism is the first truly worldwide movement. Governments, NGOs, and corporations are sharing strategies, plans, progress reports, and treaties. Ironically, even though much of this ferment takes place in the United States, the U. S. Government is not only laggard, it frequently is downright obstructive.
- It is quite predictable that any body of thought that has so many workers, and begins to turn the direction of society, will draw the skeptical attention of those committed to the currently dominant ways of public thinking. Anti-environmentalism [see Section III.2], which could hardly be expressed aloud in the 1970s and 1980s, has now found catchy formulations and new voices that launch frontal attacks on environmental thinking and action. We should expect that these attacks will continue and gain in stridency.
- Being caught up in this hugely complex societal relearning, it is difficult to get perspective on where it is going and how swiftly it is progressing. There can be no doubt, however, that it *is* moving and that it will change human society and life systems on planet earth.

—Buffalo NY, November 1995

Lester W. Milbrath is Director, Research Program in Environment and Society and Professor Emeritus of Political Science and Sociology at the State University of New York at Buffalo. He is the author of **Envisioning a Sustainable Society** [see #351] and **Learning to Think Environmentally: While There Is Still Time** (SUNY Press, Jan 1996/136p/$9.95pb.)

INTRODUCTION

by Michael Marien

Much has been written about environmental problems in recent years. A general appreciation of these interrelated problems inexorably points to the need for some sort of sustainable society. Much has been written on this topic, too. The flood of information on both environment and sustainability begs for a guide to order it in some way, much as we commonly employ guides to travel destinations and to flora and fauna in the natural world.

This guide seeks to sort things out and put these writings in context. Specifically, it has five purposes:

- To provide a bibliographic record of the important books and reports on environment and sustainability published in recent years;
- To provide a serviceable framework/taxonomy that assembles these items in an illuminating way, possibly suggesting new pathways for research and analysis;
- To provide some guidance for the beginning student as to where to start;
- To provide, for those who generally agree that humanity faces a multitude of environmental problems, a map of who says what about the most pressing issues;
- To provide, for those who do not agree with environmentalist worries, a convincing array of arguments that justify these environmental concerns and visions of sustainability.

In all, a total of 450 abstracts are presented here—mostly books and think tank reports. These selected items first appeared in *Future Survey*. Virtually all of these items were published between 1991 and 1995, and all are English-language publications. Many of the authors are Americans, but Canadians, Europeans, and some non-Western thinkers are also represented here, in addition to numerous United Nations reports. The authors include many who call themselves futurists and/or environmentalists (the line is a fuzzy one: some futurists also call themselves environmentalists and vice versa, but many futurists—especially those with a narrow outlook—are not environmentalists, and many environmentalists—notably those with a narrow view in time and space—are not futurists). Many of these items are focused on American issues; however, in most instances, both the problems and the proposed remedies may well be of great relevance to other countries.

The 450 abstracts are grouped in three broad categories: **I. Environmental Issues** (in which the primary focus is on a particular problem or set of problems); **II. Toward Sustainability** (in which the primary focus is on a sustainable society or some important aspect of it); **III. Politics and Planning** (in which the focus is on the political context, environmental critics, environmentalist groups, and actions to be taken at the global, national, regional, and local levels).

A total of 170 environment-related periodicals are also identified, including about two dozen general interest publications, about four dozen publications with a general environmental focus, and about 100 specialized environmental publications, such as those devoted to soil conservation, waste disposal, environmental education or environmental politics. The divisions between general and specialized environmental publications are often ambiguous—hence the approximate count.

Special thanks are due to Michael Lerner and the Jennifer Altman Foundation (who catalyzed this project), to Ted Trzyna of the California Institute of Public Affairs and the IUCN Commission on Environmental Strategy and Planning (serving as publishing partners in this project), to Edward Cornish and the World Future Society (who have supported the monthly publication of *Future Survey* since 1979), to Albert Sasson and the UNESCO Bureau of Studies and Programming in Paris (who supported an earlier bibliographic essay in 1992 on the literature of environmental problems and sustainable futures—see item #021), and to Lane Jennings and Monica Hahn-Koenig who ably assisted in the preparation. All opinions expressed in this guidebook are those of the editor, and are not necessarily held by the above-mentioned individuals and organizations.

I. ENVIRONMENTAL ISSUES

"Environmental Issues" cover a broad range of concerns. They are categorized here according to: Global Overviews that address all or most major issues; Yearbooks and Guides that address all or many major issues and appear periodically; Key Regions and Nations (notably the environmental problems of Russia and China); and specific areas of concern: Population Growth, Climate Change, Oceans and Coastal Areas, Freshwater, Biodiversity, Forests, Land Use, Pollution, and Toxic and Solid Waste.

A more detailed look at key items in each category:

1) Global Overviews. **The World Environment 1972-1992,** edited by Mostafa K. Tolba of UNEP and Osama A. El-Kholy, still provides an excellent overview of global environmental issues, their causes and consequences, and necessary responses (001). A broad but somewhat less imposing overview of **The Human Impact on the Natural Environment** is offered by geographer Andrew Goudie of Oxford University (002). Other noteworthy general statements include the ***World Scientist's Warning to Humanity*** (005), a volume from the US National Research Council on the human dimensions of global environmental change (006), and an interesting assortment of lists by futurist Joseph F. Coates (007).

2) Yearbooks and Guides. The biennial **World Resources** guides to the global environment (010/011), prepared by the World Resources Institute in collaboration with UNEP and UNDP, provide an authoritative source for data and analysis. The widely-distributed annual **State of the World** volumes (013/014) from Lester R. Brown and his Worldwatch Institute colleagues, always offer fresh insights on major issues. The new **Vital Signs** series from Worldwatch (015/016) provide a useful and reader-friendly focus on key global indicators.

3) Key Regions and Nations. **Planet Management** (022) provides an excellent overview of environmental problems in specific regions and nations, clearly illustrating that all regions suffer from serious problems, albeit in different mixes. Notable among them is the former Soviet Union (023/026), where problems of past neglect are especially severe, and the People's Republic of China (028/032), where a huge and still-growing population and a strong desire for economic development pose massive problems for the future.

4) Population Growth. World population will continue to grow, although projections vary, from 8.96 billion in 2025 (041), to 8.31 billion (042) and 8.5 billion in 2025 (043), and 8.0 billion in 2023 (044). The 1994 UN Conference on Population and Development stimulated several good books explaining this problem and what to do about it (046/049).

5) Climate Change. Global warming induced by human activity appears likely to be the biggest and most widely visible of all environmental problems (052/053), and various signals strongly suggest that this warming is underway (055/056). The future is likely to involve more climate chaos (057), environmental refugees (059), disrupted food supplies (060/061), many species extinctions and shifting ranges (062), and coastal land loss (063).

6) Oceans and Coastal Areas. The world's oceans are increasingly polluted (067/068) and their fisheries depleted from overfishing (069/070). In response, concepts of ocean governance (071/072), environmental security of oceans (073), sustainable coastal areas (075), and integrated coastal management (076) are proposed.

7) Freshwater. Scarcity and misuse of freshwater resources is a rapidly emerging global issue (077/080), leading to proposals for water-demand management (081/082), equity in decisionmaking (084/085), river conservation (087), ground water cleanup (088), a national aquatic restoration strategy (089), comprehensive statewide wetlands strategies (090), and control of toxic discharges and agricultural runoffs (091).

8) Biodiversity. Eloquent statements are made about threats to the world's biodiversity in general (092/094) and to China's beloved pandas in particular (095). Resulting proposals include an inventory of the world's flora and fauna (092), a Tropical Forests Fund (093), conservation breeding in zoos (094), the **Global Biodiversity Strategy** of WRI, IUCN, and UNEP (096), the companion marine biodiversity strategy (097), incentives for sound stewardship (098), meta-lessons for endangered species recovery (099), new management goals (100/101), and biodiversity prospecting agreements (103). The connection between biodiversity and the waning diversity of traditional human cultures (105/107) is also being noted.

9) Forests. Continuous global destruction of forests is a major factor in warming of the earth (108), and deforestation is also an expression of social injustice (109), the expansion of cattle production (111), and modern scientific forestry (112). Remedies include targets for global forest cover (108), multiple-use management (110), prices that tell the ecological truth (113), a concept of sustainable forestry (114), equitable participation by all stakeholders (116), and redefining wilderness (117).

10) Land Use. Topics include conserving mountain ecosystems and cultures (119/120), the shifting paradigm for natural resources management (121/122), preserving national parks (123/124), ethical land-use policy (125/126), public/private land conservation partnerships (127), a market of landscape visions (128), and reducing the environmental damage of mining (130/131).

11) Pollution. Two very thorough guides to toxic pollution (132/133) should be of help not only to Americans but to citizens worldwide. Other pollution problems include industrial decision-making (134), environmental justice (135), the threat of deteriorating nuclear power plants in Europe (136), inadequate planning for nuclear power plant accidents (137), air and water pollution regulation (138), air pollution abatement in Los Angeles (139), widespread pollution from lawn care (140), powerboat pollution (141), indoor air pollution (142), and pollution of near-earth space (143).

12) Toxic and Solid Waste. Solid and toxic waste are worldwide problems (144/145). Germany has taken the global lead in national solid waste policy (146). In the US, there are plenty of proposals for garbage (147/151), toxic waste siting (152/153), and handling nuclear and low-level radioactive waste (154/158). The Superfund approach to cleaning up toxic waste is widely considered to be a failure (159/162), and new strategies are proposed.

II. TOWARD SUSTAINABILITY

A general appreciation of the range and intensity of environmental problems leads to the conclusion that human population growth and the industrial way of life cannot continue indefinitely. Humanist alternatives to industrialism have been suggested for many years, perhaps beginning with **Issues in Post-Industrialism** edited by Ananda Coomeraswamy and Arthur J. Penty (London: George Allen & Unwin, 1914). The environmentalist critique of industrialism began after World War II, with such titles as **Road to Survival** by William Vogt (William Sloane Associates, 1948), **The Limits of the Earth** by Fairfield Osborn (Little, Brown, 1953), **Resources and the American Dream: Including a Theory of the Limit of Growth** by Samuel H. Ordway Jr (Ronald Press, 1953), and the widely-distributed classic, **The Challenge of Man's Future** by Harrison Brown (Viking, 1954).

The notions of "sustainable society" and "sustainable development" arose in the 1970s, notably in **The Sustainable Society** edited by Dennis Pirages (Praeger, 1977), **Building a Sustainable Society** by Lester R. Brown (W.W. Norton, 1981), **Quest for a Sustainable Society** edited by James C. Coomer (Pergamon, 1981), and **The Management of Sustainable Growth** edited by Harlan Cleveland (Pergamon, 1981). The concept received a significant boost with the publication of **Our Common Future** by the World Commission on Environment and Development (Oxford University Press, 1987), also known as the Brundtland Report. Many others have used the concept of sustainability in recent years; its great virtue is that it is sufficiently vague so as to offer a broad umbrella for a variety of meanings, ranging from sustainable economic growth (seen as an oxymoron by some) to strict no-growth (seen as utopian by others).

The 1987 Brundtland Report led to the 1992 United Nations Conference on Environment and Development in Rio de Janeiro, and AGENDA 21, the resulting action plan for the 1990s and beyond (163/169), available in several forms. The UNCED meeting serves to introduce a wide variety of "sustainability" overviews published in recent years. In addition to overviews, "sustainability" is a direct or indirect concern in ten specialized categories, generally reflecting key sectors of society: philosophy and ethics, economics, business and industry, agriculture, technology, energy, transportation, cities and buildings, health, and education.

A more detailed look at some of the key items in each category:

1) Overviews. In addition to AGENDA 21 and related reports (163/169), an important "Strategy for Sustainable Living" (170) was published in 1991 by IUCN, UNEP, and WWF, as an update of the 1980 **World Conservation Strategy**. The "2050 Project," examining a sustainable world in 2050, was initiated in 1993 by the improbable collaboration of the World Resources Institute, The Brookings Institution, and the Santa Fe Institute (171). Other individual and collective statements on how to save the planet and create a sustainable society have been recently issued by Mostafa K. Tolba of UNEP (172), Lester R. Brown and Worldwatch colleagues (173), Jonathon Porritt (174), Helen Caldicott (175), Paul and Anne Ehrlich (176), then-Senator Al Gore (177), Ernst von Weizsäcker (179), Shridath Ramphal (180), Jim MacNeill (181), Donella H. Meadows *et al.* (182), the World Bank (183), the Commission on Developing Countries and Global Change (184), the International Institute for Sustainable Development (185), the (US) National Commission on the Environment (190), the IUCN Commission on Environmental Strategy and Planning (191), and the Global Tomorrow Coalition (193). Even the editor of this guide has contributed to the cacophony (187), albeit belatedly!

2) Philosophy and Ethics. Some argue that a green worldview or ethical scheme is necessary, and many candidates are offered: a co-evolutionary view of transition to sustainability (194), a "universe story" of transition to the Ecozoic era (195), deep ecology (197), the 66 basic principles of an ecological world-view (198), applying the wisdom of native peoples (200/201), developing concern for future generations (202), a Public Philosophy of Sustainability (203), eco-religion (205), eco-philosophy (206), gaiasophy (207), the Earth Covenant citizen's treaty (209), Christian stewardship and responsibility (211/212), eco-socialism (213), radical ecology (214), a new "steady state" world view (215), "ecological postmodernism" (217), ecopsychology (218), green sociology (221), nature-compatible design (222), regenerative design (223), ecological engineering (224), and changing our worldview of "national security" (225-227).

3) Economics. A new view of economics is essential, and progress in this realm is reflected by the rapid growth of the International Society for Ecological Economics (228) and various books explaining ecological economics (228/230), green economics (231), real-life economics (232), and alternative economic indicators (233)—notably the new Index of Sustainable Economic Welfare (234/235). Other specific applications of green economics result in ecological tax reform (236/238), questioning consumption (240), long-overdue economic valuation of the environment (241), sustainable investment (244/245), and fostering sustainable livelihoods (246).

4) Business and Industry. Closely related to green economics are the applied notions of Paul Hawken's restorative economy (247), the Business Council for Sustainable Development (248), the EarthEnterprise initiatives of IISD (249), costing the earth (251), companies aiming for "environmental excellence" (252), green employment (253), environmental strategies (255), corporate environmental reporting (257), EcoManagement (258), pollution prevention programs (260), a sustainable materials economy (261), the industrial metabolism perspective (263), industrial ecology (264/266), greener marketing and advertising (267), sustainable world trade (268/271), and eco-tourism (272/273).

5) Agriculture. "Alternative Agriculture" has been endorsed by the US National Research Council (274), and various statements have been issued extolling sustainable agriculture (275/276), the need for soil conservation (278), sustainable agriculture in the humid tropics (279), resource-conserving production (280), reduced pesticide use (283/287), resisting "cattle culture" (288), promoting "microlivestock" (289), and greening agricultural policy (291/292). However, "sustainable agriculture" can have a wide variety of meanings (293).

6) Science and Technology. As noted in several recent reports, science and technology can be steered toward sustainable development and environmental quality (294/297), and there are immense opportunities in pursuing what is variously described as "environmentally advanced technology" (298/299), "environmentally critical technology" (300), or simply "green gold" (301). Japan has initiated a long-range "New Earth 21" program of innovative technology (302/303), and a World Engineering Partnership for Sustainable Development has been formed (304).

7) Energy. One important area of sustainable technology is energy, as illustrated by an encyclopedic MIT conference (305) and overviews of the potential for renewable energy sources (306), a sustainable electricity industry (307), energy efficiency policies (311/313), and building energy efficiency (314). Important energy sources in the future may well include biomass technologies (317), wind energy (318), and ocean thermal energy conversion (319).

8) Transportation. Another major realm of sustainable technology that overlaps with energy considerations is the transportation sector, notably growing use of motor vehicles with the extensive costs not reflected in the prices we pay. Remedies include higher fuel efficiency (320,326) "polluter pays" policies (321), comprehensive transportation policymaking (323), hydrogen vehicles (322), electric vehicles (324), ultralight hypercars (325), and greater use of rail (327) and bicycles (328/329).

9) Cities and Buildings. Traffic concerns flow into broader thinking about mega-city growth (330) and sustainable cities in general (331/335). This leads to specific proposals such as better water management and waste reduction (334/ 335), better land use (336), garden cities (337), greenways (338), and sustainable design of buildings (339/341).

10) Health. Environmental degradation is related to health in many ways (342); indeed, the industrial lifestyle—despite the wonders of modern medicine—may be generally detrimental to our health (343). Specific concerns include toxic wastes (344/345), the emergence of new infectious diseases as a result of ecological disruption (346/348), and the negative health impacts of global warming (349/350).

11) Education. As noted by Lester Milbrath, transition to a sustainable society will require considerable social learning (351). But only a few educators have thought about education for sustainability, notably David W. Orr (352/354), C.A. Bowers (354), and futurist David Hicks (355). Two textbooks on sustainability offer useful tools for teachers who are willing and able to use them (358/359).

III. POLITICS AND PLANNING

The glut of literature on environmental problems and sustainability would seem to make a compelling case for the direction that societal evolution must take. This is especially true for the long-run of the 21st century, and even the next few decades. But in the short-run, the environmentalist message has not pervaded the policy thinking of most societies. In part, this is due to conflicting signals and differing views on which policies to pursue. But two other factors warrant special sections in this guide: attention being paid to other global and domestic problems, and the anti-environmentalist backlash. Other key concerns in the transition to sustainability involve Global Action, National and Regional Action, Local Action, and effective leadership.

1) The Global Context. Many environmentalists understandably view their concerns as central, and several have even argued that an environmental consciousness and a quest for sustainability should guide all policy. But there are many other non-environmental problems, and, justifiably or not, they often take precedence in the consciousness of policy-makers and citizens. Even the United Nations, after holding the blockbuster Rio Earth Summit conference in 1992, shifted its focus to a Vienna Human Rights Conference in 1993, the International Conference on Population and Development in Cairo in 1994, the World Summit for Social Development in Copenhagen in 1995, and the Habitat II in Istanbul in 1996. Although population/resource/environment issues played a role in all of these conferences, they were not the central focus.

To give some appreciation as to how the environment fits into the broader global context a selection of ten recent overviews is provided here: the UN **Report on the World Social Situation** (360), the Boutros-Ghali **Agenda for Development** (361), an overview of ten key challenges in today's changing world (362), Paul Kennedy's excellent overview on **Preparing for the Twenty-First Century** (363), a related almanac on trends related to global development (364), John Peterson's integrated framework on the forces that are driving global change (365), Gerald Barney's **Global 2000 Revisited** (366), Ervin Laszlo's guide to global issues (367), the 1991 Club of Rome report on the global revolution (368), and a computer simulation model of international futures by Barry Hughes (369). The same is true for agendas of domestic issues in various nations: environment is a component, but not a dominating force on the entire program.

2) Environmentalist Critics. In addition to major competition from other problems, environmentalism has recently (and unexpectedly) been assaulted by a variety of critics. Some of these critics are self-described environmentalists and/or centrists: Gregg Easterbrook (370), Martin Lewis (371), Wallace Kaufman (372) and Richard D. North in the UK (373). But most of the critics are clearly anti-environmentalist, motivated by conservative, free-market, anti-government ideology. They include Ronald Baily (374/375), Julian L. Simon (376), the libertarian Cato Institute (377), the Pacific Research Institute for Public Policy (378), humorist P.J. O'Rourke (380), the late Dixy Lee Ray (381), Michael Fumento (382), Dennis T. Avery of the Hudson Institute (384), climatologist Patrick J. Michaels (386), and the late Aaron Wildavsky (387). **The Greenpeace Guide to Anti-Environmental Organizations** (389) is an excellent introduction to a wide variety of US groups.

Although this anti-green backlash appears to have strength only in the US, it could very well continue to grow and be influential elsewhere. For sustainability to seriously move forward, these critics may require a response. Yet, in all the literature surveyed here, a strong anti-environmentalist backlash is neither anticipated nor acknowledged. Many of the critics' arguments and forecasts can be readily countered. But not all. Confronting the valid points will be painful, but perhaps necessary. It is beyond the scope of this guide to analyze the critics, except sporadically, at the end of certain abstracts (notably #379, which summarizes the recurring signs of the "conservative" worldview).

3) Environmentalist Groups. The **World Directory of Environmental Organizations** (390) provides an excellent guide to over 2,600 organizations and programs in over 200 countries. Also see a more popularized global guide to projects and people (391), analyses of international environmental NGOs (392) and European environmental groups (393), an overview of US Green politics (394) and an explanation of the decline of Green parties (395), various analyses of environmentalism in the US (397/399), and environmentalism as a potential major ideology (400). Practical problems of environmental politics are also discussed for the UK (401) and the US (402/403).

4) Global Action. Three levels of action are needed, and all three should be respected. At the global level, Lynton Keith Caldwell describes global conferences and programs (404), Oran Young considers various forms of international environmental governance (405), Nazli Choucri focuses on framing global norms to guide action (406), Peter M. Haas *et al.* suggest how effective international environmental institutions can be developed (407), and Harlan Cleveland advances the concept of the "Global Commons" (408). Other writers propose a Global Commons Trust Fund (409), transforming UNEP into a UN Environment Agency (410), a "global bargain" strategy to collaborate on sustainable development (411), improving environmental treaty-making (414), and strengthening international environmental law (415/416).

5) National/Regional Action. Despite the dismal performance of green politics in recent years, effective green plans are being developed in several nations, notably The Netherlands, New Zealand, and Canada (418/419). The UNCED Earth Summit encouraged a total of 169 environmental reports from national governments (420), which could stimulate some planning activity. IUCN and IIED have distilled lessons learned from 14 years of comprehensive national strategies for sustainable development (421) and various examples of regional planning seem to offer useful models (422/423). In the US, the National Research Council has outlined necessary research to protect and manage the environment (425), and the EPA recommends a process of environmental foresight (426).

6) Local Action. The third level of action may well be equally important to the national and global levels—perhaps even more so, as suggested by reports on the process of "primary environmental care" (427) and case studies of Canadian and US grassroots groups (428/430). A variety of guides advise on community activism (431), conducting a community environmental inventory (432), local energy planning (433), community-based conservation (434), rural environmental planning (435), local reduction of toxic hazards (436/437), and community garbage source reduction (148). At the individual and family level, a guidebook advises on choosing environmentally-correct consumer products (441).

7) Leadership. All of this can only happen with environmental leadership at global, national, regional, and local levels. This final section offers books on increasing the effectiveness of environmental leaders (442/443), how to be an effective volunteer (444), tackling public problems in a shared-power world (445), the practice of mobilizing people to tackle tough problems (446), enhancing negotiated partnerships for sustainable development (447), finding common ground (448), coping with conflict (449), and how to publicize environmental causes (450).

* * * * *

Some conclusions from this survey:

1) This Guide is Both Too Long and Too Short. To do justice to recent thinking on matters relevant to environmental issues and sustainable futures, a minimum of 450 items have been selected from recent issues of *Future Survey* and **Future Survey Annual**. To many readers, this may seem exhausting and overwhelming—yet, it represents only *some* of the English-language literature published in the last few years—hopefully the more important contributions. In an age of instant global communication, we are still very far from a comprehensive and timely information system reporting on important environment/sustainability thinking in all countries.

2) Environmental Problems and Sustainable Futures are Complex and Multi-Dimensional. This statement will be perfectly unsurprising to many seasoned environmentalists. Still, this guide underscores the many facets to a wide variety of environmental problems (Part I), the many aspects to building a sustainable society (Part II), and the many levels of necessary environmental action, as well as the necessity of coping with environmental critics and obstructionists (Part III). Moving toward sustainability will be no easy matter.

3) Environmentalism is Highly Diverse and Fragmented. This is amply illustrated not only by the 450 abstracted items, but also by the listing of 170 periodicals. Several dozen of these periodicals cover general environmental problems. But no single publication stands out as the leading magazine or journal that covers all environmental issues and sustainable futures thinking. It is thus difficult to command the attention of the "environmental community" in the US, let alone the wider world.

4) Environmentalism Will Return to the Limelight. In the wake of UNCED, a period when many of the big guns of environmentalism were fired in the cause of saving the planet, there was a lapse in strong pro-environment statements, underscored by a backlash of generally ill-informed critics and attention to other problems. This condition will change, perhaps in just a few years (when millennial awareness rekindles long-range thinking) or perhaps not for a decade or so (when palpable global warming and/or other eco-disasters restore environmentalism to the forefront of public issues).

5) Futurists and Environmentalists Should Be Closely Connected. Any "futurist" without an appreciation of environmental problems is simply blindered and poorly-informed (the E/S guide seeks to correct this glaring deficiency found among far too many futurists). Any environmentalist who fails to take a broad and long-term view of environmental issues and the transition to sustainability will not be an effective leader. Environmentalists can benefit from the best of futurist thinking; futurists can benefit from the best of environmentalist thinking. If any of the best recent thinking of both futurists and environmentalists is not represented here, please let the editor know what you think is missing. This guide makes no claim to comprehensiveness; it is a beginning effort at mapping E/S ideas and people, perhaps with updated and improved editions to follow.

— LaFayette NY, January 1996

Michael Marien is founder and editor of *Future Survey*, published monthly since 1979 by the World Future Society (see back page for subscription information).

WHERE TO START:
A Short List of 21 Recommended Items for the 21st Century

Readers of this guide will doubtlessly approach it with a wide range of interests and many levels of sophistication. For the beginning student who seeks an overview of the best introductory titles, the following are recommended:

*001 **The World Environment: 1972-1992** (Mostafa Tolba/Osama El-Kholy, eds.; Chapman & Hall/Routledge, 1992)

*010 **World Resources 1994-95** (WRI/UNEP/UNDP; Oxford, 1994)

*015 **Vital Signs 1995** (Lester Brown *et al.*; Norton, 1995)

*022 **Planet Management** (Michael Williams, ed.; Oxford, 1993)

*053 **Global Environmental Issues: A Climatological Approach** (David Kemp; Routledge, 1994)

*079 **Last Oasis: Facing Water Scarcity** (Sandra Postel; Norton, 1992)

*096 **Global Biodiversity Strategy** (WRI/IUCN/UNEP; WRI, 1992)

*164 **Agenda 21: The Earth Summit Strategy** (Daniel Sitarz, ed., EarthPress, 1993)

*170 **Caring for the Earth: A Strategy for Sustainable Living** (IUCN/UNEP/WWF; IUCN, 1991)

*179 **Earth Politics** (Ernst von Weizsäcker; Zed, 1994)

*180 **Our Country, The Planet** (Shridath Ramphal; Island, 1992)

*182 **Beyond the Limits: Confronting Global Collapse, Envisioning a Sustainable Future** (Donella Meadows *et al.*; Chelsea Green, 1992)

*231 **The Gaia Atlas of Green Economics** (Paul Ekins *et al.*; Anchor/Doubleday, 1992)

*247 **The Ecology of Commerce** (Paul Hawken; HarperBusiness, 1993)

*274 **Alternative Agriculture** (National Research Council; National Academy Press, 1989)

*301 **Green Gold...The Race for Environmental Technology** (Curtis Moore/Alan Miller; Beacon, 1994)

*320 **Car Trouble** (Steve Nadis/James MacKenzie; Beacon, 1993)

*331 **The Gaia Atlas of Cities** (Herbert Girardet; Anchor/Doubleday, 1993)

*363 **Preparing for the Twenty-First Century** (Paul Kennedy; Random House, 1993)

*374 **The True State of the Planet** (Ronald Bailey, ed., Free Press, 1995)

*418 **Green Plans** (Huey Johnson; U of Nebraska, 1995)

Abstracts Selected from *Future Survey* and Future Survey Annual

[Asterisked numbers * indicate especially recommended items]

I. ENVIRONMENTAL ISSUES

1. Global Overviews

*001

The World Environment 1972-1992: Two Decades of Challenge. Edited by Mostafa K. Tolba (Executive Director, UN Environment Programme, Nairobi) and Osama A. El-Kholy (Cairo U, Egypt). London & NY: Chapman & Hall (div of Routledge), Dec 1992/884p/$89.95;$29.95pb.

Reviews trends over the past 20 years, with progress in some areas offset by accelerating destruction in others. **The Issues**: 1) air pollution (perception has broadened tremendously in the past two decades); 2) ozone depletion (once a matter of scientific curiosity, it has become an urgent policy question); 3) climate change (an issue raising important questions of international equity); 4) fresh water (due to growing population and per capita use, "it is highly likely that water, like energy in the 1970s, will become the most critical resource issue in most parts of the world by the late 1990s and the early 21st century"); 5) coastal and marine degradation (pressures on coastal zones are leading to continuing degradation, likely to be exacerbated by climate change and sea level rise); 6) land degradation ("if the process of desertification is not arrested soon, the world shortage of food will increase dramatically"; meanwhile, the cost of anti-desertification measures is escalating); 7) deforestation and habitat loss (increasing recognition of the economic value of forests and wetlands has added impetus to their conservation, and new global initiatives are underway after UNCED); 8) biodiversity (current measures to slow losses are valuable but insufficient); 9) environmental hazards (frequencies of major natural disasters have increased); 10) toxic chemicals and hazardous wastes (proliferation in many areas).

Causes and Consequences: 11) agriculture and fisheries (sustainable agriculture requires institutional mechanisms to give correct signals of emerging scarcities of land and water); 12) industry (the challenge is in restructuring for sustainable production); 13) energy (reviews production, consumption, and environmental impacts); 14) transport (one of the major contributors to environmental degradation; a coordinated global strategy for this sector is outlined); 15) tourism (now the second largest item in world trade; the concept of sustainable development should be applied to it worldwide); 16) population pressure on resources (caused by increased numbers, poverty, overconsumption, and unsustainable production technologies); 17) settlements (urbanization in poor countries and counter-urbanization in OECD countries); 18) health (infant and child mortality rates have declined steadily and life expectancy has increased, but the overall picture is still alarming in many respects); 19) peace and security (continuing military expenditure implies a lack of conviction for disarmament; a planned conversion process is needed to reverse environmental degradation).

The Response: 20) understanding the environment (all branches of science deepened tremendously since 1972, but new knowledge is not quickly applied in policies); 21) perceptions and attitudes (public concern is much greater than in 1972); 22) national responses (environmental policy is now a concern of virtually all nations and has become increasingly cross-sectoral and integrated, but laws and programs have grown haphazardly; governments are increasingly joining in international agreements and committing scarce resources to their implementation); 23) international responses (models for cooperation within the UN and between UN agencies and other bodies).

Concludes that the overriding message of this review is that human beings and human institutions wait too long before taking action. "On nearly every issue we have failed to apply the precautionary principle when it was appropriate—indeed prudent—to do so. We have had more than enough talk about environmental problems... Despite the very real achievements and advances of recent years, the state of the world environment and the living conditions of many of its people have continued to get worse... we no longer have the luxury of picking and choosing what to do next: the state of the world environment demands that we take action simultaneously on a broad front, with no further delay... It is no exaggeration to say that the ability of the biosphere to continue to support human life is now in question." [**NOTE**: A massive, authoritative overview, well-organized and clearly-written. Better than **State of the World** and **World Resources** (items #10 and #15 in this guide), both of which are very good. Especially valuable for Chapters 20-23 on "The Response."]

(environmental trends, impacts, and responses)

002

The Human Impact on the Natural Environment (4th Edition). Andrew Goudie (Prof of Geography, Hertford College, U of Oxford). Cambridge MA: MIT Press, March 1994/454p/$40.00;$19.95pb.

Trends in human manipulation of the environment: 1) the ways in which humans affect the environment are proliferating; 2) environmental issues that were once locally confined have become regional or even global problems; 3) the complexity, magnitude, and frequency of impacts are probably increasing; 4) compounding the effects of rapidly expanding human population is a general increase in per capita consumption and environmental impact [see discussion of I=PAT equation, #176]. Chapters describe the human impact on vegetation (deforestation, fires, air pollution), animals (domestication, dispersal, extinction, expansion, contraction), soil (salinity, erosion, acidification), water (pollution, river flow, lake levels), geomorphology (new landforms, accelerated sedimentation and coastal erosion, river-channel changes), and climate and the atmosphere (pollution, urban climates, the CO_2 problem). A concluding chapter on the future discusses the spatial patterning of change, consequences of global warming, projected sea-level rise, and environmental uncertainty. [**NOTE**: Offers background on what has happened and is happening, but no guidance on the human response.] **(human impact on environment)**

003

Population, Resources and the Environment: The Critical Challenges. Norman Myers (Oxford UK). NY: United Nations Population Fund (220 E 42nd St), Dec 1991/154p/$24.75pb.

The unprecedented rapid growth in human numbers, plus their distributional patterns and urbanization trends, induces critical constraints on the development process, as well as unsustainable burdens on the environmental resource base that underpins much economic activity. Based on the latest UN projections, world population is expected to grow from 5.3 billion in 1990 to 6.3 billion in 2000, 7.2 billion in 2010, and 8.5 billion in 2025. (These figures are rather higher than those projected until recently, due to a slowing in the decline of fertility in a number of countries.) The medium-level projection for eventual world population in 2100 is 11.3 billion, with a high projection of 14.2 billion and a low projection of just under 8 billion. Although the global rate of population growth has declined from a high of 2.1% in the late 1960s to 1.7% today, the number of people added each year is higher than ever, and is projected to keep on climbing, reaching an annual peak of 97 million by 2000. In 1989, the number of people estimated to be living in absolute poverty totalled 1.2 billion, or 23% of world population.

The growth in human numbers and in consumption, along with growth of environmentally adverse technology, combine to precipitate a downturn in the capacity of environmental resources to sustain human communities. The problem of land shortages, accompanied by farmland fragmentation, is becoming widespread in many if not most developing countries. Water deficits are already a serious constraint on development in 88 developing countries with 40% of world population. Chapters discuss environment and the quality of life, and seven case studies of population carrying capacities (Pakistan, India, Philippines, Ethiopia, Kenya, Mexico, and El Salvador).

(population/environment challenges)

004

Is Humanity Suicidal? Edward O. Wilson (Prof of Science, Harvard U), *The New York Times Magazine*, 30 May 1993, 24-29.

From the great diversity of large animals, *Homo sapiens* has gained intelligent control of Earth. Unlike any creature that lived before, we have become a geophysical force. The human species has doubled to 5.5 billion in the past 50 years, and is scheduled to double again in the next 50 years. But our species retains hereditary traits that add greatly to our destructive impact. We are tribal and aggressively territorial. In our search for food, we have reduced animal life in rivers, lakes, and now, increasingly, the open ocean. Everywhere we pollute the air and water, lower water tables, and extinguish species. With people everywhere seeking a better quality of life, the search for resources is expanding even faster than population. It is accelerated further by a parallel rise in environment-devouring technology.

Opinions on the human prospect fall loosely into two schools. *Exemptionalism* holds that, since humankind is transcendent in intelligence and spirit, our species has been released from the iron laws of ecology that bind other species, and that civilized human beings will somehow find a solution. The opposing idea of reality is *environmentalism*, which sees humanity as a biological species tightly dependent on the natural world. At the heart of this worldview is the conviction that human physical and spiritual health depends on sustaining the planet in a relatively unaltered state. Wilson argues that we are smart enough and have time enough to avoid an environmental catastrophe. But a redirection of much of science and technology will be required, as well as a reconsideration of our self-image as a species. There are reasons to believe that we have entered what will someday be called the Century of the Environment. The human hand is now upon the physical homeostat, with actions having been taken to restore the ozone layer and a good prospect of slowing global warming. But the human hand is not upon the biological homeostat. "Humanity is now destroying most of the habitats where evolution can occur." Mass extinctions are being reported with increasing frequency in every part of the world, and the reports are probably only a very small percentage of the actual number.

(Century of the Environment under way)

005

World Scientists' Warning to Humanity. Coordinated by Henry Kendall (Chairman, UCS). Washington: Union of Concerned Scientists (1616 P St NW, Suite 310), Nov 1992/2p (202/332-0900).

A statement signed by 1,575 scientists from 69 countries and sent to 160 national leaders. Signers include 99 of the 196 living scientists who are Nobel laureates, as well as senior officers from many of the most prestigious scientific academies worldwide.

The warning emphasizes that "human beings and the natural world are on a collision course." If harmful human activities are not checked, the future will be put at serious risk. Critical stress is apparent in the atmosphere, water resources (some 80 countries with 40% of world population have serious shortages), the oceans (total marine catch is now at or above maximum sustainable yield), soil, tropical and temperate forests, and living species. "Much of this damage is irreversible on a scale of centuries or permanent... we are fast approaching many of the earth's limits." A great change in our stewardship of the earth and the life on it is required if vast human misery is to be avoided. Five linked areas must be addressed simultaneously: 1) bringing environmentally damaging activities under control; 2) managing resources crucial to human welfare more effectively; 3) stabilizing population; 4) reducing and eventually eliminating poverty; 5) ensuring sexual equality and guaranteeing women control over their reproductive decisions. Developed nations are the largest polluters in the world today and must greatly reduce their overconsumption while providing aid and support to the developing nations. Developing nations must recognize the grave threat of environmental damage and unchecked populations. A new ethic is required in discharging our responsibility for ourselves and the earth. We need the help of scientists, business and religious leaders, and the world's people.

(1,575 scientists warn humanity)

006

Global Environmental Change: Understanding the Human Dimensions. Edited by Paul C. Stern (Study Director), Oran R. Young (Dartmouth College; Chair), and Daniel Druckman (Study Director). Washington: National Academy Press, Jan 1992/308p/$29.95.

A report of the National Research Council's Committee on the Human Dimensions of Global Change, which is separate from the international Human Dimensions of Global Change Programme or HDGCP [**FS Annual 1991** #10286]. The earth has entered a period of hydrological, climatological, and biological change that differs from previous episodes of global change in the extent to which it is human in origin. Thus to explain or predict the course of present global changes, one must understand the human sources, consequences, and responses. This book examines what is known about the human dimensions. Driving forces include population growth and standard of living (an average North American uses about 35 times as much energy as an individual living at India's average standard), economic growth, technological change, political-economic institutions (the international political economy, as well as government structure and policies, can have significant environmental consequences), and beliefs and values related to material possessions and the relationship of humanity and nature. Concludes with recommendations for a comprehensive national research program of five major elements: 1) a substantial increase in NSF support for unsolicited research on human dimensions of global change, including a category of small grants; 2) programs of targeted research supported by NSF, other Federal agencies, and private funding sources (some suggested topics: energy intensity of various societies, factors that change systems of land use and food production, valuing consequences of environmental change not well-reflected in market prices); 3) a national program to ensure that appropriate data sets for research are routinely acquired and made readily available; 4) a prestigious nationwide fellowship program to induce students and researchers to enhance their knowledge of global change issues and to interact across disciplines; 5) establishing five national centers for research on the human dimensions of global change over the next 3-5 years.

(human dimensions of global change)

007

Factors Shaping and Shaped by the Environment: 1990-2010, Joseph F. Coates (Coates & Jarratt, Inc, Washington), *Futures Research Quarterly*, 7:3, Fall 1991, 5-55.

A melange of various lists, including: **1) Negative Drivers of Global Environmental Change**: population growth (most important), economic development, urbanization, disease, war and collective violence; **2) Positive Drivers of Change**: globalizing multinational corporations (which tend to bring advanced nation standards to LDCs), developments in science, social and political value changes reflected in environmental groups, international public administration; **3) Environmental Drivers in the US**: the greening of business, the environmental movement, health concerns, monetizing the environment; **4) The Environmental Agenda in the US**: toxics, solid waste, clean air and water, preservation (the leading edge of the advanced nations' agenda); **5) Sources of New Problems**: areas that may rival or exceed present fears include immunological diseases, brain function disruptions, reproductive capability, genetic defects, trace materials in the body, electromagnetic effects; **6) Policy Recommendations for the US**: centralize environmental responsibility in some executive body, begin a series of long-range environmental policy studies aimed at new strategic approaches, create a task force to recommend revisions in GNP data, create a long-range vision of the environment for the US and the global community, evaluate experience of other nations in environmental management, certify best business practices of MNCs in host countries; **7) Potential Contributions of Science and Technology**: research frontiers in ecology, remote sensing from space, biochemistry, biotechnology, oceanography, geology, and the social sciences; **8) Scenarios for World Regions**: the world is splitting into four groups of nations, ranging from countries that have environmental problems relatively well in hand to desperate countries such as Bangladesh that are in devolution. Concludes that "Global environmental prospects for the next 20 years are on net balance in decline." We do not have the technology, the politics, the policy, or the will, on the time scale and framework required, to arrest adverse developments. Western Europe has the brightest prospects of any region, and the situation is good in other advanced nations. But Brazil, much of Sub-Saharan Africa, and much of South Asia "are on the road to environmental disaster." [**NOTE**: Some fresh overview perspectives in a jumbled format.]

(environmental problems: overview)

008

Nature, Technology, and Society: Cultural Roots of the Current Environmental Crisis. Victor Ferkiss (Prof Emeritus of Govt., Georgetown U). NY: New York U Press, March 1993/341p/$40.00.

Author of **Technological Man** (1969) and **The Future of Technological Civilization** (which proposed "ecological humanism" in 1974) discusses early views of nature and technology (the Greeks, Romans, Jews, and Early Christians), how technology flourished during the renaissance (a movement toward a scientific-technological conquest of nature), the industrial revolution and the beginnings of technocracy, views of nature and technology (in America, in Marxist socialism, in Islam, and in the Orient), technological cornucopianism (Arthur C. Clarke, Herman Kahn, R. Buckminster Fuller, Gerard O'Neill), contemporary critics of technology (Jacques Ellul, Murray Bookchin, Garrett Hardin, Lewis Mumford), ecofeminism and ecotheology, deep ecology thinking (Arne Naess, Gary Snyder, Ernest Callenbach), and the global green movement. Concludes that the drive to conquer nature through technology is virtually universal, although the modern Western world has been the leader. But technology is not irresistible, and human cultures can choose which technologies to embrace and which to resist. New, less-polluting technologies can be encouraged or discouraged by government policies. We must also radically reevaluate what our self-interest really is. The world needs a spiritual revolution—a basic change in values, and a new concept of progress. We must end the male (or at least "macho") quest to dominate nature, and rethink our place in nature.

(technology and environmental crisis)

009

A Green History of the World: The Environment and the Collapse of Great Civilizations. Clive Ponting (Honorary Research Fellow, University College, Swansea, UK). NY: St. Martin's Press, April 1992/432p/$24.95.

The opening chapter describes "The Lessons of Easter Island," which had a population peaking at 7,000 people in 1550. But the society suddenly collapsed as a result of massive environmental degradation brought on by deforestation. Subsequent chapters are much broader in time and space, describing ecosystems as the foundations of history, gathering and hunting as the way of life for 99% of human history, the first great transition to agriculture, the spread of European settlements, the second great transition to industrialization (marking a massive leap in the process of using more of Earth's resources to support more people), human thought about the relationship of people to nature, the rape of the world (e.g., the vogue for big-game hunting in the 19th and early 20th centuries, and slaughter of the huge bison herds on America's Great Plains), changing human disease patterns, population growth and urbanization, and global pollution. Concludes that "for thousands of years humans have lived with consequences of various forms of environmental degradation without necessarily incurring social breakdown." The problems now facing the world stem from a variety of pressures that have developed over long periods of time. Given the highly unequal distribution of the world's wealth and power, it is extremely difficult to deal with problems that cut across national boundaries, and with those that entail significant financial and social costs.

"The fact that a breakdown has not so far occurred does not guarantee that it will not happen." It is clearly far too soon to judge whether industrialized societies and the rapidly rising human population in the rest of the world are ecologically sustainable. [**NOTE**: Useful background with a cautious conclusion.] **(green history of world)**

2. Yearbooks and Guides

*010

World Resources 1994-95: A Guide to the Global Environment. World Resources Inst (Allen L. Hammond, Editor-in-Chief), in collaboration with the UN Environment Programme and the UN Development Programme. NY: Oxford UP, May 1994/400p(8x11")/$35.00;$21.95pb.

Sixth edition of a biennial series, with editions in eight languages. **Part I**, in support of the 1994 UN Conference on Population and Development in Cairo, has three special chapters that highlight different aspects of people and the environment: 1) natural resource consumption trends in the North and South, and their consequences; 2) the complex interactions between population growth and environmental degradation; 3) the essential role of women in sustainable development (traditional market-oriented development policies have failed to consider gender equity, the environment, and human development). **Part II**, as in previous editions, focuses on a particular region, in this case China and India. **Part III** reports on basic conditions and trends regarding food and agriculture (with a focus on pesticide use and biotechnology), forests and rangelands, biodiversity, energy, water (with a focus on marine fisheries), atmosphere and climate change, sustainable practices for industry, international institutions (with a focus on multilateral funding for sustainable development), and national policies and institutions (with a focus on strategy for sustainable development in The Netherlands, environmental planning in Chile, and the nascent planning process in Madagascar). **Part IV** provides 52 data tables for the countries and regions of the world. [**NOTE**: A massive collaborative effort; the authoritative source for data and analysis of global environmental issues. ALSO SEE: the **World Resources** Database, **Teacher's Guide to World Resources**, and **Global Trends Presentation Set** (50 color slides or 50 transparencies for instructional use), all from WRI in Washington (202/662-2596).]

(world resources guide: 6th edition)

011

World Resources 1992-93: A Guide to the Global Environment. World Resources Institute (Allen L. Hammond, Editor-in-Chief), in collaboration with the UN Environment Programme and the UN Development Programme. NY: Oxford U Press, March 1992/385p (8x11")/$29.95;$15.95pb.

Fifth edition of a biennial series, with editions in eight languages. **Part I**, in support of the UNCED meeting, focuses on sustainable development with four special chapters. The overview chapter, providing an extensive definition of sustainable development, concludes that it requires simultaneous progress along at least four dimensions: economic, human, environmental, and technological. The three other chapters provide case studies on what sustainable development might mean in industrial countries, poor countries (breaking the cycle of poverty and environmental degradation), and rapidly industrializing countries. **Part II** focuses on the severe problems faced by Central Europe. (The fourth edition focused on Latin America; the third edition on Asia.) **Part III** reports on basic conditions and trends, key issues, and recent developments. **1) Population and Human Development**: declining growth rates and increasing numbers; human development measures; children's health; urbanization; **2) Food and Agriculture**: despite disincentives, many farmers are finding it profitable to invest in sustainable agriculture; agro-chemical taxes should be imposed to control off-site impacts; **3) Forests and Rangelands**: an area about the size of China and India combined—almost 11% of the Earth's vegetated surface—has suffered moderate to extreme soil degradation in the past 45 years; **4) Wildlife and Habitat**: biodiversity is more threatened now than at any time in the past 65 million years; up to 15% of Earth's species could be doomed over the next 25 years; **5) Energy**: global energy production has increased by about 50% over the past two decades, with fossil fuels accounting for over 90% of production; **6) Freshwater**: resources are under severe and increasing stress; by 2000, withdrawals for agriculture will increase slightly and industrial withdrawals will probably double; "shortages could reach crisis proportions in the Middle East and North Africa"; **7) Oceans and Coasts**: only a small fraction of the world's wastewater is treated; trends show rising coastal pollution and accelerated destruction of coastal marine habitats; **8) Atmosphere and Climate**: includes estimates of emissions for the major greenhouse gases by source and country (the US accounts for 18.4% of global emissions, followed by the USSR at 13.5%, China at 8.4% and Japan at 5.6%); **9) Policies and Institutions**: discusses evolution of NGOs and grassroots organizations. **Part IV** provides 52 data tables for the countries and regions of the world. Concludes that "the world is not now headed toward a sustainable future, but rather toward a variety of potential human and environmental disasters... Sustainable development will require a fundamental change in existing policies and practices."

(sustainable development defined)

012

The 1994 Information Please Environmental Almanac. Compiled by the World Resources Institute (Allen Hammond, Editor-in-Chief). Boston: Houghton Mifflin, Nov 1993/704p/$11.95pb.

Third annual edition, with contents in four sections as before: **1) State of the Planet**: green taxes and accounting, various news reports, environmental highlights of the year; **2) Close to Home**: toxics in the home, water supplies, state energy report card, state goals for waste recycling (ranging from 25% to 60%), air pollution trends, grassroots activities in selected states, access to information (on-line services, a sampling of organizations, abstracts of 20 books); **3) National View**: discussion of public lands, pollutants in the food chain, environmental justice, green metro areas (San Antonio ranks best of 75 areas; Houston and St. Louis are the worst), state and province profiles and comparisons; **4) Global View**: discussion of population, marine ecosystems, atmosphere, wildlife and biodiversity, country comparisons on selected indicators, and 316 pages of individual country profiles (nearly half of the book). [**NOTE**: Popularized version of WRI's biannual **World Resources**. A useful companion to the Worldwatch Institute's **State of the World Report** (perhaps reflecting a lack of cooperative spirit, neither the Institute nor **SOTW** are mentioned by WRI). The **Almanac** provides valuable charts and data, but gets lost in the bits and pieces; **SOTW** is far better in focusing on key actions for sustainability.] **(environmental almanac)**

013

State of the World 1995: A Worldwatch Institute Report on Progress Toward a Sustainable Society. Lester R. Brown *et al.* NY: W. W. Norton, Jan 1995/255p/ $23.00;$11.95pb.

The 12th annual assessment, presently translated into 27 languages [yet, not reflected in any bestseller list]. The initial essay by **Lester Brown** focuses on three of the *earth's natural limits* that are already slowing growth in world food production: ***1) Oceanic Fisheries***: fish catch in the past four years has fluctuated between 97 and 99 million tons, dropping the catch per capita by 8%; all 177 oceanic fisheries are now being fished at or beyond capacity, and pollution and overharvesting are killing many inland seas and coastal estuaries (the annual harvest of edible oysters from the Chesapeake Bay has dropped from nearly 100,000 tons at the turn of the century to less than 1,000 tons in 1993); aquaculture is no panacea, and is also a major cause of coastal habitat destruction; ***2) Fresh Water***: some 300 cities in China are reportedly short of water—100 of them seriously short; irrigated agriculture in many Southwestern US locations is being eliminated to supply booming cities; many aquifers are being depleted, which will reduce irrigation; ***3) Fertilizer Use***: for nearly four decades, steadily rising fertilizer use was the engine driving the record growth in world food output, but the physiological limit has been reached and the era of substituting fertilizer for land came to a halt in 1990 (in the last four years, fertilizer use has fallen by some 15%, largely due to the former Soviet Union withdrawing subsidies); applying more nutrients simply has little or no effect. "With thresholds for sustainable yields now being crossed for so many resources, relationships that have been stable for centuries or millennia are becoming highly volatile."

Other topics: ***1) Protecting Oceanic Fisheries and Jobs***: a brief version of Worldwatch Paper 120, **Net Loss** (#069), on too many fishers chasing too few fish with ever more effective technology; ***2) Sustaining Mountain Peoples and Environments***: mountains span one-fifth of the landscape and are home to one-tenth of humanity, with an additional 2 billion people living close-by and depending on their water and resources; mountain ecosystems are threatened by intensive resource extraction and mass recreation; ***3) Harnessing the Sun and the Wind***: adapted from **Power Surge** (#308), on realizing the potential of solar and wind energy; ***4) Creating a Sustainable Materials Economy***: brief version of Worldwatch Paper 121, **The Next Efficiency Revolution** (#261), on the need and the potential for improving materials efficiency; ***5) Making Better Buildings***: on the environmental and economic advantages of rehabilitation over construction from scratch (buildings consume one-sixth to one-half of the world's physical resources); ***6) Facing China's Limits***: it appears likely that China's total grain production will fall by at least one-fifth between 1990 and 2030, even as new population growth and more affluent lifestyles increase demand; this will leave a shortfall of 216 million tons (more than the world's entire 1993 grain exports), and the import deficit could reach nearly 380 million tons; the prospect of one-fifth of humanity suddenly entering the consumer age will force industrial countries to face up to their unsustainable practices; ***7) Leaving Homes***: on the world's 23 million refugees (up from 15 million in 1989), 27 million people displaced in their own countries, and at least 10 million illegal immigrants; ***8) Budgeting for Disarmament***: a brief version of Worldwatch Paper 122 on coping with the legacy of war; ***9) Forging a New Global Partnership***: on global environmental treaties, the need for a dedicated funding mechanism such as a tax on currency transactions to finance transition to a sustainable society, and putting Agenda 21 and the World Population Plan of Action into practice. **(natural limits to growth)**

014

State of the World 1994: A Worldwatch Institute Report on Progress Toward a Sustainable Society. Lester R. Brown *et al.* NY: W. W. Norton, Jan 1994/265p/ $22.00;$11.95pb.

The 11th annual assessment with a first printing in English of 100,000 copies. The theme of this year's *SOTW* is carrying capacity. The initial essay by **Sandra Postel** states that, as a result of population growth, consumption patterns, and technology choices, "we have surpassed the planet's carrying capacity. This is plainly evident by the extent to which we are damaging and depleting natural capital." If current trends in resource use continue, and world population grows as projected, by 2010 per capita availability of rangeland will fall by 22%, irrigated land by 12%, cropland by 21%, forestland by 30%, and the fish catch by 10%. The days of the frontier economy are over: we have entered an era in which global prosperity increasingly depends on using resources more efficiently, distributing them more equitably, and reducing consumption levels overall. "Unless we accelerate this transition, powerful social tensions are likely to arise from increased competition for the scarce resources that remain."

Building on this chapter, **Lester R. Brown** warns that we are facing an era of food insecurity. Whereas per capita grain production rose by 40% in the 1950-1984 period, grain output per person fell 11% in the 1984-1993 period. "At present, there is nothing in sight to reverse the worldwide decline in grain output per person." There are many opportunities for expanding food production, but all the identifiable ones are small; no large jumps are in prospect, comparable to those following expanding use of fertilizer or the hybridization of corn. Since WWII, economic policies dominated the evolution of the global economy; during the next four decades ("the environmental era"), environmental limits interacting with population growth will have far more influence. In the past, national security was defined by the Cold War; in the next 40 years, food and job security will dominate, often driving hungry and jobless people across national borders.

Eight other chapters, largely adaptations of recent Worldwatch papers, discuss redesigning the forest economy, safeguarding oceans, reshaping the power industry into a service-oriented industry committed to demand-side management, reinventing transport to emphasize less destructive modes, using computers for the environment, assessing environmental health risks, destroying or dismantling military weapons and materials in an acceptable manner, and building a new World Bank to serve today's world (by measuring success in terms of quality of lending and not quantity, reorienting lending away from large infrastructure projects toward smaller efforts, reducing debt burdens, revamping the criteria for success, more widespread use of natural resource accounting, and more accountability to the public). [ALSO SEE: **Mortgaging the Earth: The World Bank, Environmental Impoverishment, and the Crisis of Development** by EDF attorney Bruce Rich (Beacon Press, Feb 1994/376p/ $29.00).] **(food insecurity ahead)**

*015

Vital Signs 1995: The Trends That Are Shaping Our Future. Lester R. Brown, Nicholas Lenssen, and Hal Kane (Worldwatch Institute). NY: W.W. Norton & Co, June 1995/176p/$19.95; $10.95pb.

The fourth in an annual series, now appearing in 14 languages, with an additional 3 translations (including Chinese) in prospect. The 50 key indicators are discussed in signed two-page essays. Among the 15 new entries this year are declining use of lead in gasoline (down 75% between 1970 and 1993), the global decline of amphibian populations, falling water tables in many areas, expanding access to safe drinking water, and growing global proliferation of both television sets and computers. In the introduction, Lester Brown comments on the accelerating pace of history as human demands collide with the earth's natural limits. "With natural constraints beginning to impinge on the expansion of food production in many countries, food security may soon become the principal preoccupation of many governments. Perhaps more than anything else, spreading food insecurity will symbolize the widespread failure of population and environmental policies." Some other key trends identified by Brown: 1) the 3% growth of world output in 1994 marked the end of a protracted slowdown, but this conventional measure of economic progress fails to reflect the loss of ecological capital; 2) energy use is growing, notably wind power generation (up 22% in 1994) and photovoltaics (factory shipments up 15% in 1994); in the "not too distant future, the closing of older nuclear power plants will exceed the dwindling number of new completions, leading to a decline in nuclear power"; 3) following the cooling effect of the Mt. Pinatubo eruption in 1991, the warming trend in the earth's temperature resumed in 1994; 4) food supplies are tightening, as world grain consumption exceeded production for the second year in a row; 5) the estimated number of people with the HIV virus increased from 22 to 26 million in 1994, and those with full-blown AIDS climbed from 6.9 to 8.5 million; 6) the human demand for fresh water exceeded the sustainable yield of rivers and aquifers in more and more areas. [ALSO SEE: **Environmental Indicators: A Systematic Approach to Measuring and Reporting on Environmental Policy Performance in the Context of Sustainable Development** by Allen Hammond *et al.* (World Resources Institute, May 1995/43p/$13.50), on how governments benefit from using such indicators, and **Index of Environmental Trends: An Assessment of 21 Key Environmental Indicators in Nine Industrialized Countries over the Past Two Decades** by Gar Alperovitz *et al.* (National Center for Economic Alternatives, 2040 S St NW, Washington 20009; April 1995/68p/$10.00), finding that, overall, the nine countries have rarely achieved stated policy goals, and have not kept up with most new environmental problems.]

(key indicators on sustainability)

016

Vital Signs 1994: The Trends That Are Shaping Our Future. Lester R. Brown, Hal Kane, and David Malin Roodman (Worldwatch Institute). NY: W. W. Norton & Co, June 1994/160p/$22.00;$10.95pb.

The third in an annual series, looking at "how we are doing in creating an environmentally sustainable economy." Key indicators are discussed in signed two-page essays: ***1) Food***: 1993 world grain harvest down by nearly 5%, world soybean crop down almost 4%, meat production up slightly, fish catch stable, aquaculture output in 1991 almost double that of 1984, grain stocks projected to drop sharply; ***2) Agricultural Resources***: world grain yield down from 1992 high, world fertilizer use in 1993 down 14% from 1989 high, slowing expansion of irrigated areas; ***3) Energy***: oil production flat since 1989, wind power up 13% from 1992, total installed nuclear generating capacity up 3% to an all-time high, slowing growth in shipments of photovoltaic cells, natural gas (now providing 21% of world primary energy) at record high production, coal use declined from peak in 1989, sales of compact fluorescent lamps quadrupled in the past five years; ***4) Atmosphere***: CFC production continues to drop from 1988 high, global temperature rises slightly, carbon emissions unchanged from 1992; ***5) Economic Trends***: world economy expanded by 2.2% over 1992, Third World debt up to all-time high of $1.77 trillion, world trade up 3.3% from 1992, global paper production up 1% to all-time high, gold production (and associated wastes) at record high due to cyanide heap leaching technology; ***6) Transportation***: bicycle production at a record high (outstripping auto output by nearly 3 to 1), world auto production essentially flat for the past decade; ***7) Environment***: resistance to pesticides growing ("at least 520 insects and mites, 150 plant diseases, and 113 weeds have developed resistance to one or more pesticides meant to control them"), sulfur and nitrogen emissions resume rise after a 1990 drop; ***8) Social Trends***: total population increase drops slightly, world cigarette production up (but per capita smoking is down from a peak in 1988), HIV/AIDS cases rise steadily, immunization rates soar, refugee flows at an all-time high of 19 million; ***9) Military***: worldwide arms exports declined 36% in 1991, spending for UN peacekeeping up 69% over 1992, nuclear warheads declined by 6% in 1993.

In the introduction, Lester Brown interprets and clusters some of the indicators, observing that: 1) of the 9,600 species of birds, populations of 6,600 species are in decline ("of all the available indicators, this may give the clearest sense of the earth's health"); 2) stratospheric levels of free chlorine appear to be peaking in the mid 1990s; if CFC production is phased out as scheduled, free chlorine levels will increase until around 2000 and then slowly diminish; 3) three food supply systems are under mounting pressure: rising demands on *fisheries* are becoming excessive worldwide, *rangelands* are being grazed at or beyond their sustainable yield almost everywhere, and *croplands* are also becoming unsustainable in many countries, leading to soil erosion; 4) stocks of all grains have declined; 5) water scarcity is becoming common; 6) "the world is entering a new phase in the era of rapid population growth, one where fishers and farmers can no longer expand output fast enough to keep up with population"; 7) the ratio of military spending to UN peacekeeping was about 2000 to 1 a few years ago, but is now closer to 250 to 1. [**NOTE**: Fascinating and important. The mix of indicators varies each year, and some (such as bicycle production and UN peacekeeping expense) can be found nowhere else.] **(key indicators on sustainability)**

017

Green Globe Yearbook 1993: An Independent Publication on Environment and Development from the Fridtjof Nansen Institute, Norway. Edited by Helge Ole Bergesen (U of Oslo) and Georg Parmann (Norwegian U Press). NY: Oxford U Press, June 1993/271p/$49.95.

The second in an annual series on progress in addressing environment and development problems, the main obstacles to effective international solutions, and what needs to be done to overcome these hindrances. **Articles**: on international efforts to combat marine pollution, NGOs at UNCED, biological diversity as common heritage vs. private property, NGOs in the Third World, trade and the environment, accelerating the greening of the World Bank, obstacles to sustainable forestry, and the partial greening of business (with suggestions for building an eco-industrial policy). **International Agreements**: brief outlines of 40 conventions on nature conservation, atmosphere, marine environment, hazardous substances, and nuclear safety. **Intergovernmental Organizations (IGOs)**: profiles of 18 organizations such as FAO, ILO, UNICEF, UNDP, UNESCO, UNEP, OECD, and the World Bank. **Non-Governmental Organizations**: profiles of 23 NGOs, including Earth Council, Earthwatch, Friends of the Earth, Greenpeace, the Commission on Environment of the International Chamber of Commerce, IUCN, Sierra Club, SID, and World Wide Fund for Nature. **Green Globe Yearbook 1992** (Oxford, July 1992/303p/$45.00) offers 11 articles on indigenous people and sustainability, foreign influence on Brazilian environmentalism, protecting the global climate, democracy and development, trade with endangered species, the global challenge of AIDS, ozone depletion, the international trade in hazardous wastes, energy for sustainable development, policy dimensions of population and environment, and protecting the Antarctic. Sections on International Agreements, IGOs, and NGOs are essentially the same as in 1993. [**NOTE**: About one-third anthology of high quality essays, and two-thirds reference information that does not need repeating every year.] **(Green Globe Yearbook #2)**

018

The Green Encyclopedia. Irene Franck and David Brownstone. NY: Prentice Hall General Reference, Nov 1992/486p/$35.00;$20.00pb.

Entries on acid rain, biotechnology, Rachel Carson, Department of Energy, Paul Ehrlich, forests, global warming, hazardous waste, indigenous peoples, James Bay hydropower, Joseph Wood Krutch, Land Trust Alliance, Man and the Biosphere Program, National Audubon Society, offshore drilling, pesticides, radiation, sea-level rise, toxic chemicals, Union of Concerned Scientists, Valdez Principles (ten guidelines for corporate responsibility proposed by the Social Investment Forum), water pollution, Yellowstone National Park, and zero population growth. Major topics include a boxed "information and action guide" leading to relevant references and organizations. Special appendices on animal rights and welfare, ecotourism, endangered and threatened species, international wetlands, World Natural Heritage Sites, toxic chemicals, acronyms ("ecobabble" explained), and a "green bookshelf" of about 200 titles. [**NOTE**: Useful introductory guide. The authors are encyclopedists who have written more than 60 reference books.]

(green encyclopedia)

019

Sourcebook on Sustainable Development. Winnipeg, Manitoba: International Institute for Sustainable Development/Institut International du Développement Durable, May 1992/134p(8x11" spiral)/C$20.00.

The IISD, founded in 1990, is a non-profit corporation established and supported by the governments of Canada and Manitoba. Its mandate is to promote sustainable decision-making in government, business, and the daily lives of individuals, and to encourage formation of partnerships to achieve creative new approaches to the problems we face. The **Sourcebook** has four sections: **1) Organizations**: a selection of 70 international institutions such as the African Centre for Technology Studies, Centre for Our Common Future, UNEP's Earthwatch, Global Resources Information Database, Greenpeace, International Academy of Environment, IUCN, The Japan Committee for the Global Environment, The Third World Network, UNDP, UNU, WRI, WWF; **2) Publications**: listing of about 450 books, reports, and articles in seven categories (information on sustainable development, economy and ecology, law and public policy, social and community issues, environment and natural resources, investment, regional and country reports); **3) Audio-Visual**: international TV series and documentaries, UNDP videos, regional and national productions, cable TV, independent producers; **4) Computer Networks, and Online Databases** (e.g. the APC Networks, NGONET, GeoNet, BITNET, Compu-Serve, Biobusiness, Textline). **Appendices** include a glossary, Association for Progressive Communications affiliated networks, APC conferences, networks in the TCN Electronic Village, and BITNET mailing lists related to sustainable development. [ALSO SEE: **The IISD EarthEnterprise Tool Kit** (#249).]

(sustainable development sourcebook)

020

The Island Press Bibliography of Environmental Literature. The Yale School of Forestry and Environmental Studies (Joseph A. Miller *et al.*). Washington: Island Press, Jan 1993/396p/$48.00.

The 3,084 briefly annotated books (published through 1991) and journals are arranged in two parts of seven sections each. Within each section, items are divided into groups of 10 to 40 references. **Natural Environment**: 1) whole earth systems and conditions (global resources, ecology, etc); 2) air and atmosphere; 3) water and hydrosphere; 4) land; 5) plants and agriculture; 6) forests and forestry; 7) animals. **Human Environment**: 8) society and culture (country studies, population, movements); 9) ethics and arts (philosophy, religion, nature writing and thought); 10) law and government; 11) economy and business; 12) health and medicine; 13) engineering and technology (solid waste, recycling, hazardous waste, water engineering); 14) science and research (information sources, reference books, journals, modeling, statistics). Includes an author/title index, a subject index, and a Quick-Reference index of general topics. [**NOTE**: Treats each item equally, with no indication of outstanding works.]

(environmental literature bibliography)

021

Environmental Problems and Sustainable Futures: Major Literature from WCED to UNCED, Michael Marien (Editor, *Future Survey*), *Futures*, 24:8, Oct 1992, 731-757. (Also in *FUTURESCO* #2-3, July 1992.)

A bibliographic essay on environmental literature reported in *FS* during the five years bounded by the 1987 report of the World Commission on Environment and Development and the 1992 UNCED Earth Summit. Attention is focused on a framework for assembling the literature, based on two major categories: **1) Environmental Problems**: global overviews (planet earth in peril, regions and nations, human population growth), global climate change, endangered resources (overviews, biodiversity, deforestation, protected land areas, freshwater, wetlands, oceans), pollution (air and water, other), waste and hazards (solid and toxic waste, nuclear waste and radiation, natural hazards); **2) Toward a Sustainable Future**: new environmental thinking (ecophilosophy), elements of a sustainable society (general overviews and agendas, ecological economics, green business, ecological agriculture, green technology and transportation, energy, cities and communities, other sectors), actors and environmental movements, government, and information frontiers. In all, 312 items are cited in 255 footnotes, each with *FS* or **FS Annual** reference numbers. Concludes that green infoglut is a neglected problem. [**NOTE**: A precursor to this guide, with many earlier references not listed here.]

(green infoglut underappreciated)

3. Key Regions and Nations

*022

Planet Management. Edited by Michael Williams (School of Geography, Oriel College, U of Oxford). NY: Oxford U Press, April 1993/256p(9x12")/$45.00.

Lushly illustrated book, with maps of polluted or threatened areas (worldwide and within regions), explanatory diagrams, and 247 color photos. Contents provide brief essays in two sections: **1) Managing the World's Environments**: people in the environment (the gathering pace of change, controlling nature, the rise of environmental concern), major environmental issues (atmospheric pollution, global warming, water pollution, waste and waste disposal, marine pollution, deforestation and erosion, ecosystem disruption), prospects for management (alternative energy sources, combating pollution, sustainable agriculture, reforestation, and global agreements). **2) Regions of the World**: *Canada / Arctic* (acid rain and soil erosion, but there is strong local and national environmental awareness), *US* (acid rain, desertification, waste disposal), *Central America / Caribbean* (widespread soil degradation and deforestation, severe air pollution in urban areas, heavy impact of tourism), *South America* (rainforest destruction, soil erosion, urban pollution), *Nordic Countries* (air and water pollution largely from outside the region; problems are being addressed), *British Isles* (pollution, loss of wetlands, soil erosion), *France* (potential hazards from nuclear power, pressures on land use, relatively little acid rain), *Netherlands / Belgium* (persistent pollution that has prompted concerted action), *Spain / Portugal* (deteriorating air quality, water supply problems, soil degradation), *Italy / Greece* (air and water pollution, tourism pressures, water supplies stretched to the limit), *Germany / Austria / Switzerland* (air pollution and acid rain, severe forest die-back, Alps harmed by tourism), *Eastern Europe* (stricken by state-directed industrialization, deforestation, and intensive agriculture; costly cleanup needed), *Former USSR* (widespread ecological destruction; high pollution levels of air, water, and land), *Middle East* (large land areas now waterlogged and made infertile by salt build-up; population explosion, inadequate drinking water, desertification), *Northern Africa* (land degradation, water quantity and quality), *Central Africa* (population explosion, disease, famine, fuelwood shortage), *Southern Africa* (land degradation, population explosion), *Indian Subcontinent* (prone to large natural disasters, population explosion, pollution, land degradation), *China* (a huge and growing population, air and water pollution, soil degradation, desire for rapid economic growth), *Southeast Asia* (population explosion, air and water pollution, deforestation, war damage), *Japan / Korea* (acute pressure on low-lying land, severe air and water pollution), *Oceania* (degraded soils, desertification risk).

[**NOTE**: Not much on planet management, but of great value for the detailed maps of environmental problems and as an introductory overview to different problems in 22 world regions. Clearly demonstrates that all regions have problems, ranging from bad to disastrous. ALSO SEE: **Embracing Earth: New Views of Our Changing Planet** by Payson R. Stevens and Kevin W. Kelley (San Francisco: Chronicle Books, Oct 1992/176p/$39.95), an even larger book (10x14") of satellite photos (some showing the widespread effects of pollution) and portentous commentary ("a warning and an inspiration for the third millennium").]

(global distribution of environmental problems)

023

Ecological Disaster: Cleaning Up the Hidden Legacy of the Soviet Regime. Murray Feshbach (Research Prof of Demography, Georgetown U). A Twentieth Century Fund Report. NY: The Twentieth Century Fund Press (41 E 70th St; 800/275-1447), Feb 1995/155p/$9.95pb.

Co-author of **Ecocide in the USSR** (BasicBooks, 1992; **FS Annual 1993 #11606**) updates his argument that "no other industrial civilization so systematically and so long poisoned its land, air, and people" with much new data—yet it still may "seriously underestimate the problem." ***1) A Nuclear Plague***: many unofficial nuclear dump sites have been found in the last two years; large-scale illegal trading in nuclear material does not yet exist, but "this situation could change"; ***2) Chernobyl***: 8,000 people died from the disaster, and the Ukranian minister of the environment estimates that the health of two-thirds of Ukraine's 52 million people may have been harmed by the accident; such an accident could recur elsewhere in the former Soviet Union (FSU); ***3) Oceans and Seas***: radiation pollution threatens the Arctic Ocean, the Sea of Japan, and the Baltic Sea; ***4) Freshwater Bodies***: the Aral Sea is one of the world's worst ecological catastrophes, the Black Sea is dying, the Caspian Sea and the Sea of Azov are badly polluted, and most of the rivers of the FSU are hazardous to the health of populations along their banks; ***5) Chemical and Biological Warfare Facilities***: the total number of such polluted facilities could be in the hundreds or higher; ***6) Air and Land Pollution***: an average of two major oil and gas pipeline spills occur every day; toxic waste dumping has spread throughout urban areas; some 30% of all food was contaminated in 1990, with 10% actually toxic to the consumer; ***7) Forests***: "the denuding of the Siberian forests has the potential to contribute more to global climate change than the assault on the Amazon basin."

The Russians don't know where to begin the cleanup and they have no money, and helpful Western governments may well do too little, too late. An international mechanism is needed for reaching common decisions and coordinating a comprehensive attack on the problems: ***1) Prepare for the Long Haul***: build ecological and health considerations into long-term aid strategies, launch an immediate program to prevent future Chernobyls (27 aging reactors are of primary concern), prevent further degradation of the Arctic Ocean and Baltic Sea, turn the Red Army into the principal source of manpower for clean-up, rebuild the Russian forestry and national park service; ***2) Treat the Health Crisis as a Major Human Catastrophe***: initiate a multinational immunization campaign, help provide clean water to medical facilities; set up 25 Regional Diagnostic and Health Training Centers; ***3) Build Health and Ecology Alliances***: support an environmental survey to obtain better data; strengthen local institutions and political support.

(Russia's environment/health crisis)

024

Ecocide in the USSR: Health and Nature Under Siege. Murray Feshbach (Research Prof of Demography, Georgetown U) and Alfred Friendly Jr (former Moscow bureau chief, *Newsweek*). Foreword by Lester R. Brown. NY: BasicBooks, April 1992/376p/$24.00.

In many ways, this book is a product of the information revolution that preceded and propelled the disintegration of the Soviet Union. But for *glasnost*, the sources on the disastrous consequences of health and environmental policies on the Soviet people would have been inaccessible. "When historians finally conduct an autopsy on the Soviet

Union and Soviet Communism, they may reach the verdict of death by ecocide." No other great industrial civilization so systematically and so long poisoned its land, air, water, and people. By endangering the health of its people, the productivity of its soil, and the purity of its air and water, the Soviet Union has endangered prospects for economic recovery, even if sweeping systemic reforms take hold in successor states. Some of the problems: 1) to make up for inefficient farming practices, Soviets resorted to massive but untutored use of toxic agricultural chemicals (25 million acres of cropland are still overloaded with DDT); 2) the volume of the Aral Sea, once larger than Lake Huron, shrank by two-thirds as contributory rivers dried up; 3) due to contaminated drinking water, mothers in the Aral region cannot breast-feed their babies without running the risk of poisoning them; 4) 70 million of the 190 million Soviets living in urban areas are in danger of respiratory and other life-shortening diseases from air that carries five or more times the allowed limit of pollutants; 5) almost three-fourths of surface water is polluted; one-fourth is completely untreated; 6) average life expectancy of Soviet males dropped from 66.1 years in 1964-65 to 62.3 years in 1980-81, making only a small recovery to 63.8 years in 1989 [**NOTE**: Between 1991 and 1995, male life expectancy plunged from 63.8 to 57.3 years (*NY Times*, 2 Aug 1995, p.1)]; 7) by 1989, nearly half of the 1.5 billion acres of cultivated land in the USSR was seriously imperiled by salinity, erosion, or waterlogging. Chapters discuss repressed data and self-deception in making the USSR an industrial giant, agriculture and industry, pollution, wasting of energy resources, the military as polluter, the backwardness of Soviet health care, the greening consciousness of the Soviet public, and the price of cleanup (estimated at 390 to 485 billion rubles over 15 years). [**NOTE**: This massive tale of woe, largely based on anecdotal evidence, is very similar to **China on the Edge** (#029), which employs much better data. Both countries illustrate extreme instances of the consequences of ignoring and suppressing negative information. But neither book points to worldwide use of industrial-era economic ideas that continue to ignore environmental matters. ALSO SEE: ***Poisoning Russia***, *U.S. News & World Report* Cover Story, 13 April 1992, 40-51.] **(USSR's environmental disaster)**

025

Troubled Lands: The Legacy of Soviet Environmental Destruction. D. J. Peterson (Graduate Fellow, RAND/UCLA Center for Soviet Studies). A RAND Research Study. Boulder CO: Westview Press, March 1993/276p/$55.00;$19.95pb.

Among the most disturbing revelations resulting from *glasnost* are numerous reports of environmental damage inflicted throughout the former USSR during 70 years of communist rule. Material in this book, based primarily on recent Soviet sources, discusses: **1) Air Pollution**: in the 1980s, pollution in over 100 cities across the former Soviet Union periodically exceeded air quality standards by a factor of ten (emissions have fallen since 1989, largely due to the decline in economic output); 2) **Water Pollution**: large and small rivers have been choked with sewage, petroleum products, phenols, heavy metals, and agricultural runoff; 90% of the Black Sea has been declared dead; stocks of sturgeon in the Volga River have plummeted; **3) Land**: Soviet agricultural policy emphasized massive investment in mechanization and chemicals that proved catastrophic; broad swaths of Russia and Kazakhstan have become infertile desert; the Aral Sea (once the world's fourth largest lake, the size of West Virginia) is fast becoming the Aral desert; **4) Solid and Hazardous Waste**: little provision was made for solid waste; over half of the almost 6,000 official landfills did not meet public health norms in the 1980s; radioactive contamination far greater than that of Chernobyl is widespread.

Chapters also consider environmental policy in the post-Soviet era and Green movements in Soviet successor states (in a July 1991 Russian survey three-quarters of respondents rated environmental conditions in their hometown as "intolerable"). Concludes that successful economic reform, if not mediated by effective environmental protection, "could spell a second onslaught against nature" due to increased energy consumption and higher emissions. In the mid-1980s, the Soviet economy accounted for 10% of the world's CFCs and 18% of emissions. "It is in everyone's interest that these emissions not be sustained." [**NOTE**: Extensively documented and clearly written. Better than the much-touted **Ecocide in the USSR** (above).] **(environmental disaster in former USSR)**

026

The Geography of Survival: Ecology in the Post-Soviet Era. Ze'ev Wolfson (Hebrew U of Jerusalem). Armonk NY: M. E. Sharpe, June 1994/160p/$35.50;$17.95.

Author of the widely circulated 1979 *samizdat* book, **The Destruction of Nature in the Soviet Union,** written under the pseudonym of Boris Komarov, points to ecological disaster areas in the former Soviet Union and elsewhere, and how such deteriorated areas can have an effect on "normal" ones—the *eco-domino effect*. The eco-domino process acquires a geopolitical significance since it has a long-distance impact. The belt of forests and taiga in Siberia has heretofore stabilized the highly vulnerable northern ecosystems. With this mighty forest belt shrinking from massive felling, Siberia represents a danger zone for the entire continent. On the southern front, desertification is advancing. Perestroika in the USSR not only failed to give rise to an "eco-perestroika," but was accompanied by a marked increase in the pace of environmental destruction. "Russia is in the grips of an environmental crisis that involves huge areas and has led to physical and psychological deterioration of the people inhabiting those areas... Today's Russia is a disintegrating society which is not only powerless to curb environmental disintegration but, on the contrary, effectively contributes to the headlong advance of the destruction." The disintegration of a society's political and economic structures, along with its physical environment, is unprecedented. No country can create an "island" of stability and safety from the eco-domino effect. From now on the geography of survival will involve not only socioeconomic factors inside each country but also neighborly goodwill, requiring a policy of co-operation and aid to weaker neighbors. The idea of joining a "zone of ecological interest" may become an attractive option. [**NOTE**: A Russian Rachel Carson? For updates on this unfolding disaster, Wolfson/Komarov edits *Environmental Policy Review (The CIS and Eastern Europe)*.] **(Russia's environmental disaster growing)**

027

Environmental Problems in Eastern Europe. Edited by Francis W. Carter (London U) and David Turnock (Leicester U). London & NY: Routledge, July 1993/249p/$55.00.

Essays on environmental conditions in Albania, Bulgaria, Czechoslovakia, Hungary, Poland, Romania, and the former Yugoslavia (by Barbara Jancar-Webster). The editors summarize such topics as: ***1) Air Pollution***: a deadly pall of sulphurous smoke hangs over Europe, mainly due to heavy reliance on lignite (brown coal); its noxious fumes have damaged many building and forests and have caused serious health problems; ***2) Water Pollution***: the damage from water mismanagement is "far more awesome than anything in Western Europe"; many rivers and lakes suffer eutrophication, the Mediterranean basin is endangered, and the Black Sea is on the verge of catastrophe; ***3) Contamination of Land and Forests***: soil degradation from overfarming and mining is often severe; half the forests in Czechoslovakia are said to be nearly extinct; ***4) Noise Pollution***: there is a legacy of neglect, compounded by continuing use of antiquated equipment, poor abatement controls, and inferior building standards; ***5) Landscape***: evidence of pollution damage to scenery in all its forms can be seen throughout East-Central Europe.

Environmental Action in Eastern Europe: Responses to Crisis, edited by Barbara Jancar-Webster of SUNY-Brockport (M.E. Sharpe, Oct 1993/238p/$49.95; $19.95pb), offers essays on environmental problems and movements in Poland, Hungary, the Czech Republic, Slovakia, the former Soviet Union, the Baltic states, and Slovenia. Jancar-Webster concludes that, despite a positive attitude in Eastern Europe toward Western pollution control efforts, and passage of new legislation, there is a strong sense of inadequacy and uncertainty, and doubts that the former communist countries can realize the Western model. "All are skeptical of any immediate possibility of improvements in environmental management. All stress the enormity of the problems."

(environmental disasters in Eastern Europe)

028

China's Environmental Crisis: An Inquiry into the Limits of National Development. Vaclav Smil (Prof of Geography, U of Manitoba). Armonk NY: M. E. Sharpe/ East Gate, April 1993/257p/$42.50;$16.95pb.

Author of **Energy, Food, Environment** (Oxford, 1987; **FS Annual 1988-89** #9037) stresses that China's future "is obviously circumscribed both by its inherent environmental limitations and by the environmental consequences of its bold, belated quest for modernity." Chapters discuss China's population and its one-child policy (during the 1990s, China will add at least 125 million people, the equivalent of today's Japan or half of the US), water supplies, deforestation and false afforestation claims from 1950 to 1990, quality-of-life measures, food supply, extensive rural energy shortages, air pollution (from largely uncleaned coal, burned with little or no pollution controls), increasing frequency of respiratory morbidity and mortality in all large urban areas, inefficiency of energy use, China's large contribution to global CO_2 generation (by 2020, China could be the clear number one emitter, producing about one-fifth of global CO_2 from fossil fuels, compared to 17% for the US and 13% for Russia), unsustainable farming practices and a steadily diminishing area of arable land (intensified fertilization is necessary: "no other choice can produce more food from less land for more people"), and expected farmland losses from 1989 to 2000 (ranging from 3.5-6 million hectares, largely due to construction and industrial activities).

"Assessed against this background, China's modernization plans guarantee further extensive environmental deterioration, destruction, and pollution." There are no solutions within China's economic, technical, and manpower reach that could halt and reverse these degradative trends in the next two decades. "For the world, China's environmental debacle will be yet another intractable destabilizing factor." The best practical possibilities for the 1990s (price reforms, technical and managerial innovations, better law enforcement) "are not even for stabilizing current losses, declines, and emissions, but merely for reducing the rates of their growth." *Wu ji bi fan*—when things are at their worst, they begin to mend—goes a Chinese saying, but in this unfolding drama, nothing guarantees that crisis will end in catharsis.

(environmental crisis worsening in China)

029

China on the Edge: The Crisis of Ecology and Development. He Bochuan (Philosophy Dept, Sun Yat-sen U, Guangdong). San Francisco: China Books and Periodicals Inc (2929 24th St), Dec 1991/208p/$29.95;$16.95pb.

English translation of a widely-circulated book in the People's Republic of China, with over 430,000 copies in print since first publication in 1988 (printing was abruptly halted by decree from above after the Tienanmen Square incident). Chapters discuss: 1) overpopulation as the heart of China's environmental and development crisis (a problem of numbers, distribution, and education); 2) the ecological crisis (one-seventh of the land has become desert, deterioration of forests due to bad management and blatant looting, worsening soil loss now at 5 billion tons per year, decline in cultivable land from 0.44 acres per capita in 1955 to 0.30 acres in 1970 and 0.21 acres in 1990) 3) rapidly deteriorating water quality, and looming water shortages in the cities and countryside (88% of the 289 major cities have periodic problems with water supplies); 4) pollution damage to China's marine fishing grounds "almost to the point of annihilation" (most of China's rivers are polluted, and the catch of freshwater fish fell by half between the 1950s and 1970s, and continues to decline); 5) air pollution as a major cause of disease and decreased productivity (environmental protection regulations are vague and lack the force of law; "indeed, there was general silence about environmental problems in China until the mid-1980s"); 6) the questionable economic goal of quadrupling output by 2000 (20 flaws are pointed out, such as capital shortage and the need for twice as much energy); 7) energy and transportation as two weak links (China faces a severe energy shortage, and suffers from low efficiency in use, unreasonably low energy prices, and poor transport); 8) problems in industry and agriculture (too much priority to heavy industry, lack of innovation, the low educational level of most peasants, inadequate efforts in agricultural extension); 9) poor management of imported technology; 10) low investment in education, with enormous waste of resources and talent. Concludes that "there is nothing to indicate that things will improve in the near future... Most people are only concerned with immediate material benefits and very few look at long-term and far-ranging effects... because they are afraid of failure, they prefer to avoid attempts to change. And they have no concept of the notion of a crime against nature." [**NOTE**: Harsh criticism combined with a remarkable orchestration of charts and data. In the Foreword, Lester Ross suggests that the author paints an "unduly bleak" picture, tending in places to project past trends into the future. But in his Preface to the English-language edition, He Bochaun states that China "has no effective or feasible measures to reverse the trend in the foreseeable future... China has gone past the point of no return."] **(China's environmental disaster)**

030

Fueling One Billion: An Insider's Story of Chinese Energy Policy Development. Yingzhong Lu (Director, Institute for Techno-Economics, Tsinghua U, Beijing). Washington: Washington Institute Press (1015 18th St NW, #300), Oct 1993/266p/$29.95;$19.95pb.

China's economy grew 10 times from 1953 to 1989, and its energy consumption has increased 18 times. Severe energy shortages before 1980 encouraged a scientific energy policy during the past decade of economic reform, and the energy situation has been greatly improved. China seeks to quadruple production in the 1980-2000 period, using only its own resources, so ensuring future energy supply remains a daunting task. Environmental concerns are also growing, and an adequate policy of reducing energy intensity, improving the energy structure, and mitigating CO_2 emissions will be of great importance for China and the world. Chapters discuss how the scientific approach to energy policymaking was born, the four paradoxical "coexistences" of China's energy-economy (abundance and shortage, vastness and diversity, modernization and backwardness, central planning and market economy), energy conservation potential (by 2000, the level of energy intensity for major industrial products should reach the 1980s level of the advanced countries), "walking on two legs" (depending on both conservation and production), the great progress in rural energy policy since 1980, energy-environmental issues, forecasts of energy demand through 2050 (population scenarios range from 1.0 billion to 1.8 billion; GNP per capita ranges from $6,700 to $14,500 in 1984 US dollars), the medium/long-term energy-economic model to forecast supply and demand for 2000 and beyond, the dissemination of 100 million fuel-saving woodstoves in Chinese rural households (1983-1988), trends in electricity supply, rural electrification (about 30% of rural households are still not served), and a comparison of Chinese and Indian energy consumption and supply. [**NOTE**: Extensive detail, including about 300 tables and figures. Will be of great interest to anyone concerned with China's long-term economic and environmental impact. This "insider's" account, by an expert who supervised much of the planning described here (according to Alvin M. Weinberg in a brief Foreword), suggests that China is coping satisfactorily with its energy and environmental problems. But an "outsider's" account, **China on the Edge:** (above) also offers abundant data and sharply questions China's ambitious economic goals. Which of these accounts should we believe?]

(China's energy to 2050)

031

All the Coal in China, Nicholas Lenssen, *World Watch*, 6:2, March-April 1993, 22-29.

Carbon emissions in China have increased 65% in the past decade, largely due to a sharp rise in burning coal. This boosted the country's share of global carbon emissions to 11%. Although China still emits only half as much carbon as the US, and only one-ninth as much per capita, its fast-growing economy is driving CO_2 emissions upward. According to a UN panel studying climate change, "China is expected to emit more CO_2 by 2025 than the current combined total of the US, Japan, and Canada." Redirecting China's energy economy may be as important to the global atmosphere as similar changes in the US and Europe. But China's leaders don't share this view, arguing that the industrial West has caused the accumulation now hanging over us. They may discover, however, that a concerted move toward energy efficiency could reduce exorbitant costs and lead to greater employment and health. Only one-third of all fuel burned in China produces useful energy—far below the 50-60% efficiency of the US and Japan. Building one $7.5 million compact fluorescent light bulb factory, for example, would eliminate the need for $4.9 billion worth of coal-fired power plants and transmission facilities. **(China as leading polluter in 2025?)**

032

China's Demographic Dilemmas. H. Yuan Tien (Professor Emeritus of Sociology, Ohio State U) et al., *Population Bulletin*, 47:1, June 1992/44p/$8.00 single copy.

Author of **China's Strategic Demographic Initiative** (Praeger, 1991) and colleagues in China discuss China's demographic transition from high to low fertility since 1950, one of the fastest—if not the fastest—in history. Chinese fertility fell nearly 50% between 1970 and 1979, from 34 to less than 18 births per 1,000 population. But no further large drops have occurred, and the 1991 birth rate estimate was 19.7 births per 1000. The ideal of only one child proved controversial, and its implementation provoked strong resistance in rural areas. The population ceiling of 1.2 billion originally set for the year 2000 thus remains elusive. The official target has been raised to 1.294 billion, but even this will be difficult to achieve. If fertility cannot be further reduced from the 1990 rate, "the Chinese population will rise above 1.3 billion by 2000, top 2 billion in 2050, and continue to grow unabated for decades thereafter." It would not be prudent to venture an opinion on when the Chinese population will stop growing. The demographic transition in many European nations with below-replacement fertility evolved under a much longer time table, and under different social, economic, and political conditions. Socioeconomic conditions in China's vast countryside do not resemble the situation in developed countries now practicing minimal reproduction. **(2 billion Chinese by 2050?)**

033

The "Second India" Revisited: Population, Poverty, and Environmental Stress over Two Decades. Robert Repetto (Vice President and Senior Economist, WRI). Washington: World Resources Institute, Aug 1994/95p (8x11")/$14.95pb.

In the late 1960s, when India's population stood at about 500 million, the forward-looking *Second India Study* asked how India could cope with demands for more food, land, water, jobs, schooling, and energy, as its population doubled by the year 2000. With the creation of a "Second India" nearly completed, and projections of a Third and a Fourth India before population stabilizes, a research team of Indian experts along with Repetto compared the study's forecasts with the actual effects of a rapid population doubling. By and large, developments in the past 20 years have fallen within the range of scenarios the authors thought plausible, although aggregates mask an enormous diversity of experience within India's regions and populations. There were external shocks that the study could not clearly foresee (oil prices, the collapse of the Soviet Union, and China's switch to market-oriented growth). The study anticipated some problems that have not materialized (it greatly overestimated the resource requirements of economic development by discounting the impact of new technologies). And problems have materialized that were not foreseen (despite extensive agricultural research, India has been unable to demonstrate continuing yield improvements for major crops; the Green Revolution didn't fail, but neither has it produced a quantum increase in foodgrains). "Overall, the record of the *Second India Study* teaches both the value of serious efforts to look ahead at the implications of massive population expansion, and the need for a certain humility in making predictions." However, the study gave little attention to environmental problems, which "will undoubtedly be more prominent in the coming decades, as India's economy and population expand." **(forecasts for India revisited)**

034
The Fragile Tropics of Latin America: Sustainable Management of Changing Environments. Edited by Toshie Nishizawa (Prof of Geography, Tokyo Seitoku U) and Juha I. Uitto (UNU). Tokyo: United Nations U Press, March 1995/325p/$35.00.

The Amazon region is an important part of the global environment. The Amazon River alone accounts for about one-fifth of all fresh water discharged into the world's oceans, and about one-third of the world's tropical rain forests are located in Amazonia, mostly in Brazil. Essays discuss managing rich and poor ecosystems, the Brazilian Amazon, the Peruvian Amazon, and the semi-arid North-East of Brazil. Gilberto C. Gallopin (Cali, Columbia) and Manuel Winograd (Grupo de Análisis de Sistemas Ecológicos, Buenos Aires) provides two scenarios for the next 40 years. ***1) Reference Scenario***: moderate economic growth, unchanged pattern of development, expanding influence of transnational corporations, diminished state regulation, additional pressures on the environment, weak environmental policies with little enforcement, growing preponderance of the export sector, general increase in intensity of land use and inputs, increasing fuelwood deficit in dry areas, growing scarcity of agricultural land; ***2) Sustainable Scenario***: implementation of national and regional environmental policies and socio-economic reforms, technological pluralism (complementary use of traditional, modern, and high tech), protecting/managing large-scale ecological functions, productive pluralism (co-existence of different types of agriculture), innovations to create a sustainable peasant agriculture. The latter scenario is desirable and feasible, but more recent trends in the region favor the former.

(sustainable management in Amazon region)

035
The Green Cathedral: Sustainable Development of Amazonia. Juan de Onis (IPHAE, Rondonia, Brazil). A Twentieth Century Fund Book. NY: Oxford U Press, July 1992/280p/$24.95.

Former Rio bureau chief for *The New York Times* writes that Amazonia has become a primary scenario for humanity's new concerns over protection of the environment. If the political and social forces at work in Amazonia produce an ecological debacle, the whole world will pay a price. The environment is threatened because a new breed of native settlers and exploiters entered this habitat without a thought to the sustainability of their behavior. This has happened rapidly, with little time to adapt: more than half of the 16 million people now living in Amazonia have migrated there in the past three decades, attracted by cheap land made accessible by new roads, airports, and telecommunications. The settlement that has taken place cannot be reversed. Practical solutions to provide a livelihood for these settlers: 1) differences in soils, rainfall, topography, and temperature require differentiated strategies for ecosystem management in each specific area; 2) land-use zoning is essential to encourage productive investments; 3) militant vigilance by international and local NGOs will remain a major task; 4) donor governments and development banks need to fund small projects with the participation of the people directly affected in design and implementation; 5) a more secure, stable existence for remaining indigenous Indian communities is an important part of any strategy; 6) the lack of social fabric and community pride are among the causes that make settlers so wasteful of their natural environment; 7) the greatest single challenge is to help small farmers develop technologies for the humid tropics that will make them stable landholders. Developing this new frontier will be a slow process, like constructing a great cathedral. If ecology is a new religion, as some say, what better place to build a "Green cathedral" than Amazonia, the greenest part of planet Earth?

(sustainable development in Amazonia)

036
Clearing the Air: Environmental Reform in Mexico, Stephen P. Mumme (Prof of Political Science, Colorado State U), *Environment*, 33:10, Dec 1991, 6ff.

Mexico's rapid economic growth since WWII occurred with little concern for the environment. The unregulated proliferation of industries and human settlements have exacted a severe toll on the quantity and quality of Mexico's water, e.g. 77% of river basins are so badly polluted as to jeopardize public health. Beaches at Acapulco and Puerto Vallarta are fouled with raw sewage. Forests have been harvested at rates far exceeding sustainability. Mexico City, with nearly 20 million inhabitants in 1990 (one-quarter of Mexico's population), has major air pollution and garbage problems. President Salinas is the first of a new generation of leaders compelled to reckon with environmentalism as a political force. The government has begun to address various problems, but there is mismanagement, corruption, and poor enforcement of ordinances. Although the ruling Institutional Revolutionary Party (PRI) is not threatened, a number of environmental organizations emerged in the 1980s, such as the Mexican Ecology Movement, the Pacto de Ecologistas, and Alianza Ecologista. **(Mexico faces environmental issues)**

037
After the Greening: The Browning of Australia. Mary E. White. Kenthurst NSW: Kangaroo Press, Aug 1994/288p(8x11")/$49.95 (dist in US by Seven Hills, 49 Central Ave, Cincinnati OH 45202; 513/381-3881).

In the prehistoric past, Australia was part of the southern supercontinent Gondwana, and was attached to Antarctica. The two lands shared a common flora and fauna, and both lands were well-watered and supported forests of conifers. White, a paleobotanist, describes the long-term evolution of the Australian continent: how it broke away from Gondwana, the changing climate and vegetation, the drying of the continent, and the "devastating" impact of European settlement in the past 200 years. "Management of the continent is steadily turning more of it to unproductive desert and making the rest more fire-prone . . . the environmental degradation we have caused has reached crisis proportions." Feral herbivores (rabbits, goats, pigs, horses, donkeys, buffalo, camels) are literally eating Australia to death, with attendant damage to soils and impact on native fauna. All these feral animals are desert-makers, adding to sheep and cattle doing the same thing on marginal grazing lands. Native wildlife is under siege. "Much of the change we have to accept because we have no option and it is too late." In some places we still have a chance to hold back the tide, and even to win back a little lost ground. [**NOTE**: Includes many maps and sumptuous color photos of a strange land becoming stranger.] **(degradation of Australia)**

038
Global Change and Canadians. Royal Society of Canada, Canadian Global Change Program. Ottawa K1G 5J4: Royal Society of Canada (Box 9734), 1993/57p (8x11")/Can$5.00.

Global change is important to Canada because it is a large country still highly dependent on natural resources.

With less than 0.5% of the world's population, Canada has 6.7% of the world's land surface area, 3% of its cropland, 9% of its fresh water, 9% of its forests, and 13% of its wilderness. Forests cover 45% of Canada's land surface, and almost half of this wood is considered commercial. Canada's geography and location make it particularly susceptible to certain types of global change, in that ozone depletion and global warming are predicted to have a greater impact in the higher northern latitudes. Chapters discuss: **1) Causes of Global Change**: population growth, energy and resource consumption, global economic forces that lead to over-exploitation, military expenditures, technology, culture; **2) Consequences of Global Change**: likely impacts on Canada and the world from climate change, ozone depletion, air pollution, soil degradation, biodiversity loss; **3) How Global Change Will Affect Society**: requirements for adaptation and for addressing the root causes of global change (concerning population, food, shelter, production and consumption, and international action); **4) Canadian Responses to Global Change**: decision-making despite uncertainties (the most prudent course is the precautionary principle), actions by individuals, actions by corporations (realizing that environmental considerations have become a fundamental business concern), possible actions by Canadian governments (a carbon tax, an increase in fuel efficiency standards, educational programs, foreign aid directed at the causes of global change instead of symptoms). Concludes that "we have more than enough information already to demonstrate the need for action, and the initial solutions are both obvious and simply applied. What is still lacking is the will to act." [**NOTE**: More reflection should be devoted to why the "will to act" is lacking. Will this clear, concise, and authoritative message reach its intended audience—and make any difference?]

(Canada and global environmental change)

039

America's Renewable Resources: Historical Trends and Current Challenges. Edited by Kenneth D. Frederick and Roger A. Sedjo (both RFF). Washington: Resources for the Future, Oct 1991/296p(8x10")/$34.95;$19.95pb.

Essays on how the quantity and quality of the US resource base has changed over the last century: **1) Water**: future demand will increase faster than supply in most areas, and costs will rise (if costs are borne by users, there will be incentives to conserve); **2) Forests**: the nadir of US forests was 1920; by many criteria, present forests are in their best condition since the beginning of this century; long-term effects of global warming on forests may be negative; **3) Rangelands**: environmentalists and private ranchers are "miles apart" in judging range condition and trends, and many studies are flawed; rangeland management does need improvement; **4) Cropland and Soils**: facing the challenge of developing new technologies and management practices that maintain economic competitiveness while protecting environmental values; **5) Wildlife**: a drastic decline in the numbers and variety of North American wildlife continued until 1900; since then, growing concern began to be effective and populations of many bird and mammal species have staged remarkable recoveries; looking ahead, expectations for wildlife populations are mixed, depending on who makes the projection for what species; **6) The Growing Role of Outdoor Recreation**: a category of uses that cuts across several different resources; ample capacity exists in the US to meet future demand. Concludes that many earlier concerns about the demise of renewable resources have faded (e.g. threats of an impending timber famine), and US economic development has become much less dependent on consuming renewable resources. Less relative emphasis is now placed on the direct economic benefits of putting resources to productive uses, but greater weight is being given to protecting resources for their contribution to the quality of life, and there is growing concern about the health of ecosystems due to pollution. The challenge is to balance competitive and changing demands on resources, while ensuring that they are managed on a long-term sustainable basis. "Continued growth of population and of per capita consumption of the world's resources makes the concept of sustainability increasingly attractive as a guide for future development policies." [**NOTE**: Useful historical perspective, with restrained and balanced commentary on challenges; many may find the analysis overly restrained.]

(US renewable resources: long-term view)

040

California's Threatened Environment: Restoring the Dream. Edited by Tim Palmer (Planning and Conservation League Foundation, Sacramento). Washington: Island Press, Feb 1993/305p/$29.95;$15.95pb.

The coming years will bring unprecedented change in California, leaving environmental impoverishment in its wake. California's population grew from 24 million in 1980 to 30 million in 1990, and is projected to be 36 million in 2000 and 39 million by 2005. Chapters consider air pollution, energy use, the costs of automobile dependency, water supply and quality, the troubled coastline, land use and growth management, forests, solid and toxic wastes, pesticides, parks and recreation, wildlife, and protecting native landscapes. Proposed actions: **1) Population**: assist the Federal government in enforcing existing immigration laws, eliminate incentives that encourage reproduction; **2) Urban Problems**: develop public transportation, impose a fee on fossil fuels; **3) Resources**: employ a guiding principle of sustainability, strict water conservation for agriculture; **4) Waste**: favor reuse over creation and disposal. **(sustainability for California)**

4. Population Growth

041

The Future of World Population. Wolfgang Lutz (Director, Population Project, IIASA, Austria). *Population Bulletin,* 49:1, June 1994/47p/$7.00 single copy.

Brief version of a forthcoming report, **The Future Population of the World: What Can We Assume Today?**, based on a late 1992 meeting of demographers at the International Institute of Applied Systems Analysis. Current trends are discussed in the three components of population change—fertility, mortality, and migration—and scenarios are presented on these components for 12 world regions. *Three major certainties* are apparent: world population will continue to grow (increasing at least 50% and perhaps even doubling by 2030), developing countries will account for a greater share of world population (rising from 78% in 1990 to 85-87% by 2030), and the average age in all world regions will increase under all scenarios (the share of people age 60 and over will rise from today's 9% to 13-17% in 2030). Some other projections: 1) world population under the IIASA central scenario will grow to 8.96 billion in 2025 [contrast with lower PRB projection of 8.38 billion, below] and 12.56 billion in 2100; 2) the high migration scenarios through 2030 assume annual net migration gains of 2 million in North America, 1 million in Western Europe, and 350,000 in the Japan/Australia/New Zealand region; 3) by 2030, under the central scenario,

Western Europe will increase in population by 10%, North America by 35%, China by 50%, South America by 75%, and Sub-Saharan Africa by 298%; 4) AIDS may not reduce population growth significantly if fertility remains high; in sub-Saharan Africa, a three-year decline in life expectancy per decade is the worst case and a three-year-per decade increase is the best case; 5) life expectancy in Japan, Europe, and North America will increase one year per decade through 2030 (low scenario), two years per decade (central scenario), or three years per decade (high scenario). [**NOTE**: Excellent array of scenarios and discussion of uncertainties and assumptions. The upper limit of mortality improvement reflects the cautious view of demographers, not of biologists on the frontiers of aging research.] **(world population at 12.6 billion in 2100)**

042

1995 World Population Data Sheet. Carl Haub and Machiko Yanagishita (PRB). Washington: Population Reference Bureau, May 1995/one page (20x28")/$4.00.

Demographic data and estimates in 14 categories for the countries and regions of the world. ***1) World Population:*** 5.70 billion in mid-1995, with a projection of 7.02 billion in 2010 (up 23.2%) and 8.31 billion in 2025 (up 45.8%); ***2) Regional Population Growth (1995, 2010, 2025):*** Africa from 720 million to 1.07 and 1.51 billion; Latin America/Caribbean from 481 million to 601 and 706 million; Asia from 3.5 billion to 4.2 and 4.9 billion; Europe from 729 million to 743 and 743 million; ***3) Growth in Selected Countries (1995, 2010, 2025):*** US from 263 million to 300 and 338 million; Canada from 30 million to 34 and 37 million; Mexico from 94 million to 118 and 137 million; Brazil from 158 million to 194 and 225 million; China from 1.22 billion to 1.39 and 1.52 billion; India from 931 million to 1.18 and 1.38 billion; Pakistan from 130 million to 188 and 252 million; Russia from 148 million to 150 and 153 million; Nigeria from 101 million to 162 and 246 million[!!]; ***4) World Population Doubling Time:*** at the current rate of growth, population will double in 45 years (the doubling rate was 43 years in 1994, 41 years in 1992, and 39 years in 1990); ***5) Regional Population Doubling Time:*** 24 years in Africa, 29 years in Western Asia, 30 years in Central America, 38 years in South America, 66 years in East Asia, 105 years in North America, and 741 years in Western Europe; ***6) Annual Change:*** some 88.4 million people were added to world population in 1994-95, with the less-developed world accounting for 97.9% of the increase. [**NOTE**: Continues to be an essential reference on the basic "facts of the future," prepared in a handy format and offered at a modest price. Don't think about the future without it. AIDS is not discusses, but seems to be making scarcely a dent in Africa's population growth, still expected to double in 24 years!] **(world population up 24% by 2010)**

043

The State of World Population 1994: Choices and Responsibilities. United Nations Population Fund (Nafis Sadik, Executive Director). NY: UNFPA (22 E 42nd St), Aug 1994/66p(8x11").

World population in mid-1994 is 5.66 billion. It will be six billion by 1998. Although overall fertility rates have fallen, annual increases are at 94 million a year, the highest in history. After 1997, annual additions will start to decline, falling to 85 million by 2020. Some other key points: ***1) Alternative Projections***: the latest UN medium projections indicate a population of 8.5 billion by 2025 and 10 billion by 2050; the high projection is 12.5 billion by 2050; the low projection is 7.8 billion; ***2) Migration***: urban population will be 50% of the world total by 2000; by that time, 300 cities in developing countries will have populations over one million, compared to 125 in 1993; ***3) AIDS***: as of early 1992, some 2.7 million individuals worldwide had developed AIDS, 90% of whom had already died; by 2000, 1.8 million people will die of AIDS each year; ***4) Poverty***: the numbers of the poor are rising: some 1.1 billion people (62% in South Asia) are now living on about $1 a day; ***5) Gender Security***: a guarantee of gender security (freedom from discrimination, physical and sexual abuse, and nutritional deficiencies; access to reproductive health information) is needed to preserve women's rights; it is a realistic and necessary development goal enabling attainment of other goals; ***6) Funding Priorities***: based on lessons learned in the most successful countries, donors must re-examine their funding priorities and direct them toward goals of human development, including equal access to education and health services. **(world population and gender security)**

044

World Population Projections, 1994-1995 Edition: Estimates and Projections with Related Demographic Statistics. Eduard Bos *et al.* (World Bank). Published for the World Bank. Baltimore: Johns Hopkins U Press, Oct 1994/521p/$25.95pb.

The sixth edition in a biennial series that projects population by sex and age group through 2150 for more than 200 countries and territories, and for various geographic regions, fertility status groups, and income groups. *Overall world population* passed the 5 billion point in 1987, and is projected to pass 6 billion in 1999, 7 billion in 2011, 8 billion in 2023, 9 billion in 2039, 10 billion in 2060, and 11 billion in 2103. Under the assumption of *standard fertility decline*, world population will be 9.6 billion in 2050 and 11.0 billion in 2100. Assuming *slow fertility decline*, it will be 10.1 billion in 2050 and 12.2 billion in 2100. Assuming *rapid fertility, decline*, it will be 8.6 billion in 2050 and 9.4 billion in 2100. *Total US population* will rise from 250 million in 1990 to 276 million in 2000, 323 million in 2025, 335 million in 2050, and 344 million in 2100. *Total Canadian population* will rise from 26.5 million in 1990 to 29.5 million in 2000, 34.2 million in 2025, 34.4 million in 2050, and 34.5 million in 2100. [**NOTE**: John Kettle's **Population Projections for Canada 1991-2031** (*The Future-Letter*, 208 Winona Ave, Oshawa Ont L1G 3H5, June 1993/$95.00), offers nine different forecasts based on fertility and immigration. The low fertility/low immigration forecast for 2026 is 31.7 million; the high fertility/high immigration forecast for 2026 is 45.1 million. Both the World Bank and Kettle projections are on diskette. In **How Many Americans?** (Sierra Club, 1994; #051), Bouvier and Grant assume greater migration and project a US population of 350 million in 2025, 397 million in 2050, and 492 million by 2100—a whopping 148 million more than the World Bank projection!]

(World Bank population projections)

045

Population Debate: The Premises Are Changed, Barbara Crossette, *The New York Times*, Wed, 14 Sept 1994, A3. (Also see accompanying article on the 20-year U.N. Program of Action, A2.)

The International Conference on Population and Development concluded its deliberations on September 13 with a plan of action adopted by more than 150 nations, seeking to stabilize world population at about 7.27 billion by 2015. Many experts agreed that the Conference reflected a change in the premises for dealing with the crisis of too many people on earth. The phrase "population control" is now gone from official vocabulary, and with it several decades of policies that relied on goals that often went unmet because women in poor countries had little say in family planning. The Conference accepted that many women and men, if given adequate information, general health care, and access to a wide choice of contraceptive methods, will very often make having fewer children a priority. Among the societal changes influencing the tone of the Cairo conference were an expansion of the middle class and a rise in the number of women in the paid labor force in many developing nations, and the growing propensity in Catholic and Muslim countries to disregard religious rules. The Vatican endorsed the principles of the 113-page declaration, as well as chapters on the family as the basic unit of society, gender equality, and empowerment of women (the Vatican claimed a victory in shaping the language of the text so that it did not enshrine any universal right to abortion). Ellen Chesler (Director, International Women's Health Coalition, NYC) viewed the Conference "as ending 2,000 years of ecclesiastical authority and jurisdiction over marriage and women's lives. Medicine and science, not religion and belief, will govern family planning." **(population stabilization by 2015)**

046

Population and Environment: Rethinking the Debate. Edited by Lourdes Arizpe (National Autonomous U of Mexico), M. Priscilla Stone (Program on Environmental Change, Social Science Research Council), and David C. Major (PEC/SSRC). Boulder CO: Westview Press, Sept 1994/352p/$29.85pb.

The debate about the global relationship between population and the environment has progressed relatively little in the past 20 years. Much of it remains polarized between two extreme positions, each based on macro data and a partial scientific understanding of the problem. One view, represented by Julian Simon, holds that human creativity will improve technology to expand the planet's carrying capacity; the other, represented by Paul and Anne Ehrlich, holds that population is the principal threat to the environment because of the planet's finite resources. The aim of this volume is to assist in rethinking the debate from the perspectives of the micro data of the social sciences, along the lines of new and more complex propositions, asking what social/political/economic factors shape the use and management of natural resources. Essays discuss world population trends (by Wolfgang Lutz), gender and environment, population and deforestation, the social context of desertification, and urbanization and the environment. "The most important lesson of the work reported here is that the terms and models of the debate on population and the environment have been oversimplified and decontextualized." The starting point for research should be to determine use of local resources by local people—the "user's perspective." To achieve sustainable livelihoods and a sustainable pattern of resource use and population growth, a much deeper understanding of the interactions of population and resource consumption as mediated by technology, culture, and values must be developed. **(population/environment reconsidered)**

047

Full House: Reassessing the Earth's Population Carrying Capacity. Lester R. Brown and Hal Kane (both Worldwatch Inst). Worldwatch Environmental Alert Series. NY: W.W. Norton, Aug 1994/261p/$21.95;$8.95pb.

The world is entering a new era in which it is far more difficult to expand food output. Fisheries and rangelands are being pressed to the limits of their carrying capacities. Future growth in food demand can be satisfied only by expanding output from croplands, but there are problems in doing so. The authors assume that there will be no dramatic technological breakthroughs leading to quantum jumps in world food output. They project grain production and consumption in the four largest food-producing nations. By 2030, the US will have a surplus of 82 million tons, but the other three countries will have deficits (-25 million tons in the former Soviet Union, -45 million tons in India, and -216 million tons in China). Nine other populous countries will also have grain deficits by 2030. In sum, rapid population growth, which once slowed the rise in living standards, is now lowering them for large segments of humanity. "Seldom has the world faced an unfolding emergency whose dimensions are as clear as the growing imbalance between food and people." Given the limits to the carrying capacity of each country's land and water resources, "every national government now needs a carefully articulated and adequately supported population policy." World leaders must speak out, urge small families, and ensure universal access to family planning. At the national level, more and more governments will have to choose between more people and better diets. And, as the food/population balance becomes more precarious, concern about potential climate instability will escalate. We need a new vision of an environmentally sustainable economic system to guide us. In April 1994, the UN Population Fund argued for stabilizing world population at 7.8 billion by 2050. "There is no reasonable alternative to such an effort."

(growing food/population imbalance)

048

All Our People: Population Policy with a Human Face. Klaus M. Leisinger (Prof of Sociology, U of Basel) and Karin Schmitt (Ciba-Geigy Foundation for Cooperation with Developing Countries). Foreword by Robert S. McNamara. Washington: Island Press, Sept 1994/267p/ $45.00;$24.95pb.

World population passed the 5.6 billion mark in 1994, up from 1 billion in 1800 and 3 billion in 1960. About 90 million people were added to the human race in 1993: more than 250,000 per day and 10,000 per hour. "While you are reading this sentence, world population will grow by about 20 people." Of the 3.2 billion people to be added to world population between 1990 and 2025, at least 3 billion (94%) will live in developing countries that are in the poorest position to cope with increased numbers. Chapters discuss population growth in industrial and developing countries, urbanization, religious and sociocultural norms, determinants of high fertility, the negative effects of rapid population growth, the demographic transition theory assuming that modernization will lead to lower birth rates (the historical model is too crude to predict the future of developing countries), moral issues in population policy, evaluating programs for controlling population growth, the role of birth control techniques, and population policy with a human face (the two most important elements are satisfaction of basic needs, and equal rights and opportunities for women). Although prime responsibility for creating basic conditions for human development lies in the developing nations, action by all members of the international community (particularly the rich) is indispensable. Insofar as sustainable development for the whole planet, at least as much will be gained from a life-style change by the rich minority as from reducing birth rates among the poor majority. **(population policy)**

049

Critical Masses: The Global Population Challenge. George D. Moffett (*The Christian Science Monitor*). NY: Viking, Aug 1994/353p/$26.95.

Former Middle East bureau chief of the *Monitor* discusses optimistic and pessimistic views on population growth, the explosion of cities, feeding the world, environmental degradation, family planning, the importance of educating and empowering women, the influence of the world's religions, and the international policy response. After many interviews, he finds a large measure of common ground on what needs to be done and what can be done: 1) the vast majority of experts believe that any prudent strategy for dealing with the future must include measures to slow population growth; 2) broad strategic agreement also exists that slowing this growth requires a combination of effective family planning and development plans that include increased educational opportunities for girls. The end of the Cold War offers an unprecedented opportunity to accelerate the dramatic progress that has already been made in reducing global birthrates. Many men and women worldwide increasingly seek access to family planning services. If this opportunity is not seized in the 1990s, it will be far harder in the next century to stabilize population growth at a sustainable level. If political leaders act now, "the population problem could be solved within two generations." [**NOTE**: Fair-minded and readable.]

(world population: strategy consensus)

050

World Access to Birth Control. Edited by Sharon L. Camp (VP, Population Crisis Committee). Washington: PCC (1120 19th St NW; 202/659-1833), Oct 1992/1 page (25x22")/$5.00.

A wall chart with analysis on the reverse side, rating the availability of family planning information and services in 124 countries. Several hundred individual country experts contributed information on over 120 measures of family planning program strength. This was consolidated into ten indicators related to the range of birth control choices available (60% of total score), the competence of those providing family planning services (20% of total), and the convenience of services and the amount of information available to contraceptive users. Countries are scored from 0 to 100. Top countries were Denmark (97 out of 100), The Netherlands (95), Sweden (95), and Norway (93). China rated 90 and India 73. The US was given a total score of only 75. Canada rated 71 and Japan 43. Seven countries received a 0 score: Gabon, Iraq, Kuwait, Liberia, Libya, Saudi Arabia, and Somalia. Overall, the 1992 assessment showed "a dramatic increase in access to birth control around the world, reflecting the expansion of organized family planning programs in developing countries and much better program implementation." Of the 87 developing countries studied in both the 1987 and 1992 editions, 57 improved their scores by 10 points or more (of these, 33 gained 20 points or more). But some 12 countries appear to have moved backwards with lower 1992 scores (including Brazil, Columbia, Jamaica, and the Philippines). In 22 countries scoring less than 25 points, couples still have virtually no access to birth control information or services. If the world is to reach population stabilization in the next century at less than double the current 5.4 billion people, the proportion of couples practicing contraception worldwide must increase from roughly 50% to 75% or more.

(access to birth control increases)

051

How Many Americans? Population, Immigration and the Environment. Leon F. Bouvier (Adjunct Prof of Demography, Tulane U) and Lindsey Grant (former Deputy Asst Secretary of State). San Francisco: Sierra Club Books, Sept 1994/174p/$18.00.

The US population was 261 million in mid-1994, and will grow to 279 million by 2000. Given current demographic trends, US population will reach 300 million around 2012, 350 million in 2025, 397 million by 2050, and 492 million by 2100. The direct and indirect effects of immigration between 2000 and 2050 will account for 72% of total growth. The costs of doing nothing are estimated for energy, agriculture, forests, biodiversity, social infrastructure, and urban problems. If the US remains on the present course, we will pay high taxes for a declining quality of life in 2050, with the gap between rich and poor growing ever wider. Alternative demographic futures are considered: 1) low fertility and annual immigration of about 1 million (the expected rate in the mid-1990s), resulting in a US population of 337 million in 2050 and 298 million in 2100; 2) low fertility and annual immigration of 500,000, resulting in a population of 301 million by 2050 and 233 million by 2100; 3) the recommended scenario of low fertility and annual immigration of 200,000, resulting in a population of 279 million in 2050 and 193 million in 2100. Although the American public favors reductions in immigration levels, all recent legislation has been geared to raising levels because of ethnic pressure groups, the erroneous business view that increasing labor supply is valuable, and liberals feeling that the US has a duty to take in the poor surplus populations of the world. Population growth must end, the sooner the better. No other single issue is so fundamental, or so deeply intertwined with other issues. [ALSO SEE: **Peaceful Invasions** by Bouvier and **Elephants in the Volkswagen** by Grant (**FS Annual 1993** #11845/11846).]

(US population near 400 million by 2050?)

5. Climate Change

052

Climate Change: The IPCC Response Strategies. Intergovernmental Panel on Climate Change, Working Group III (Frederick M. Bernthal, Chairman). Washington: Island Press, May 1991/272p(8x11")/$55.00;$34.95pb.

The IPCC was established in 1988 as a first attempt to address the threat of global warming on an international scale. Working Group I (Science) and Working Group II (Impacts) concluded that "We are certain emissions resulting from human activities are substantially increasing the atmospheric concentrations of the greenhouse gases." These increases will enhance the greenhouse effect. Based on current model results, "business-as-usual" emissions will increase global mean temperature during the next century by 0.3°C per decade (with an uncertainty range of 0.2°C to 0.5°C per decade), and the average rate of global mean sea level rise will be about 6 cm per decade over the next century (with an uncertainty range of 3-10 cm per decade). In many cases, the impacts will be felt most severely in regions already under stress, mainly the LDCs.

The main findings of Working Group III (Response Strategies): 1) "effective responses (will) require a global effort that may have a considerable impact on humankind and individual societies"; 2) a major part of emissions affecting the atmosphere presently originates in industrialized countries; 3) emissions from LDCs are growing and are likely to be an increasingly significant part of global emissions; 4) sustainable development requires proper concern for environmental protection as the necessary basis for continuing economic growth; "it is imperative that the right balance between economic and environmental objectives be struck"; 5) limitation and adaptation strategies must be considered as an integrated package, and should complement each other to minimize net costs; 6) response strategies are justified immediately, even in the face of significant uncertainties; 7) a well-informed population is essential to promote awareness of the issues and provide guidance on positive practices. The Working Group identified short-term measures (improved energy efficiency, cleaner energy sources, better forest management, etc) and long-term measures (accelerated and coordinated research, development of new technology, expanding the global ocean observing and monitoring systems, etc). [**NOTE**: A new IPCC report is due in Dec. 1995.]

(intergovernmental panel on climate change)

*053

Global Environmental Issues: A Climatological Approach (Second Edition). David D. Kemp (Prof of Geography, Lakehead U, Thunder Bay, Ontario). London & NY: Routledge, Dec 1994/224p/$59.95;$19.95pb.

An introductory, multi-disciplinary textbook with chapters on human activities and the atmosphere, characteristics of the atmosphere, drought/famine/desertification, acid rain, particulate matter, the ozone layer threat, and the greenhouse effect. Some conclusions on the "current state of the issues": ***1) Atmospheric Turbidity***: evidence has begun to appear indicating that increased fouling may contribute to atmospheric warming, but this currently raises little concern among the general public; ***2) Nuclear Winter***: interest in this issue has waned from a peak in 1983-1985; investigation since 1983 produced increasing evidence that the climatic impact of nuclear war was overestimated in the original study; ***3) Drought and Desertification***: desertification accelerated by human interference has become a major problem in some arid countries; attempts at its prevention and reversal have met with only limited success; ***4) Acid Rain***: the study of acid rain "has declined remarkably since the early 1980s"; although it has not disappeared, abatement programs "have met with some success" and "rain in many areas is measurably less acid than it was a decade ago"; there are indications (albeit disputed) that "lakes and forests are showing some signs of recovery in the most vulnerable areas of North America and Europe"; in less developed areas, "the full impact of acid rain may still be in the future"; ***5) Ozone Depletion***: there is already evidence of a decrease in the growth rate of atmospheric halons and certain CFCs; even with such developments, the ozone layer "will take time to recover" [time span not indicated]; ***6) Global Warming***: the size and complexity of the problem is such that there has been little development of replacements for the products and processes releasing greenhouse gases. The technology exists to tackle most if not all of these problems, but the gap between theory and practice is immense, due to political intransigence and government budget constraints. Despite an increasingly intensive research effort, global environmental problems are still not completely understood; as better models are developed, "there will be surprises."

(overview of global climate issues)

054

Between Earth and Sky: How CFCs Changed Our World and Endangered the Ozone Layer. Seth Cagin (Telluride CO) and Philip Dray (NYC). NY: Pantheon Books, May 1993/430p/$30.00.

The invention of chlorofluorocarbons in 1928 by a General Motors scientist was celebrated as a boon to humanity, making possible the mass proliferation of air conditioning and refrigeration. By the 1950s, CFCs found further applications as propellants in spray cans and as vital industrial solvents. In 1974, F. Sherwood Rowland and Mario Molina of the University of California at Irvine theorized that these wonder substances had altered the fundamental chemistry of the atmosphere and had begun to erode the ozone layer—the protective shield of life on earth. In the early 1980s it was discovered that stratospheric ozone over Antarctica was severely depleted each springtime. In 1987, a NASA expedition brought back irrefutable evidence that CFCs were responsible for the ozone hole. Sunblock, sunglasses, hats, and staying in the shade have become daily facts of life in many southern-latitude communities. In 1992, scientists observed that severe ozone depletion was occurring not only above Antarctica, but over heavily populated portions of the northern hemisphere as well.

Between 1974 and 1988, global atmospheric levels of chlorine rose from 1.8 parts per billion to 3.6 ppb. With the current phaseout agreements (the 1990 London Agreements, superseding the 1987 Montreal Protocol), it is believed chlorine levels will peak at about 4.5 ppb by 2000, then slowly decline to about 4 ppb by mid-century (some scientists believe chlorine levels may go as high as 6-8 ppb). This ozone layer stabilization by the mid-21st century depends on strict adherence to the CFC phaseout schedule now in place. But one particularly worrisome scenario is that a large volcanic eruption could inject huge amounts of sulfuric acid particles into the atmosphere, reacting with chlorine and triggering a large-scale global depletion of ozone. [**NOTE**: A thorough journalistic account, based on interviews with scientists, environmentalists, lawyers, and industrialists.]

(ozone layer to stabilize by mid-21st century)

055

Signals from Earth, Jessica Mathews (Senior Fellow, Council on Foreign Relations), *The Washington Post*, Sunday, 19 March 1995, C7.

Some recent news bulletins from the planet: 1) warming is changing the shape of the Antarctic Peninsula by breaking off an iceberg larger than Rhode Island; 2) glaciers are in full retreat and snow cover and Arctic sea ice are declining; 3) the weather pattern known as El Niño is occurring more frequently and lasting longer; 4) coral reefs from the Caribbean to the South Pacific are sick, apparently from warmer water; 5) a 50,000-square-mile region of the Pacific Ocean has lost 80% of its zooplankton, coinciding with a four-decade rise in water temperature; 6) the US has seen a steady stream of multi-billion dollar weather disasters in the last 10 years (consistent with the prediction of models for more frequent floods and more severe droughts). None of these proves that global warming is underway, but the accumulation of suggestive evidence leaves fewer doubts. "For more and more scientists, the question is not if, but when and in what form, the greenhouse smoking gun will appear." It is unlikely, however, that nations will negotiate binding targets and timetables for cutting emissions until there is conclusive evidence of global warming. [ALSO SEE: ***Most Precise Gauge Yet Points to Global Warming*** (*New York Times*, 20 Dec 1994, C4), reporting new data on rising sea levels from the Topex/Poseidon satellite, a joint French-American project.]

(global warming: suggestive indicators)

056
Global Warming Unchecked: Signs to Watch For. Harold W. Bernard Jr (Climate Institute). Bloomington IN: Indiana U Press, Oct 1993/186p/$29.95;$12.95pb.

A weather forecaster describes the warning signs of global warming in the 1980s, climate models, and the expected warmer world of the 1990s. Signs to watch for: 1) rising global temperature trends by the mid-1990s, after the temporary cooling due to Mt. Pinatubo's eruption; 2) record temperatures in the US, with readings to 120°F in the Great Plains and 110°F or more in some states east of the Mississippi; 3) severe to extreme dryness in the Great Plains as a harbinger to a major drought (the Dust Bowl era of the 1930s is used as an analog); 4) a few years where the number and strength of hurricanes is greater than in the past (hurricanes Hugo and Andrew may well be harbingers). Even if we were to take drastic steps now to curtail the greenhouse effect, the models warn that we've already commited the earth to a palpable climatic warming. We seem determined to wait until the pain becomes unbearable before reacting. [**NOTE**: Very informal style—remarkably so for a university press—with no footnotes or references.] **(global warming likely in 1990s)**

057
Climate and Chaos: Societal Impacts of Sudden Weather Shifts, Lester W. Milbrath (Prof Emeritus of Poli Sci and Sociology, SUNY-Buffalo), *The Futurist*, 28:3, May-June 1994, 26-28.

Author of **Envisioning a Sustainable Society** (SUNY Press, 1989) points to vulnerability of our society rooted in our unexamined premises: the premise of continuity (that the past will continue into the future), and that nature can be controlled. Human population will continue its rapid growth in the decades ahead, along with higher throughput, producing greater amounts of industrial and consumer waste. As more and more people look to life systems for sustenance, their actions will reduce the productive capability of these systems. "As this scenario plays out, it is improbable that the climate system will not change at all or that it will gradually change to a new pattern and settle down, as is assumed in most current economic thinking. The most probable climate scenario is for even more chaos" (as suggested by the floods and hurricanes of recent years). If humans increasingly perturb the system, we could expect it to become even more chaotic. But how chaotic will it become, what kinds of chaos might we expect, and how long will it last? One climate scenario is that of frequent climate disasters interspersed into "normal" climate patterns. Another scenario suggests an extended period, perhaps a decade or two, of oscillation-type chaos in the climate system. This could result in extensive plant dieback and resulting human starvation, precipitating collapse of world financial markets and sharp economic declines. Many humans, in turn, will revert to religious fundamentalism or seek overly simple solutions. More positively, the chaotic episode will likely force an unlearning of obsolete beliefs that would have been impossible earlier. It might then become possible to build a truly sustainable society. **(climate system chaos likely)**

058
Global Climate Change: Linking Energy, Environment, Economy, and Equity. Edited by James C. White (Cornell U). NY: Plenum Press, 1992/242p/$65.00.

Proceedings of the 8th annual conference of The Center for Environmental Information (Rochester NY), held December 1991 in Washington. Topics include steps to reduce US emissions, improving energy efficiency, the costs of reducing CO_2 emissions in the US (using the Global 2100 model), "no regrets" tax reform, four scenarios of multiple benefit environmental policies, the role of markets in energy/environmental policy, local and regional policies, sectoral perspectives (government, electric utilities, auto industry, industry), scenarios of future energy consumption, strategies for international negotiations, corporate and personal responsibility, intergenerational equity, energy policy to insure against the unknown, designing a carbon tax strategy, and successful national energy strategy. In the introduction, Ray T. Oglesby (Global Environment Program, Cornell U) warns of unidirectional trends toward major changes in future climate, and points to virtual certainties: 1) accelerated inputs of greenhouse gases can be expected to continue for at least several decades; 2) warming will be greater at higher than at lower latitudes (very roughly, a mean increase of 2.5°C translates to a 1° increase in equatorial temperature and as much as 10-12° at the polar circles); 3) more warming will occur in winter than in summer, decreasing temperature differences between seasons and creating troubles for urban water supplies. **(global warming response)**

059
Environmental Refugees in a Globally Warmed World, Norman Myers (Visiting Fellow, Green College, Oxford U), *BioScience*, 43:11, Dec 1993, 752-761.

Much has been written about the physical impacts of global warming, but little attention has been directed to the human impact, notably the emergent issue of environmental refugees. According to recent estimates, there are at least 10 million environmental refugees today (roughly half of them in Sub-Saharan Africa), compared with 17 million refugees under the conventional definition. The first figure is certainly on the low side because governments often take little official account of this unconventional category. In any event, we can assume that the number of environmental refugees is likely to swell rapidly as burgeoning throngs of poor people press ever harder on overloaded environments. Some estimates are made of key world regions, assuming a global population of 10 billion people in 2050 and a sea level rise of 30cm (about one foot): **1) Bangladesh**: by 2050, expected hazards would destroy homes and holdings in an area containing 7% of today's population, affecting 15 million people; **2) Egypt**: the country could well lose 12-15% of its scarce arable land, now containing 14% of the population; sea-level rise would thus displace more than 14 million people by 2050; **3) China**: the government estimates that 30 million people would be displaced by coastal flooding (a highly conservative estimate, in that 525 million Chinese will be living in the coastal plain by 2050); **4) India**: the total of flood-zone refugees could range from 20 to 60 million; **5) Caribbean Island States**: may eventually become subject to tropical storms of increased intensity over a longer season, "sufficient to reduce greatly the suitability of parts of the islands for permanent human habitation." Overall, "the total of environmental refugees as calculated here for a greenhouse-affected world is about 150 million." This amounts to 1.5% of the world population in 2050, compared with only 0.2% of the global population today. [**NOTE**: An important speculation by a prolific author best known for his estimates of future species extinctions (**FS Annual 1991** #10517).]

(150 million environmental refugees in 2050?)

060
Climate Change and World Food Security (Special Issue). Edited by Robert S. Chen (CIESIN, Univ Center MI) and Robert W. Kates (Trenton ME). *Global Environmental Change*, 4:1, March 1994, 3-77.

In the 1990s, more than a billion people suffer from micronutrient deficiencies, more than 700 million endure chronic undernutrition, and tens of millions are highly vulnerable to famine. The possibility of global climate change poses a dual threat: that our accustomed climate will undergo significant and disruptive change, and that attention devoted to mitigating or adapting to climate change will detract from other efforts to improve food security and sustainable development. The possibility also offers a glimmer of hope: that the issue of climate change will bring to the forefront the need to improve food security. These five essays draw on material presented at a July 1993 U of Oxford workshop. Who benefits and who loses in the long run—up through 2060? Various scenarios indicate that effects on crop yields in mid- and high-latitude regions will be positive, or less adverse than those in low-latitude regions. As a result, overall world cereal production will decrease by 1-7%, and net imports of cereals into LDCs will increase by 20-50%. This loss of production in developing countries is likely to increase the number of people at risk of hunger by 5-50%. Implications for food security of policies to prevent climate change are also assessed: mandatory reductions in use of fossil fuels will make energy more expensive, and afforestation on the requisite scale will compete with agriculture for good land.

(food security and climate change)

061

Climate Change and World Food Security (Special Issue). Edited by Thomas Downing and Martin Parry (both Environmental Change Unit, U of Oxford). *Food Policy*, 19:2, April 1994, 99-208.

Additional essays from the July 1993 workshop at U of Oxford, on supply and demand in global agriculture over the next 40 years (emphasizing that the supply of knowledge can compensate for insufficient supplies of land, water, and other natural resources), involuntary migration and food security, multiple threats to regional food production (arguing that, nevertheless, it seems possible to develop sustainable food production systems to feed a population of 12 billion), climate change and food security in Mexico, some likely impacts of climate change in Southern Africa (higher ambient temperature, aridification, reduction in biodiversity, impairment of hydroelectric generation, expansion of disease vectors), and a normative scenario of a *food-secure world* in 2060 by **Robert S. Chen** and **Robert W. Kates**. Such a world will be warmer, more crowded, and more connected. To avoid famine, it must not only have plenty of food, but also a widespread recognition of the human right to food, and effective mechanisms to prevent famine and alleviate its root causes.

(food-secure world in 2060?)

062

Global Warming and Biological Diversity. Edited by Robert L. Peters (Global Change Program, Conservation International) and Thomas E. Lovejoy (Smithsonian Inst.). New Haven CT: Yale U Press, July 1992/386p/$45.00.

Papers derived from an October 1988 World Wildlife Fund conference held in Washington, where biologists made projections based on a climate scenario of warming by 3°C. "The most general conclusion to be drawn from the conference is that many ecological systems will be dramatically changed by warming." Among the projections are large shifts in the ranges of species, hundreds of kilometers toward the north in the north temperate zone, causing ecological communities to break up and reassort. Also, many species extinctions are likely. Specific attention is given to vegetation, soil processes, tropical forests, wildlife, host-parasite and disease-vector relationships, Arctic marine ecosystems, Arctic tundra, forests in Northwestern North America, ecosystems in California, and fauna in Florida. It is pointed out in the Foreword that rapid global climate change could not happen at a worse time, because of human population growth and the rending of habitats into fragments due to logging, highway construction, damming, overhunting, overgrazing, acid rain, and eutrophication. This makes adjustment of species more difficult. The solutions to ecological upheaval are not yet clear. But one of the best things that can be done in the short term is to minimize further encroachment of development on existing natural ecosystems. [ALSO SEE: **FS Annual 1992** #11135 for a detailed EPA study on the potential effects of global climate change on the US.]

(ecological disruption from climate change)

063

The Regions and Global Warming: Impacts and Response Strategies. Edited by Jurgen Schmandt and Judith Clarkson (both Center for Global Studies, Houston Advanced Research Center, The Woodlands TX). NY: Oxford U Press, Sept 1992/337p/$59.95.

Essays by the finalists in the 1991 competition for the Mitchell International Prize for Sustainable Development. The theory of global climate change is widely accepted in the scientific community, but there is equally widespread uncertainty about the extent, timing, and regional implications. Contributors focus on mitigative and adaptive strategies that can be implemented regionally. Topics include coastal land loss in the Chesapeake Bay region, expected sea level rise in Bangladesh (a rise of 1 meter will inundate about 16% of its total area), impacts in Mexico (current climate models indicate significant increases in temperature and changes in precipitation), impacts on human settlement patterns in West and Central Africa, implications for South Asia (also for Northeast Brazil and Northeast North America), the possible aridization of the continents, methodology for assessing regional economic impacts, global warming and forests of the Great Lakes states (First Prize Winner by Daniel B. Botkin, proposing a new management strategy using nature as a guide to change), the impact on rice production in Thailand, assessing regional damage costs (by Robert U. Ayres, on environmental accounting and the problem of valuation), the contribution of LDCs to total CO_2 emissions (now 43% and likely to increase because of growth in energy consumption), policies for resilient regional response, and policy responses to climate change in Southeast Asia. The editors conclude that "climate change will be most painful in regions that are poor, that suffer from conventional pollution, that are equipped with outdated infrastructure, and that are least able to deal with additional stress."

(global warming and world regions)

064

Shaping National Responses to Climate Change: A Post-Rio Guide. Edited by Henry Lee (Director, Environment and Natural Resources Program, JFK School, Harvard U). Washington: Island Press, April 1995/ 303p/$48.00;$24.95pb.

In June 1992, Earth Summit negotiators from more than 150 countries signed what is now known as the Framework Convention on Climate Change, and more than 100 countries have ratified the convention. In the US, the Climate Change Action Plan, announced by President Clinton in October 1993, contained 47 initiatives. All other countries that ratify the Framework Convention will prepare and submit similar plans. This international process, along with domestic processes of developing national action plans, guarantees that design, development, and implementation of mitigation responses will be on the public policy agenda for many years. This volume, culminating the four-year Harvard Global Environmental Policy Project and involving some 30-40 scholars, seeks to inform these processes. The eight chapters explain overcoming obstacles to a successful climate convention, negotiating environmental treaties in an unequal world, improving compliance with the climate change treaty, trading in greenhouse permits, a global carbon tax strategy, economic effects of a carbon tax, lessons from development experience applied to technological cooperation, and environmental policy in the US, EC, and Japan.

Lee summarizes the principal lessons of the chapters for forging a workable international agreement: 1) no significant worldwide consensus for action now exists; it must be forged by governments, the expert community, and NGOs; 2) indirect and flexible strategies will be more successful than highly visible and aggressive responses (success should be measured by results, not by paper mandates, rhetorical positions, or the number of protocols signed); 3) a treaty that demands the same obligations from all nations will be impossible to negotiate, or unsustainable once negotiated; 4) the most difficult challenge in achieving meaningful global reductions of greenhouse gases will be to structure a portfolio of incentives and policies to encourage developing nations in emissions-reduction strategies; solutions that only involve the developed world will be unsuccessful. Lee also notes that designing policies to mitigate global warming will be complex and costly, and that scientific understanding of global climate change will be constantly evolving. "Over the next two decades, countries will go through five or six cycles of debating, developing, and implementing responses to the problem of global warming—each cycle triggered by new scientific information." It is likely that this science will take a nonlinear path of zigs and zags; policy responses thus must be flexible. **(climate change plans)**

065

National Case Studies of Institutional Capabilities to Implement Greenhouse Gas Reductions (Special Issue). Edited by Steve Rayner (Pacific Northwest Laboratory, Washington DC). *Global Environmental Change*, 3:1, March 1993, 7-128.

The Framework Convention on Climate Change requires signatories to develop National Action Plans with information on their net greenhouse gas emissions, as well as efforts to mitigate or adapt to climate change. But the Convention does not offer specific guidance on the form or content of such plans While technological feasibility and economic costs will be prominent concerns, the institutional aspects of such plans, and their political feasibility and societal sustainability, are likely to receive scant attention, despite their key role to any success. These five essays examine what might realistically be expected in terms of initiative, leadership, cooperation, and compliance: **1) The US**: deep divides exist between those who seek to rely on the marketplace and environmentalists who seek more drastic and immediate action (the ambiguous rhetoric of "sustainable development" may temporarily conceal this conflict, but has yet to transcend it); the US economy has deep dependence on fossil fuels, and "major changes to break this addiction would be a momentous political task"; **2) India**: the term "sustainable development" thinly disguises a fundamental cultural cleavage over equity; **3) Indonesia**: the economy is dominated by a dense network of public monopolies controlled by a tiny elite; industry is putting on a green face, but Indonesia has little ability to follow through on its international commitments; **4) China**: the leadership asserts its dedication to environmental issues and public opinion is aroused, but the environmental problems are enormous and resources to confront them are few; "China will continue to develop and consume coal as its primary source of energy well into the next century; there are simply no other choices, given its desire to expand its economy"; **5) European Community**: efforts to set EC policy on global warming have coincided with growing conflict over moves to greater monetary and political union; a proposed carbon tax is likely to be favored as leaving the least footprint of central authority. [**NOTE**: Useful case studies of "soft" factors.

(greenhouse gas reduction barriers)

066

Policy Implications of Greenhouse Warming. Policy Implications of Greenhouse Warming—Synthesis Panel (Daniel J. Evans, Chair). Washington: National Academy Press, April 1991/127p/$14.95pb.

The Committee on Science, Engineering, and Public Policy (a unit of the NAS, NAE, and IOM) conducted a study on global warming involving nearly 50 experts on four panels. This report of the Synthesis Panel combines assessments of the Effects Panel, the Mitigation Panel, and the Adaptation Panel with additional materials. Overall, it is concluded that, despite considerable uncertainties in knowledge, greenhouse warming from various human-induced gases in the atmosphere "poses a potential threat sufficient to merit prompt responses." Americans could probably adapt to likely changes associated with global warming, but the costs would be substantial. Investment in mitigation measures acts as insurance protection against great uncertainties and possible dramatic surprise. Substantial mitigation can be had at modest cost: "in other words, insurance is cheap." Recommendations for mitigation: 1) study "full social cost pricing" of energy production and use, with a goal of gradually introducing such a system, to be coordinated internationally; 2) enhance energy conservation and efficiency; 3) make greenhouse warming a key factor in planning for future energy supply; 4) continue the aggressive phaseout of CFCs and other halocarbon emissions; 5) reduce global deforestation and remove incentives contributing to US deforestation. Other recommendations concern enhancing adaptation (plan margins of safety in long-lived structures, make water supply more robust, slow present losses in biodiversity), improve knowledge for future decisions, evaluate geoengineering options (which is not a recommendation that such options be undertaken now), and resume full US support of programs to slow world population growth. **(greenhouse insurance recommended)**

6. Oceans and Coastal Areas

067

The Wealth of Oceans. Michael L. Weber and Judith Gradwohl (both Washington DC). NY: W.W. Norton, May 1995/256p/$25.00.

An adjunct to "Ocean Planet," a major traveling exhibition opening at the Smithsonian Institution in 1995, with chapters on the history of ocean exploration, climate change and the oceans, threatened life in the seas, economics of ocean resources, coastal areas, pollution, depleted fisheries, population growth and fish consumption, and the fading of past visions of future riches from the seas. Foremost among major trends is continued growth in the number of people, their consumption of natural resources, discharge of their wastes into rivers and coastal waters, global warming, sea-level rise, and new technology for exploiting fisheries. Positive trends include growing public awareness and concern about the oceans, new satellite technology aiding compliance with conservation measures, and the new field of ecological economics questioning current accounting for ocean/coastal resources. [**NOTE**: Nice popular overview. ALSO SEE: **Sea Change: A Message of the Oceans** by marine biologist Sylvia A. Earle (G.P. Putnam's Sons, April 1995/361p/$25.95), an equally popularized but more personalized call for protecting the oceans.] **(ocean problems)**

068

Abandoned Seas: Reversing the Decline of the Oceans. Peter Weber (Research Associate, Worldwatch Institute). Worldwatch Paper 116. Washington: Worldwatch Institute, Nov 1993/66p/$5.00. (Brief version in **State of the World 1994**, 41-60; #014.)

Public outcries over oil spills, fouling of beaches, and killing of whales have prompted some promising actions to protect the marine environment. But these high-profile issues are not the largest problems. Less dramatic, but more pervasive and ultimately destructive, are the slow and persistent incursions of coastal habitat destruction, the relentless push to increase the global fish catch (the potential yield of this essential food source has been decreased by as much as 20%), and the dispersed sources of pollution that ultimately end up in the sea. Vital processes of the oceans are heavily concentrated in coastal waters; disruptions there have a disproportionate effect on the whole ocean ecosystems. Coastal areas are also where human activity is concentrated: over half the people in the world live within 100 km of the coast. About half of the world's saltmarshes and mangrove swamps have been cleared, drained, diked, or filled. Some 5-10% of world coral reefs have been eliminated, and another 60% could be lost in the next 20-40 years. Even beaches are threatened, with 70% eroding worldwide.

Oceans must become a substantive part of the global agenda for sustainable development. Three areas have the highest priority: ***1) Fishing***: depletion of fish stocks must be halted by re-establishing basic tenets of local fisheries management once widely observed in traditional Pacific Island and coastal cultures; trawlers could increase their long-term catch by using nets with larger mesh (the good news is that driftnetting has fallen off sharply); governments should end subsidies to fishing fleets (saving some $54 billion a year) and collect adjustable rents based on status of fish stocks for use of fishing grounds (earning $25 billion a year); ***2) Coastal Development***: eliminate government-sponsored insurance and funding for ocean-altering roads, dikes, and dams; control excessively invasive tourism; ***3) Inland Pollution***: improve sewage treatment; reduce soil erosion and use of pesticides and fertilizers. **(ocean decline, how to reverse)**

069

Net Loss: Fish, Jobs, and the Marine Environment. Peter Weber (Research Associate). Worldwatch Paper 120. Washington: Worldwatch Inst., July 1994/76p/$5.00pb.

After decades of buying bigger boats and more advanced hunting technologies, fishers have nearly fished the oceans to the limits. The catch has fallen in 13 of the world's 15 major fishing regions (in 4 of them, the catch has shrunk by more than 30%). In the last few years, over 100,000 fishers have lost their source of income. Some 10 million of the world's 15-21 million fishers could be out of work in the coming decades, in that fishers possess about twice the capacity needed to fish the ocean. Rather than contributing to a solution, governments have generally promoted the overexpansion of the fishing industry. Worldwide, they subsidize national fleets on the order of $54 billion annually to catch $70 billion worth of fish. Fish, fishers, and consumers would all benefit from programs to protect and rehabilitate marine fisheries. Fishers could potentially increase their catch by 25% if they allowed fish populations to rebuild. To make the transition to sustainable fisheries, governments and fishers must move beyond the current state of deadlock. In virtually all cases, a combination of government oversight and community-based management promises the best solution. Examples are provided. [ALSO SEE: **Folk Management in the World's Fisheries: Lessons for Modern Fisheries Management**, edited by Christopher L. Dyer and James R. McGoodwin (University Press of Colorado, 1994/347p/$24.95).] **(fishery depletion)**

070

Crisis in the World's Fisheries: People, Problems, and Policies. James R. McGoodwin (Associate Prof of Anthropology, U of Colorado). Stanford CA: Stanford U Press, Dec 1990/235p/$35.00.

All around the world there is a crisis in fisheries: quite simply, too many people chasing too few fish. The world's total annual fish catch has steadily risen from 2 million tons in 1850 to 55 million tons by the end of the 1960s, with little indication of reaching an upper limit. But the total catch abruptly leveled off at about 70 million tons in the early 1970s, and throughout the 1970s world per capita fish catch declined, as well as catch per unit of fishing effort and per dollar invested. Reacting to these first signs of crisis, many fishing fleets diversified in the 1980s, ranging more widely, using more effective technology, and targeting a greater variety of species. The world catch rose to over 90 million tons by the late 1980s, but world per capita catch and catch per unit of fishing effort continued to decline. A few fisheries experts are optimistic that the annual world catch can eventually stabilize at about 100 million tons, but only with effective environmental protection and fisheries management. Without such efforts, "there is every likelihood that the world's total catch may soon begin to decline, and some stocks will eventually be wiped out permanently."

McGoodwin attempts a fresh reconsideration of fisheries management, emphasizing fishers and fishing peoples, especially small-scale fishers—the overwhelming majority of all fishers worldwide. Recent work in the social sciences, particularly marine anthropology, indicates that the widely-known "tragedy of the commons" model of Garrett Hardin has many defects. It ignores important social institutions and their roles in managing the commons (local fishers often develop their own means of coping with overfishing), it erroneously assumes that private property will always be more diligently conserved by its owners, and it takes a cynical view of the mentality and character of fishers—that they are ignorant and greedy. A more humanized and sustainable process of fisheries management must promote three broad activities: incorporating indigenous means of management, applying more social analysis into the policy-making process, and providing for more extensive participation by the people concerned (while shifting away from autocratic management modes). The oldest and most successful regimes of cooperative co-management between experts and fishers are found in Japan. Just as the Green Revolution made matters worse for many poor people, the Blue Revolution (capital-intensive industrial fishing for export markets) has spelled disaster for many small-scale fishers in LDCs. Meeting local subsistence needs should be a first priority. [**NOTE**: Scholarly but very readable.]

(humanized sustainable fisheries management)

071

Ocean Governance: Sustainable Development of the Seas. Edited by Peter Bautista Payoyo (Dalhousie U). Tokyo: United Nations U Press, Dec 1994/369p/$50.00pb.

Proceedings of Pacem in Maribus XIX, held in Lisbon in November 1991 (PIM, paraphrasing Pope John XXIII's encyclical, *Pacem in Terris*, is the annual conference of the International Ocean Institute, founded by Elizabeth Mann Borgese). PIM XIX involved 75 participants from 35 countries. The 19 papers discuss such topics as the UN Convention on the Law of the Sea, the role of indigenous peoples in ocean governance, new structures for decision-making in integrated ocean policy, environmental accounting and valuation in the marine sector, ocean policy in India and Japan, the Regional Seas Programme to integrate environment and development, joint management zones, regional centers for marine sci/tech, information and communication on the oceans, the question of financing, and the proposal for an Ocean Assembly or World Ocean Forum. Some conference proposals: 1) ocean governance should be examined as a possible pattern for governance of other global concerns such as energy, food, atmosphere, outer space, and sci/tech; 2) national, regional, and global institutions must be restructured to reflect the shift from a sectoral to an intersectoral approach to planning/decision-making; ocean policy must be integrated horizontally across disciplines and agencies, as well as vertically across levels of governance; 3) human resource development for ocean management must be a priority of national governments; 4) states should integrate sustainable ocean development into their general development strategy; 5) a global forum involving the economic and NGO sector should be established within the UN system, enabling periodic discussion of all ocean issues; 6) to finance assistance, new taxation systems are needed, such as levies on trade, tourism, and other uses of ocean space.

(sustainable development of oceans)

072

Freedom for the Seas in the 21st Century: Ocean Governance and Environmental Harmony. Edited by Jon M. Van Dyke (Prof of Law, U of Hawaii), Durwood Zaelke (President, Center for International Environmental Law), and Grant Hewison (Greenpeace New Zealand). Washington: Island Press, Dec 1993/504p/$55.00;$27.50pb.

The image of the limitless sea has dominated Western thought for the past millennium. Because the seas have been viewed as a frontier territory, an "anything goes" attitude has been accepted as appropriate. Legal terminology to justify this approach is "freedom of the seas"—a concept developed by the major maritime powers during the age of imperialism. This freedom has historically included the freedom to pollute, assuming that oceans have a limitless capacity to assimilate wastes. Technological developments such as drift nets on the high seas demonstrate that "anything goes" is no longer acceptable. Most of these essays, based on a December 1990 meeting in Honolulu, argue that the governing notion should be "freedom *for* the seas," whereby the key goal will be to protect ecological vitality. The 29 essays are in six parts: ***1) Ocean Governance***: principles for a comprehensive regime, the process of creating an international ocean regime (by Elizabeth Mann Borgese), potential contributions from indigenous peoples (who generally believe that the sea is not a resource only for humans to exploit), principles of a new Law of the Sea; ***2) Pacific Approaches***: five essays on how Pacific Islanders have begun to build a regime for the South Pacific area based on high environmental consciousness; ***3) Controlling Ocean Pollution***: mending the seas through a Global Commons Trust Fund, what is needed in international agreements, protected marine areas and low-lying atolls; ***4) Living Resources***: problems of unregulated fisheries, legal issues concerning drift net fishing, trade restrictions of a conservation measure; ***5) The 1982 Law of the Sea Convention***: environmental ethics and deep seabed mining; ***6) Military Activities***: security regimes for the oceans, denuclearizing the seas, naval operations, military exclusionary zones. In sum, five brands of idealism are reflected here: environmentalism, structuralism, humanism, nonstatism, and transnational bureaucratism. [ALSO SEE: **Global Marine Environment: Does the Water Planet Have a Future?** by Herman Prager of NE Louisiana U (University Press of America, March 1993/161p/$42.50;$18.50pb), on the ocean pollution crisis and the role of international organizations and INGOs in protection.]

(ocean governance)

073

The Oceans and Environmental Security: Shared U.S. and Russian Perspectives. Edited by James M. Broadus (deceased, 1994; WHOI) and Raphael V. Vartanov (IMEMO, Russian Acad. of Sciences). Washington: Island Press, Nov 1994/328p/$55.00;$35.00pb.

Environmental security—the reasonable assurance of protection against threats to national well-being or the common interests of the international community associated with environmental damage—is both a goal and a way of thinking. This collaboration between the Woods Hole Oceanographic Institute and the Institute for World Economy and International Relations (IMEMO) seeks to apply the concept of environmental security to the world's oceans and to identify problems of high mutual interest. Seven such problems are explored: land-based marine pollution (which accounts for more than 75% of total marine pollution), resource problems in the North Pacific (most likely requiring a multilateral scheme for cooperative management of fisheries), hazardous materials transport, radioactivity in the oceans, environmental protection for the highly vulnerable Arctic ecosystems, management of the Southern Ocean, and the Law of the Sea Convention. Nine cross-cutting themes: 1) the absence of clearly specified and widely recognized criteria for assigning responsibility, liability, and compensation for environmental damage; 2) the value of viewing each problem from a global perspective; 3) the growing density of coastal populations; 4) the desirability of a "net benefits criterion" for environmental policy (which implies some softening of certain rigid standards); 5) issues of monitoring, data, and information management; 6) the importance of international cooperation and coordination; 7) the importance of international organizations; 8) virtually all the problems reviewed are regrettable spillover effects of advancing technology, but technology can also help environmental security; 9) military conversion is important: some of the greatest sci/tech skills in both the US and Russia are still bottled up in their military-industrial establishments. **(environmental security of oceans)**

074

The Rising Tide: Global Warming and World Sea Levels. Lynne T. Edgerton (National Resources Defense Council-Los Angeles). Washington: Island Press, May 1991/140p/$29.95;$17.95pb.

If current trends continue, they are expected to raise the earth's average surface temperature by 1.5°- 4.5°C in the 21st century, with warming at the poles perhaps 2-3 times as large as warming at the middle latitudes. Large uncertainties remain about the timing and ultimate magnitude of climate change, but there is little dispute that there has been a significant increase in global concentrations of heat-trapping gases. The US and the world can minimize global warming by taking decisive action now, but we are probably too late to eliminate it altogether. As a minimum insurance policy, "nations should be planning for a minimum 0.7-meter global sea level rise along their coasts by 2050." This rise will produce major losses of beaches, dunes, estuaries, barrier islands, and wetlands. It will result in salinization of coastal drinking water supplies, threats to coastal infrastructure (roads, ports, waste and wastewater facilities, power plants), and potentially enormous human suffering from loss of life and property during storms. Concludes with recommendations for the Federal and state governments, such as a nationally coordinated response to sea level rise, studies of anticipated effects in all coastal states, and public education. [**NOTE**: A competent non-technical survey.]

(planning for sea level rise)

075

An Introduction to Coastal Zone Management. Timothy Beatley (U of Virginia), David J. Brower (U of North Carolina), and Anna K. Schwab (UNC). Washington: Island Press, July 1994/210p/$49.00;$29.95pb.

The US is increasingly becoming a coastal nation, with coastal areas around the country continuing to grow in absolute and relative terms. While these pressures are probably inevitable, the destructive and hazardous nature of coastal land use and development patterns need not be. Chapters discuss the special nature of coastal areas, coastal forces and processes, critical management issues (patterns of ownership and control, storm damage mitigation, shoreline erosion and sea level rise, wetland protection, controlling nonpoint pollutants, protecting the recreational commons), the fragmented management framework, the Coastal Zone Management Act of 1972, state and Federal coastal management programs, and ecosystem management. In the future, all coastal management decisions must assume sustainability as their primary and essential goals. Efforts aimed at sustainability should involve action at three levels of government: ***1) Federal***: implement a national coastal management plan, eliminate or reduce subsidies for dangerous development patterns, expand funding and technical assistance for state programs, expand acquisition of sensitive coastal areas; ***2) State***: establish minimum development standards, promote "strategic retreat" in light of predicted sea level rise (resisting the forces of nature is largely futile); ***3) Localities***: develop a sustainable community plan, limit extensive low-density development, view the community in terms of resources consumed and waste streams generated. **(sustainable coastal areas)**

076

Managing Wastewater In Coastal Urban Areas. National Research Council. Washington: National Academy Press, 1993/477p/$49.95.

At least 37% of the US population is located in counties adjacent to the oceans or to major estuaries. Waste from this population and its associated activities is a major contributor to the widely documented deterioration of ocean and coastal waters. Wastewater constituents of concern are ranked at three levels: high priority (nutrients, pathogenic viruses and bacteria, toxic organic chemicals), intermediate priority (toxic metals, oil and other hazardous materials, floatable and plastics), and low priority (oxygen depletion from organic material, solids). The NRC Committee on Wastewater Management identified specific areas for improvement: tailoring strategies to regional differences, greater attention to preventing harmful levels of nitrogen and other nutrients in coastal waters, encouraging source control of pollutants, vigorously pursuing techniques for monitoring the presence of pathogens, making management systems flexible so they can respond to new information, and developing a system of "integrated coastal management" (ICM) at the regional level. In the next 20 years, the US should implement a system of ICM for all of its urban shores. Full integration should include all sources of environmental stress, all environmental media, and principles of pollution prevention and source control. Regional ICM, in contrast to the existing top-down policy of setting national standards, is "nothing less that a new paradigm for wastewater management in the US."

(integrated coastal management needed)

7. Freshwater

077

Global Water Resource Issues. Gordon J. Young (Wilfrid Laurier U, Ontario), James C.I. Dooge (University College Dublin), and John C. Rodda (World Meteorological Organization, Geneva). NY: Cambridge U Press, Aug 1994/194p/$44.95.

The International Conference on Water and the Environment (ICWE), held in Dublin in January 1992 in preparation for UNCED, was the most all-embracing conference on global water management since the 1977 UN Water Conference in Argentina. It attracted more than 500 participants from 114 countries, 28 UN agencies, and 58 NGOs and inter-governmental organizations. It issued a *Report of the Conference* (ICWE, 1992) and *The Dublin Statement on Water and Sustainable Development*, reproduced here in Annex 1. The Statement warns that "scarcity and misuse of fresh water pose a serious and growing threat to sustainable development and protection of the environment," and calls for "fundamental new approaches to the assessment, development, and management of freshwater resources." Concerted action is needed to reverse present trends of overconsumption, pollution, and rising threats from drought and floods. Guiding principles for action at local, national, and international levels: 1) effective management of water resources demands a holistic approach; 2) water management should be based on a participatory approach, involving planners, policy-makers, and users at all levels; 3) women play a central part in the provision, management, and safeguarding of water, requiring positive policies to empower women to participate in all levels of water resources programs; 4) water should be recognized as an economic good, which will lead to more efficient and equitable use.

The authors seek to enlarge upon the two main documents produced at the Dublin conference, with chapters on the global scene, integrated water resources development and management, water resources assessment, protection of water quality and aquatic ecosystems, impacts of climate change on the hydrological cycle, water and sustainable urban development, efficient and equitable allocation, sustainable food production, sanitation, capacity building, and mechanisms for implementation at global, national, and local levels. **(global water issues)**

078

Water: The International Crisis. Robin Clarke (Consultant, UNEP). Cambridge MA: MIT Press, March 1993/193p/$30.00;$15.95pb. (First published in London by Earthscan, 1991.)

In 1940, total water use was about 1,000 cubic kilometers. Usage doubled by 1960 and doubled again to 4,138 cubic km by 1990, and is projected to be 5,189 cubic km by 2000. Chapters discuss water and climate, degradation of land and water, spreading deserts, growing water scarcity, the neglect of water issues in development planning, the enormous potential for international conflict over water, the high costs of high-tech schemes (diverting rivers, desalting the sea, moving icebergs, large dams), traditional solutions (run-off farming, flooding the fields, small-scale irrigation), new twists to old solutions (changing crops, small dams, micro-irrigation, rainwater harvesting), and water security as a key to food security.

(water crisis: global overview)

*079

Last Oasis: Facing Water Scarcity. Sandra Postel (Vice President for Research, Worldwatch Institute). Worldwatch Environmental Alert Series. NY: W. W. Norton & Co, Nov 1992/239p/$19.95;$9.95pb.

Masking scarcity is a principal aim of water development, the collection of engineering projects and technologies that give people access to and control over nature's supply. The illusion of plenty leads to overconsumption and adverse consequences for the environment and for future generations. Signs of water stress abound worldwide: water tables are falling, lakes are shrinking, and wetlands are disappearing. Today there is a large unmet demand for household water: nearly one of every three people in the developing world (some 1.2 billion people) do not have access to a safe and reliable supply for their daily needs. In many places, pollution is rapidly diminishing the usable supply. Engineers propose "solving" water problems by mammoth river diversion schemes, but they have already selected the easiest and least-costly sites for development. Over time, water projects have become increasingly complex, expensive to build, and more damaging to the environment. Desalination is technically feasible and use of the process has grown, with more than 7,500 desalting plants now operating worldwide. Still, desalination produces only 0.1% of the world's total water use, and its contribution "is likely to remain small for the foreseeable future." Both mega-diversion projects and desalination hold out the unrealistic hope of a supply-side solution.

The first and easiest step toward water security is doing more with less. "With technologies and methods available today, farmers could cut their water needs by 10-50%, industries by 40-90%, and cities by a third with no sacrifice of economic output or quality of life." Most investments in water efficiency, recycling, reuse, and conservation yield more usable water per dollar than investments in water supply projects. But they will not materialize until policies, laws, and institutions begin to foster such measures, rather than hinder them. A new water era has begun. In contrast to earlier decades of unfettered damming, drilling, and diverting, the next generation will be marked by limits and constraints. Most fundamentally, water scarcity challenges us to adopt a new ethic to guide our relationship to earth's natural systems. Measures to conserve water and use it more efficiently are now the most economical and environmentally sound options for much of the world—and they have barely been tapped. Together, they constitute our "last oasis." [ALSO SEE: ***Water Tight*** by Sandra Postel (*World Watch*, 6:1, Jan-Feb 1993), on case studies of moving toward water conservation in Mexico City, Waterloo (Ontario), San Jose (California), and Boston.]

(water conservation as "last oasis")

080

Population and Water Resources: A Delicate Balance. Malin Falkenmark (Prof Emerita, Swedish Natural Science Research Council) and Carl Widstrand (U of Linkoping & Carleton U). *Population Bulletin*, 47:3, Nov 1992/36p/$7.00 single issue.

Water is a dynamic resource with a fixed rate of renewal. The world's population lives at the mercy of the water cycle. Many of the world's poorest countries lie in the global zones most subject to drought, and many of these countries also have rapidly growing populations that must compete for limited water resources. Arid and semi-arid regions of Africa are already living on the hydrological margin. By 2025, 1 billion people worldwide will live in areas of extreme water scarcity. Technological solutions (dams, pipelines, desalinization) are expensive, especially for low-income countries suffering the worst shortages. A crucial step in averting a water crisis is to slow population growth. Promising water management solutions include storing water during dry seasons, reducing irrigation waste, reuse of irrigation runoff, and planting drought-resistant crops. [**NOTE**: Outlines a key problem in the present, but does not consider potential climate change impacts on water supply or the possibility of low-cost desalinization technology.] **(water and population)**

081

Metropolitan Water Use Conflicts in Asia and the Pacific. Edited by James E. Nickum (East-West Center, Honolulu) and K. William Easter (U of Minnesota). Boulder CO: Westview Press, Dec 1994/219p/$55.00pb.

Throughout the world, growing urban demands for water are confronting increasing costs of developing new supplies, aggravated by growing pollution of traditional water sources. These problems are particularly acute in the large, rapidly growing cities of the Pacific Rim, where water quantity problems are accelerating. Case studies examine water management in Bangkok, Beijing, Honolulu, Madras, Manila, Osaka, Seoul, and Central Japan. Concludes that the future of supply management approaches does not look promising for many of these eight areas, but demand management strategies have been used only to a limited extent so far. It will be increasingly necessary to formulate ever more sophisticated blending of supply-side and demand-side options, as many US electric power utilities already do. Water pricing, conservation, and reallocation promise to be at the forefront of future policy options. In many places, this will bring an end to the era of "cheap" water. **(urban water supply problems)**

082

Tapping Alternatives: The Benefits of Managing Urban Water Demands, Benedykt Dziegielewski and Duane E. Baumann (both Dept of Geography, Southern Illinois U), *Environment*, 34:9, Nov 1992, 6ff.

Rather than seeking new water supplies, water planners in southern California are turning to "water-demand management" as a promising alternative. By 2010, water conserved through increased efficiency of water use in southern California will constitute the largest single source of additional water. The broadening of water planners' perspectives has been brought about by increasing difficulty in finding untapped water sources,

depletion and contamination of ground water sources, the increased frequency of droughts in the last decade, and environmental concerns. New alternatives to increase water supply include: 1) more efficient utilization of existing supplies by evaporation suppression and lining reservoirs; 2) protecting existing supplies by cleaning up waste sources contaminating water and constructing barriers against intrusion of salt water; 3) using ground water aquifers for storage of excess supply of surface water (water banking); 4) regional management of water supply to improve distribution among users; 5) water-demand management programs: plumbing retrofit campaigns, conservation audits and ordinances, system efficiency programs, educational programs, landscape replacement, and conservation-oriented pricing. [**NOTE**: Parallels shifting thought about energy and transport (#12549/12551.]
(water-demand management in California)

083
America's Water: Federal Roles and Responsibilities. Peter Rogers (Prof of Environmental Engineering & City Planning, Harvard U). A Twentieth Century Fund Book. Cambridge MA: MIT Press, Nov 1993/285p/$27.50.

Each year the US uses over 500,000 gallons (2,075 tons) of water per capita. It is made available at bargain prices averaging for all uses about 19 cents per 1,000 gallons (in contrast, bottled mineral water sells for $4.00 per gallon). For more than 200 years, towns, states and the Federal government have grappled with problems of ensuring abundant high-quality water to all citizens, supplying flood protection, and securing commerce along waterways. Over this period, public goals concerning water policy have continually shifted as the nature of the problems changed. Affluence, for example, leads to a rise in quantities of water used, and later to a utilitarian change in attitude toward the environment, toward long-term sustainability. Rogers seeks to give "a balanced view" of water use in the US, avoiding the temptation to side with environmentalists (who see the resource in dire straits) or with prodevelopment forces (who see no need to restrain use). No major crises are presently seen in either water quantity or quality. "In every case examined, whether for irrigation, domestic supply, or wildlife support, there are good alternatives available to us, usually at small cost." We must always guard against unnecessary fears, but we must also maintain vigilance by building flexible institutions. Still, "The US has all the water it needs for all conceivable uses well into the middle of the next century (and beyond, if the current rates of per capita decline in water use continue)." The large and unexpected decline in consumption (a broad indicator of success for Federal water policy) is not directly attributable to any single policy. Overall, despite population increase, "the US is now using less total water than it did in 1975."

"There is very little new that can be said about water policy... All of the recommendations made in this book are included somewhere among the 47 Long's Peak recommendations." [See below.] Nevertheless, Rogers proposes a "Blueprint for Federal Water Policy" including such suggestions as: 1) a President's Water Council to take comprehensive view of national water programs so as to influence planning of cohesive policies throughout the US; 2) a national water information program to gather useful data about water; 3) increasing research emphasis on political and institutional arrangements, in keeping with the needed shift in emphasis from water development to water management; 4) a concerted public-private water education effort to make the US more aware of its resources, problems, and needs; 5) a series of regional forums to link Federal water administrators with state, local, and private counterparts; 6) a review of cost-effectiveness and timetables for the present technology-based standards and regulations, so as to achieve coherence among water quality programs; 7) addressing questions of social equity in water availability, in light of the need for increased application of the user-pays principle. **(water policy)**

084
America's Waters: A New Era of Sustainability. Report of the Long's Peak Working Group on National Water Policy. Boulder CO: U of Colorado Natural Resources Law Center, Dec 1992/12p (303/492-1288).

A working group of 30 national experts in water policy met near Colorado's Long's Peak in early December. In America's past, water seemed abundant and nature forgiving, Federal funding was plentiful, and extensive subsidies for development encouraged inefficient use of water. The era of major water projects has passed: neither the economy nor the environment can tolerate more. A national water policy must be based on sustainability and seek four objectives: **1) Efficient Water Use**: the efficient use of water resources must be a central aspect of all water policy, with the Federal government facilitating transfers of water from one use to another; **2) Ecological Integrity and Restoration**: watersheds should form the basic unit of analysis and activity, and prevention strategies and integrated responses should replace crisis-oriented management; **3) Clean Water**: the central objective of the Clean Water Act remains unfulfilled; many obstacles stand in the way: e.g., nonpoint discharges, fragmented regulatory responsibility, inadequate water quality standards and lax enforcement; **4) Equity and Participation in Decisionmaking**: Indian tribes have not received benefits of Federal water development in spite of treaty promises, and numerous rural communities have had their water supplies transported to urban centers; many citizens have been excluded from making water policy. Concludes with 47 specific proposals for the Clinton Administration. **(water policy for sustainability)**

085
Searching Out the Headwaters: Change and Rediscovery in Western Water Policy. Sarah F. Bates *et al.* (Natural Resources Law Center, U of Colorado). Washington: Island Press, Nov 1993/242p/$35.00; $17.95pb.

Water use in the Western US is governed by one of the most outmoded collections of rules found anywhere in American public policy. The society that created the current system well over a century ago saw the rivers of the American West as a commodity that needed to be put to use, and extravagant expenditures to impound and transport water were justified in the name of "conservation." This single-minded focus was successful in achieving the goals of encouraging settlement in the dry and inhospitable Western lands. Today, the region has become a very different place, and people worry about long-term sustainability. Water policy is under pressure to respond to new technology, new values, and changing demands. "Perhaps the single most influential development in Western water law during this century has been the rediscovery of the public nature of water." Future Western water policy should be based on three key principles: 1) conservation (demanding that the pros and cons of every water use be carefully weighed); 2) equity (treating the whole community fairly); 3) ecology. This "water ethic" can lead to balanced water use. **(water ethic for the West)**

086
River Life Through U.S. Broadly Degraded, William K. Stevens, *The New York Times*, Tues, 26 Jan 1993, C1.

"Two decades of Federal controls have sharply reduced the vast outflows of sewage and industrial chemicals into America's rivers and streams, yet the life they contain may be in deeper trouble than ever. The main threat now comes not from pollution but from humans' physical and ecological transformation of rivers and the land through which they flow. The result, scientists say, is that the nation's running waters are getting biologically poorer all the time and that entire riverine ecosystems have become highly imperiled." Few if any major river systems are unaffected by the threat to ecological integrity. Major classes of factors that affect river environments: organic material entering a stream (e.g. sediment from farm fields), chemicals and nutrients affecting water quality, destabilization of substrate and banks by erosion and increased sediment, the volume and timing of water flow (interrupted by dams), and interactions of living things through introduction of exotic species. A number of scattered efforts to restore rivers and streams are being undertaken. **(river ecosystems "highly imperiled")**

087
Lifelines: The Case for River Conservation. Tim Palmer. Washington: Island Press, Sept 1994/259p/$35.00; $16.96pb.

During the past two decades, profound changes have taken place in the way we treat US rivers, the lifelines of a continent. Proposals for large, destructive dams seldom surface anymore, and some of the foulest pollution has been stopped. But smaller dams for hydropower still threaten many rivers, and concerns for water quality have shifted from conventional industrial discharges and municipal sewage to an epidemic of toxic wastes and the blight of polluted runoff. The US has 3.6 million miles of rivers and streams of all sizes; 30% of that mileage does not meet clean water standards, and the figure would rise to 50% or more if biological criteria were also considered. According to the National Park Service, 75,000 large dams block nearly every river in the US outside Alaska, and 200,000 miles have been channelized in the name of flood control. A clear majority of river/stream frontage in most regions has been developed for industry, farming, or homes. Proposals for protection: 1) drop the remaining Federal incentives backing large dams; 2) reduce nitrate pollution by reducing use of chemical fertilizers; 3) tax pesticides and fertilizers harmful to public waters; 4) retire acreage on which irrigation causes land and water damage; 5) raise grazing fees on public land in the West and enforce the requirement that land and water not be degraded. To make any progress in river conservation, the key principle is to build broad community support. [ALSO SEE: **Rivers at Risk: The Concerned Citizen's Guide to Hydropower** (Island Press, 1989; **FS Annual 1991** #10530).] **(river conservation)**

088
Alternatives for Ground Water Cleanup. National Research Council, Committee on Groundwater Cleanup Alternatives. Washington: National Academy Press, July 1994/315p/$39.95.

America's groundwater resource, which supplies more than 50% of drinking water, is threatened not only be excessive overdrafts, but by contamination. It is estimated that 300,000 to 400,000 sites may have contaminated soil or ground water requiring some form of remediation. One 1991 report concluded that over the next 30 years the US will spend $480 billion to $1 trillion in cleanup activities, with a best guess of $750 billion, or $8,000 per household. (The magnitude of the problem may be equally significant in other industrialized countries.) Restoration of contaminated ground water to drinking water standards is possible, but cleanup is inherently complex, and will require large expense and long time periods. The inability of "pump-and-treat" systems to reach drinking water standards at many sites to date is not only due to complexity, but insufficient or inaccurate characterization of the problem. "The Committee strongly believes that existing ground water cleanup goals cannot be attained in reasonable time frames (decades) at a large number of sites with current technologies." Some sites will remain contaminated above drinking water standards for the foreseeable future, even when the best available technologies are used. Although long-term goals may remain unchanged, interim objectives are still needed in order to acknowledge current technological limitations. **(ground water cleanup goals)**

089
Restoration of Aquatic Ecosystems: Science, Technology, and Public Policy. National Research Council, Committee on Restoration of Aquatic Ecosystems (John Cairns Jr of VPI, Chairman). Washington: National Academy Press, Feb 1992/552p/$39.95.

Rapid population increases in many parts of the US have led to pollution of surface waters by fertilizers, insecticides, motor oil, toxic landfill leachates, and feedlot waste. At the same time that pollution and sewage effluents have increased, water consumption has also increased, reducing the flows for dilution of wastes. Add to this the increased sediment from urban construction, agriculture, and forestry, resulting in greater turbidity and sedimentation in downstream channels, lakes, and reservoirs, with attendant losses of water storage and conveyance capacity, recreational and aesthetic values, and quantity and quality of habitat for fish and other wildlife. America's investment in different types of environmental programs has been considerable but piecemeal, and has not always been effective because of failure to recognize interconnected ecosystems.

A large-scale national aquatic restoration strategy is recommended: **1) Lakes**: a net gain over the next 20 years of 2 million acres of restored lakes is achievable (out of the current 4.3 million acres of degraded lakes); **2) Rivers and Streams**: 400,000 miles of river-riparian ecosystems (about 12% of US rivers and streams) should be restored in the next 20 years by accelerated erosion control programs, changing grazing practices, bio-engineering techniques for bank stabilization and repair; **3) Wetlands**: an overall gain of 10 million wetland acres is recommended by 2010, largely through reconverting crop and pastureland and modifying or removing existing water-control structures (this is a 10-fold increase in the 1990 wetlands restoration target, but less than 10% of the 117 million acres total wetlands loss in the last 200 years). This new emphasis on stewardship and restoration cannot succeed without public understanding and support through various education programs. Without such an ambitious program, "our swelling population and its increasing stresses on aquatic ecosystems will certainly reduce the quality of human life for present and future generations." Restoring these ecosystems is a high-priority task at least as urgent as protecting water quality through abatement of pollution. Concludes with 13 case studies. **(aquatic restoration strategy)**

090

Statewide Wetlands Strategies: A Guide to Protecting and Managing the Resource. World Wildlife Fund (Heidi Sherk, Project Director). Washington: Island Press, 1992/268p(8x11")/$60.00;$40.00pb.

The US continues to lose 290,000 acres of wetlands each year, an area almost half the size of Rhode Island. Wetlands are important for water quality, groundwater recharge, erosion prevention, flood control, fisheries, waterfowl, biodiversity, and recreation. But programs to protect wetlands lack consistency, predictability, and timeliness, and local, state, and Federal guidelines often conflict. This guidebook proposes a "comprehensive statewide wetlands strategy" that is more inclusive and responsive to regional and local needs than current approaches. The strategy draws on all levels of government and the private sector to coordinate efforts toward the goal of no net loss of wetlands, and long-term net gain. Chapters discuss strategy development, mechanisms for protecting wetlands (acquisition, regulation, planning, restoration, creation, etc.), and wetlands data sources and collection methods. [ALSO SEE: **Wetlands: An Approach to Improving Decision Making in Wetland Restoration and Creation**, edited by Ann J. Hairston (Island Press, Dec 1992/151p/$40), based on an approach developed by the EPA Wetlands Research Program.] **(wetlands protection strategy)**

091

Turning the Tide: Saving the Chesapeake Bay. Tom Horton (*Baltimore Sun*) and William M. Eichbaum (VP, World Wildlife Fund/Conservation Foundation). Washington: Island Press, Aug 1991/324p/$22.95;$14.95pb.

A comprehensive assessment of the entire watershed of the Chesapeake Bay, sponsored by the 80,000-member Chesapeake Bay Foundation, founded in 1967 to save the Bay. Overall, as the human population has increased in the Middle Atlantic region, the health of the Bay has declined. The natural systems have been overwhelmed in recent decades by wasteful land use and energy consumption, the use of agricultural and industrial chemicals, and higher-tech overfishing by both sportsmen and watermen. Although there have been notable efforts and some successes in cleanup and restoration, "we still are winning battles but not the war." There is no reason to accept a vision of inevitable decline or holding the line at some semidegraded level. A brighter vision is articulated of a cleaner and more resilient Bay with a more livable watershed. Numerous recommendations are made concerning agricultural runoffs, the Susquehanna River (source of half the Bay's fresh water) and Pennsylvania's unmet obligations, limiting sewage discharges, eliminating toxic discharges, reducing oil spills, restoring oyster populations (reduced to 1% of what existed 100 years ago), managing waterfowl, protecting wetlands, etc. Concludes that we must learn to see the Bay whole, and, rather than correct piecemeal problems as they arise, "we must define and enforce limits on all of society's environmental impacts," and replace myths of "pollution as the price of progress" that underlie current development of the watershed. [**NOTE**: As suggested in the preface, this book can serve as a model for restoration efforts in other coastal regions worldwide.]

(Chesapeake Bay restoration)

8. Biodiversity

092

The Diversity of Life. Edward O. Wilson (Prof of Science, Harvard U). Cambridge MA: The Belknap Press, Harvard U Press, Oct 1992/424p/$29.95; W.W. Norton $14.95pb.

"Biodiversity is our most valuable but least appreciated resource." Wilson describes the processes that create new species in bursts of adaptive radiation (the spread of species into different niches), and the cataclysmic events that have disrupted evolution and diminished global diversity over the past 600 million years. The five previous natural blows to the planet (such as meteorite strikes and climatic changes) required 10 to 100 million years of evolutionary repair. Humanity has initiated the sixth great extinction spasm, rushing to eternity a large fraction of our fellow species in a single generation. We do not know the vast majority of species of organisms well (we have yet to anoint 90% of them with scientific names). In the small minority of groups of plants and animals that are well known, extinction is proceeding at a rapid rate, far above prehuman levels. In many cases the level is calamitous: the entire group is threatened. By 1989, the tropical rainforests of the world had been reduced to about 8 million square km., or slightly less than half the prehistoric cover. Surviving rain forests occupy an area about that of the contiguous 48 US states, and are being reduced by an amount equivalent to the size of Florida every year. "A 20% extinction in total global diversity, with all habitats incorporated, is a strong possibility if the present rate of environmental destruction continues." The most conservative estimate based on our current knowledge of the extinction process is 27,000 species doomed each year; "each day it is 74, and each hour 3."

The mindless horsemen of the environmental apocalypse have always been overkill, habitat destruction, introduction of exotic species such as rats and goats, and diseases carried by these exotics. In prehistory, the paramount agents were overkill and exotic animals. In recent centuries and especially in our generation, habitat destruction is the major lethal force, followed by invasion of exotic animals. With the problem only beginning to come into focus, there is little doubt about what needs to be done: **1) Survey the World's Flora and Fauna**: a complete inventory of the world's species is needed over the next 50 years, with immediate emphasis on identifying "hot spots" (habitats with many endangered endemic species); **2) Create Biological Wealth**: make bioeconomic analysis a routine part of land management policy, protecting ecosystems by assigning them future value (it can buy time against removal of entire communities ignorantly assumed to lack value); a key to this analysis is *chemical prospecting*, the search among wild species for useful chemical products; **3) Promote Sustainable Development**: the rural poor of the Third World are locked onto a downward spiral of poverty and the destruction of diversity; to break free they need work that provides basic food and housing; "the proving ground of sustainable development will be the tropical rain forests"; **4) Save What Remains**: biodiversity can be saved by a mix of programs, notably locating the world's hot spots and protecting the environment they contain; creating new species in the laboratory or resurrecting extinct species from DNA are doubtful (and the best efforts by zoos will only slow the tide of extinction); **5) Restore the Wildlands**: there is still time to save many of the "living dead" if natural habitats are not only preserved but enlarged; "the next century will, I believe, be the era of restoration in ecology." Concludes that the ethical imperative should be prudence, and that "we should judge every scrap of biodiversity as priceless while we learn to use it and come to understand what it means to humanity." [**NOTE**: Fresh, eloquent, authoritative, and very readable.]

(halting the "extinction spasm")

093

The Primary Source: Tropical Forests and Our Future. Norman Myers (Oxford UK). NY: W.W. Norton, March 1992/416p/$10.95pb.

First published in 1984 (**FS Annual 1985** #6708), this updated edition adds a new Prologue and Epilogue to chapters on biological richness, the impacts of modern man, and the unrealized potential for rational use and preservation. In the new material, Myers estimates that the rate of deforestation increased in the 1980s, and that the destruction rate could well double within the decade. Since writing the first edition of this book, a forest area 1.5 times the size of Texas has been destroyed. But half of all tropical forests that ever existed are intact, and there is much to battle for. Commercial logging and cattle ranching are a problem, but the "shifted cultivator" accounts for well over half of deforestation—a proportion that is expanding rapidly. The shifted cultivator (roughly estimated at 200-500 million people worldwide) is the peasant farmer squeezed out of traditional farmlands, who heads for the only unoccupied lands available. The survival of tropical forests depends on a range of policy initiatives (agrarian reform, extension services, credit facilities) to help the subsistence farmer make an acceptable living in established farmlands. At the planetary level, a Global Forests Convention is needed, leading to a World Forests Organization backed by a Tropical Forests Fund. This would mobilize a critical mass of intersectoral expertise in all areas related to tropical forests, and offer a forum for NGO contributions. Safeguarding all tropical forests could cost us roughly $20 billion a year—much less than the eventual price we shall have to pay for global warming. Grandscale reforestation might cost another $4 billion a year. **("shifted cultivator" and tropical forests)**

094

Last Animals at the Zoo: How Mass Extinction Can Be Stopped. Colin Tudge. London: Hutchison Radius, 1991; Washington: Island Press, March 1992/266p/$22.00.

A widely-published British zoologist/science writer argues that zoos are now an essential part of modern conservation strategy, and that of the several tasks that fall to them, by far the most important is the breeding of endangered animals or "conservation breeding." This may seem obvious, but it has not been obvious for very long, and it is not obvious to every senior zoo person now. Today, zoo populations are often far safer than those in the wilderness, and in many cases are growing while the precarious populations of the wild continue to dwindle. The ideal, of course, is to save what is left of the wilderness—to protect the wild places where animals live. But this is not always an option in the short term. Even the best zoos should be seen as part of a spectrum that extends from the intensive breeding center, to fenced reserves for one species, managed reserves with a select list of species, the national park which resembles wilderness, and wilderness itself in which nature unfolds according to its own rules.

Chapters discuss the reasons for conserving animals, the scope of the problem, the theory of conservation breeding, projects in progress, prospects for "the frozen zoo" (in which gametes and embryos are stored, perhaps augmented by cloning), and progress in understanding the whole animal. Concludes that zoos, where animals live in a protected place, are necessary. But they are not the end of the road. The aim of serious conservation breeders is to return animals to their native habitats, as soon as these habitats can be made safe again. This is already happening in some areas. "In a few centuries' time, if all goes well, the importance of zoos will diminish. For the present, their significance must grow. For the next few centuries they must be perceived as a necessary part of civilization. When they rise to this challenge they become exciting. Any lesser perception, or lesser endeavor, is a trivialization." **(conservation breeding and zoos)**

095

The Last Panda. George B. Schaller (Wildlife Conservation International Division, NY Zoological Society). Chicago IL: U of Chicago Press, April 1993/291p/$24.95.

China's giant panda has been enshrined as an icon of our environment. It serves as the distinctive logo of the World Wildlife Fund, and no animal is more popular with the zoo-going public. The panda's black and white markings personify Yin and Yang, summer and winter, life and death. But humankind has upset the balance, and the panda's existence is now shadowed by fear of extinction. Only about 1,000 giant pandas survive in the wild, dependent on a shrinking supply of bamboo and threatened by human greed. A live panda can bring over $100,000 on the Chinese black market; a panda pelt is worth $10,000 or more. Western zoos pay millions to rent the ever-popular panda for exhibit; by 1988, at least 30 American zoos and other institutions had applied for panda rentals from the Chinese.

Schaller and Chinese colleagues conducted a five-year study of the panda in the early 1980s, a project that, in retrospect, "may have harmed rather than helped the panda" because of resulting publicity that raised the price for panda rentals and trophies. Most of this book describes the study, but there are concluding commments about the moral obligation to maintain the species in the wild. There will never be enough pandas for all who clamor for them. Schaller hopes that panda management will improve. Captive breeding in China should achieve sustained success before pandas are sent on stressful world trips that disrupt their reproductive capacity. Some day China may make more money from panda tourism than from panda loans, just as visitors pay high fees to observe mountain gorillas in Africa. **(pandas loved to near-extinction)**

*096

Global Biodiversity Strategy: Guidelines for Action to Save, Study, and Use Earth's Biotic Wealth Sustainably and Equitably. World Resources Institute, The World Conservation Union (IUCN), and UN Environment Programme, in consultation with FAO and UNESCO. Washington: WRI, Feb 1992/244p/(8x11)/$19.95pb.

All life on Earth is part of one great, interdependent system, interacting with and depending on non-living components of the planet. Humanity depends totally on this community of life—this biosphere—of which we are an integral part. Economic development is essential if the millions of people who live in poverty are to achieve a quality of life commensurate with the most basic of human rights. "Development has to be both people-centered and conservation-based.... Unless we use Earth's resources sustainably and prudently, we deny people their future. Development must not come at the expense of other groups or later generations, nor threaten other species' survival. The conservation of biodiversity is fundamental to the success of the development process." Conserving biodiversity is not just a matter of protecting wildlife; it is also about safeguarding the natural systems of Earth that are our life-support systems.

Following brief chapters on the nature and value of biodiversity (by Peter Raven) and losses of biodiversity and their causes, the bulk of this report is devoted to describing 85 actions in 8 chapters on the strategy: **1) Catalytic**

Actions: adopting the international Convention on Biological Diversity, designating 1994-2003 as the International Biodiversity Decade, creating an International Panel on Biodiversity Conservation to develop priority lists, establishing an Early Warning Network, and integrating biodiversity conservation into national planning processes; **2) Establish National Policy Frameworks for Biodiversity**: abandon unsustainable practices in forestry and agriculture, adopt new accounting methods that promote conservation, reduce demand for biological resources; **3) An International Policy Environment That Supports Biodiversity**: integrate conservation into international economic policy, develop a principle of "national ecological security," strengthen the international legal framework, incorporate biodiversity values into development assistance, improve debt-for-nature swaps; **4) Conditions and Incentives for Local Conservation**: use land under cultivation more efficiently, increase incentives for local stewardship, promote the value of local knowledge and affirm rights of local people; **5) Manage Biodiversity Throughout the Human Environment**: develop new methods for bioregional development and planning, support private sector initiatives; **6) Strengthen Protected Areas**: identify priorities to promote protected areas in conservation, broaden participation in designing management plans; **7) Conserving Species, Populations, and Genetic Diversity**: strengthen the capacity of off-site conservation facilities to conserve, educate the public; **8) Expanding Human Capacity to Conserve**: help institutions disseminate needed information, produce periodic national biodiversity assessments, reconsider research priorities, increase support for training biodiversity professionals. Concludes with a listing of 46 "partner organizations" and about 800 individuals worldwide who helped in the research/preparation of this report. [ALSO SEE: **Caring for the Earth: A Strategy for Sustainability** by IUCN, UNEP, and the World Wildlife Fund (#170).] **(biodiversity strategy)**

097

Global Marine Biological Diversity: A Strategy for Building Conservation into Decision Making. Elliott A. Norse (Chief Scientist, Center for Marine Conservation, Washington). Washington: Island Press, Oct 1993/383p/ $50.00;$27.50pb.

This companion document to the **Global Biodiversity Strategy** (above) is sponsored by CMC, IUCN, WWF, UNEP, and the World Bank. It argues that conserving biodiversity in the sea has been even more neglected than on land. Despite their importance to us, humankind is destroying marine species and ecosystems, not only in estuaries and coastal waters, but in the open seas. We harm marine biodiversity by overexploiting living things, altering the physical environment, polluting the sea, introducing alien species, and adding substances to the atmosphere that increase ultraviolet radiation and alter climate. The root causes underlying these threats to marine life are too many people consuming too much, institutions that degrade rather than conserve, lack of knowledge, and insufficient valuation of nature. Chapters discuss the importance of marine biological diversity, impediments to marine conservation (national sovereignty, North-South divisions, jurisdictional gaps and overlaps, fragmented decision making, etc.), the key strategy of building conservation into decison making (best done by institutionalizing sustainability into laws and actions, and providing incentives to choose sustainability over other options), tools for conserving marine biodiversity (expanding the knowledge base, planning, regulating threats, economic tools, protecting and restoring areas), and lessons from existing institutions and instruments. Concludes with about 150 recommendations for implementing *"The Strategy,"* such as: 1) nations should sign and implement the Convention on Biological Diversity; 2) UNEP, IUCN, WRI, and WWF should produce biennial reports on the state of biological diversity; 3) the UN should establish 1995 to 2004 as the International Biological Diversity Decade and establish a Global Biodiversity Forum; 4) the UN should establish an Early Warning Monitoring Network on threats to marine biodiversity; 5) UNEP and others should establish an International Marine Conservation Network to link decision makers worldwide; 6) coastal countries should have a management agency responsible for protecting marine resources; 7) nations should push for a new, binding climate convention to replace the one signed at UNCED. **(marine biodiversity strategy)**

098

Saving Nature's Legacy: Protecting and Restoring Biodiversity. Reed F. Noss (Editor, *Conservation Biology*) and Allen Y. Cooperrider (formerly Bureau of Land Management). Washington: Island Press, May 1994/416p/ $48.00;$27.50pb.

Today we stand on the threshold of a revolution in wildlife conservation policy, due to emergence of a more holistic view of the relationships among plants, animals, and the environment. As a nation, we have tended to focus on the pieces—our favorite places and preferred species; "we now are moving increasingly toward a recognition that if we are to save the pieces, we must also protect the systems upon which they depend for their survival" (from Foreword by Rodger Schlickeisen, President, Defenders of Wildlife—the sponsor of this book). Chapters discuss the value of biodiversity, ecological processes, current status of biodiversity in North America ("ample evidence" suggests substantial impoverishment and an increasing rate of loss), past and present conservation strategies, designing biosphere reserves, managing forests and rangelands, guidelines for conserving aquatic biodiversity, monitoring programs, and barriers to continuing biodiversity (exploitative philosophies, fragmented institutions, political suppression of scientific findings, and ignorance arising from the increasing separation of people from nature). Some of the proposed priority actions for conserving biodiversity at global, national, and local levels: 1) encourage rapid stabilization of human population; 2) reduce the US rate of resource consumption to set a positive example to the Third World; 3) provide incentives for sound stewardship of private lands and land donations to conservation agencies; 4) implement a national biological survey; 5) set recovery goals for endangered ecosystems; 6) strengthen the Endangered Species Act by greater protection of habitats; 7) establish a new Civilian Conservation Corps oriented toward land restoration projects; 8) place biodiversity conservation as the highest management objective for all Federal lands; 9) create bad local publicity for those engaged in destructive activities. **(priority actions for biodiversity)**

099

Endangered Species Recovery: Finding the Lessons, Improving the Process. Edited by Tim W. Clark (School of Forestry, Yale U), Richard P. Reading (Bureau of Land Mgt), and Alice L. Clarke (U of Trondheim, Norway). Washington: Island Press, Oct 1994/450p/$48.00;$25.00pb.

Despite dedication and hard work by many people in and out of government, and vast expenditures of time and money, implementation of the 1973 Endangered Species Act has been weak. The disappointing performance is well-documented, but poorly understood. These essays, arising from a University of Michigan conference in 1993, point to the profound need and numerous opportunities to improve professional and organizational responses to the endangered species crisis. After examining nine case studies, several "meta-lessons" were distilled: 1) endangered species conservation is a multifaceted task of interacting biological, professional, sociological, organizational, economic, political, and policy dimensions; 2) there is no substitute for sound professional training well grounded in state-of-the-art theories and techniques; 3) high-performance teams should carry out the work of species restoration; 4) government agencies must be made more effective; 5) goals must be set high enough to ensure long-term species conservation, and overall progress must be continuously monitored; 6) mechanisms to promote learning should be institutionalized; 7) conservation leaders must be encouraged, trained, and empowered. In sum, "too many recovery efforts fall prey to their own narrow and inadequate problem analysis, crippling organizational arrangements, and limited learning approaches."

(endangered species recovery reconsidered)

100

Environmental Policy and Biodiversity. Edited by R. Edward Grumbine (Director, Sierra Institute, U of California Extension, Santa Cruz). Washington: Island Press, Oct 1994/415p/$45.00; $22.00pb.

As the biodiversity crisis has accelerated, debates over fundamental approaches to science and policy analysis have intensified. Calls for new environmental management goals have risen in conjunction with the loss of biodiversity and new knowledge from conservation biology. "But no policy initiatives appear to be effective at slowing down the rate of loss." The safety net of US environmental laws is being stretched thin, as society exceeds environmental limits. These essays discuss preservation of natural biotic communities, conservation biology, the design and management of nature reserves, biological reasons why the Endangered Species Act doesn't work, biodiversity law, creating knowledge for species and ecosystem conservation, paradigm shifts in land and water policy, and case studies of conservation biology (California, Great Lakes, Northern Florida, and the Wildlands Project, aka the North American Wilderness Recovery Project). Grumbine reviewed recent literature and policy documents, finding general agreement on such key themes as upholding ecological integrity over sustained yield as a new management goal, hierarchical context (noting connections among all levels of biodiversity), interagency cooperation, new research and data collection needs, monitoring actions, adaptive management, and changing public lands agencies. A new ecological model is noted that recognizes "flux" over "balance" in nature. Whereas ecologists once saw natural systems as orderly, steady state, and in equilibrium, the new model views nature as dynamic, uncertain, and episodic. **(biodiversity policy: new thinking)**

101

Reclaiming the Last Wild Places: A New Agenda for Biodiversity. Roger L. DiSilvestro (northern Virginia). NY: John Wiley & Sons, July 1993/266p/$24.95.

Former editor for the National Audubon Society describes the history of conservation and Federal land management in the US, highlighting fundamental errors. Foremost among them has been the historic practice of creating isolated pockets of wilderness exposed on all sides to human "progress." Rather than creating wildlife sanctuaries, such places become ecological prisons affording animals far from sufficient room and resources in which to thrive. The total of our protected lands equals about 9% of the US, a fragment of the once vast natural ecosystems, itself broken up into a sprinkling of tinier fragments. The best way to maintain wildlife and ecosystems is to minimize this habitat fragmentation. The US needs a national conservation plan, created by Congress as a National Biodiversity Policy Act. The act would create a new land-protection agency with the sole mandate of protecting natural biodiversity and ecosystems within the lands it manages. The new agency would create a network of connected wildlands now managed by various agencies. The conservation community should find means for making the protection of wildlands remunerative to people living near the lands, and help promote a national school science curriculum. **(new approach to biodiversity)**

102

Noah's Choice: The Future of Endangered Species. Charles C. Mann (Contributing Editor, *Science*) and Mark L. Plummer (Senior Fellow, The Discovery Institute, Seattle). NY: Alfred A. Knopf, Feb 1995/302p/$24.00.

In some ways, Noah had it easy. The materials he needed to build his Ark were at hand, the design was provided by the Lord, and the vessel was sufficient for the task. Two by two, the creatures walked aboard the Ark, and Noah fulfilled his obligations when the parade finished. There was no need to exercise judgment or agonize over tough choices; no creature was excluded. But in the role of modern Noahs, we face momentous choices. We want to load endangered species on our ark, but the task must compete for scarce resources with other worthy projects. Rather than orders from a Supreme Deity, we are without blueprints for our conduct or for building the ark. We don't know the number of potential passengers, and we can't accommodate everything. It is up to us to make these tragic choices consciously and thoughtfully. Unfortunately, the present system seems unable to do this. In 1973, Congress faced the incoming tide of extinction and commanded it to stop with the Endangered Species Act. The endangered species list has grown steadily in the past two decades, and the trend shows no sign of stopping. For every species the Fish and Wildlife Service has removed from the list since 1973, it has added more than 30 others (and most of those removed have been due to extinction or new information). Not including 134 species carried over from pre-1973, 721 species have been added to the list. The net total was 836 species in 1994, up from 563 in 1990. Only five gained enough ground to warrant removal of the Act's protections. The best that can be said about the Act is that in many cases it may have slowed the fall to extinction. Four principles are proposed in reforming the law: 1) the system for making choices must be ethical, balancing among different interests, equities, and values; 2) the system must be practicable, and the means to attain goals must be present; 3) the system must be knowledgeable (too many species are being listed too early or too late); 4) the system must be political, and not run solely by experts.

(endangered species: new principles needed)

103

Biodiversity Prospecting: Using Genetic Resources for Sustainable Development. Walter V. Reid (VP, WRI) *et al.* Washington: World Resources Institute, June 1993/341p/$29.95pb.

"Biodiversity prospecting"—the search for wild species that can yield new medicines, better crops, and raw materials for industry—can help protect the habitats of plants and animals on the edge of extinction if developing countries have more than a token stake in the profits from commercially valuable discoveries. This report analyzes the laws, contracts, and organizations needed to promote conservation and assure that profits from biodiversity are divided fairly. Special attention is given to Costa Rica's National Biodiversity Institute (INBio) and its arrangement with Merck & Co, the world's largest pharmaceutical firm. This agreement is just one of a rapidly growing number of biodiversity prospecting ventures. But much of this activity is taking place in a policy vacuum. Needed institutional arrangements include intellectual property regimes (IPR), international agreements, and intermediary organizations to facilitate access to resources and their transfer to industry. IPR can be used to protect indigenous people, farmers, and traditional healers, although extension of IPR to wild species appears unworkable at present. **(biodiversity prospecting agreements)**

104

Genes for Sale: Privatization as a Conservation Policy. Joseph Henry Vogel (Associate Prof of Economics, Facultad Latinoamericana de Ciencias Sociales, Quito, Ecuador). NY: Oxford U Press, Aug 1994/155p/$29.95.

Mass extinction is a complex problem and its solution will also be complex. Synthesizing the economics of property rights, the biology of conservation, and nonequilibrium thermodynamics, Vogel (who recently held research posts in Australia and Brazil) explores the "very counterintuitive idea" that privatization can become a conservation policy. The best way to retain useful genetic information is to create property rights over it and, at the same time, constrain this newly created property (similarly, a permit to release greenhouse gas is the creation and attenuation of the right to pollute). Ten principles for conserving genetic information: 1) replace the word "biodiversity" with the term "genetic information"; 2) legislate equal protection to artificial and natural information; 3) endorse a temporary general environmental fund (GEF) to support conservation through royalties ("under no circumstances should the World Bank assume responsibility for the GEF; it has a proven record of failure"); 4) remit part of the royalties to landowners on whose property the genetic information is stored; 5) resist giving or receiving money for environmental protection; charity is not sustainable; 6) endorse family planning on a massive scale; 7) if climate evidence is persuasive, endorse a greenhouse permit accord with harsh penalties for recalcitrant countries; 8) decouple northern finance from southern dictatorships; 9) endorse mechanisms that guarantee human rights; 10) watch for northern aggression on the pretense of human-rights abuse (various violations in Brazil invite such intervention and "resultant internationalization of the Amazons"). This policy of privatization is markedly different from that of self-constraint by individuals, cooperation among governments, and economic reform, as notably proposed by the **Global Biodiversity Strategy** of WRI/IUCN/UNEP (#096). Also includes a chapter on "genesteading" and population control in the Brazilian rain forest, with rewards tied to smaller families. [**NOTE**: An intelligent and readable provocation by a truly "unorthodox economist," with extensive footnotes and boxed asides in the text. Back cover blurbs are supplied by Jeff McNeely (Chief Officer of IUCN), Ghillean Prance, Helen Caldicott, Jane Goodall, and Norman Myers.] **(privatizing for conservation?)**

105

Biodiversity: Culture, Conservation, and Ecodevelopment. Edited by Margery L. Oldfield (Dept of Biology, Trinity U) and Janis B. Alcorn (Biodiversity Support Program, Washington). Foreword by E.O. Wilson. Boulder CO: Westview Press, Dec 1991/349p/$42.50pb.

Today, traditional cultures and biodiversity are both threatened by the continuing expansion of modern, industrialized societies. These papers from a 1988 regional AAAS conference discuss the importance of cultural and biological diversity, genetic resources and traditional agroecosystems, traditional conservation and use of biodiversity, and linking rural development with biological conservation. Alcorn concludes that two strategic actions are absolutely necessary to achieve global conservation action: 1) empowering the local, traditional conservation ethic wherever it still exists; 2) constructing a new, modern conservation ethic for the emerging global capitalist economy. Planners and decisionmakers have pursued neither of these actions, because of incorrect assumptions about biological conservation and ecodevelopment. To retain biodiversity, all actors—from rural poor to international corporations—must be kept as players by making them stakeholders in long-term conservation. [**NOTE**: Important for linking biological diversity with cultural diversity.] **(cultural and biological diversity)**

106

Guardians of the Land: Indigenous Peoples and the Health of the Earth. Alan Thein Durning (Senior Researcher, Worldwatch Institute). Worldwatch Paper 112. Washington: Worldwatch Inst, Dec 1992/62p/$5.00.

Human cultures are disappearing at unprecedented rates: "worldwide, the loss of cultural diversity is keeping pace with the loss of biological diversity." One linguist projects that half of the world's 3,000 languages, and arguably the cultures they embody, will disappear within a century. The contest between industrial societies and native peoples is so lopsided that supporting indigenous cultures may seem quixotic. But, far from disappearing, the world's indigenous people number in the hundreds of millions, and their cause is gaining ground. Two reasons why the world's dominant societies should heed their voices: 1) indigenous peoples are the sole guardians of vast, little-disturbed habitats that modern societies depend on; 2) their ecological knowledge is an asset of great value; encoded in indigenous languages and practices "may be as much understanding of nature as is stored in the libraries of modern science." Over the past few decades, hundreds of native groups have organized and joined forces to defend their resources and ways of life. They offer the world's dominant consumerist/individualist culture living examples of ancient values that may be shared by everyone: devotion to future generations, ethical regard for nature, and commitment to community. Indigenous peoples are also "the miner's canary of the human family:" their cultures are the first to suffer when the natural realm is poisoned, degraded, or exhausted. In the words of a Guarani holyman, "When the Indians vanish, the rest will follow." [ALSO SEE: **Wisdom of the Elders: Honoring Sacred Native Visions of Nature** by David Suzuki and Peter Knudtson (Bantam, 1992; #200), **Our Responsibility to the Seventh Generation: Indigenous Peoples and Sustainable Development** (IISD, 1992; #201), and **The New Resource Wars: Native and Environmental Struggles Against Multinational Corporations** (South End Press, 1993/270p/$15.00pb).]

(indigenous peoples and the environment)

107
State of the Peoples: A Global Human Rights Report on Societies in Danger. Edited by Marc S. Miller (Project Director, Cultural Survival, Cambridge MA). Boston: Beacon Press, Nov 1993/262p/$35.00;$18.00pb.

Jason W. Clay, founding editor of *Cultural Survival Quarterly* and director of a database on the languages of the world, notes that there are now more than 190 states in the world, up from only 50 before WWII and 170 as recently as 1989. By contrast, the world's 6,000 or so indigenous groups (defined by a unique language, a long-term relationship to a homeland, and one-time exercise of political control over their destinies) have been around for centuries, even millennia. At the heart of the matter is the state-building, nation-destroying process, founded on the belief that indigenous peoples and states can't coexist. "More than half the world's 15,000 known languages have disappeared; only 5-10% of the remaining 6,000 to 7,000 are likely to survive another 50 years." Despite this history of cultural destruction, indigenous peoples account for some 10-15% of the world's population, and have traditional claims over some 25-30% of the earth's land area and resources. This report summarizes overall conditions in ten world regions, and the major groups within them. It also includes the Draft Declaration on the Rights of Indigenous Peoples, created by the UN Working Group on Indigenous Populations, specifically recognizing rights to self-determination and protection against cultural genocide. When adopted by the UN General Assembly, it will form the basis for a new relationship of indigenous peoples with states and with the UN. **Julian Burger** notes that, whereas indigenous peoples were once considered a remnant of the past, destined to inexorable assimilation by mainstream society, the second World Conference on Human Rights has asked the General Assembly to proclaim a decade for indigenous peoples.

(rights of indigenous peoples)

9. Forests

108
World Forests for the Future: Their Use and Conservation. Edited by Kilaparti Ramakrishna and George M. Woodwell (both Woods Hole Research Center). New Haven CT: Yale U Press, April 1993/156p/$18.50.

Papers from an October 1991 workshop at Woods Hole MA. Woodwell discusses the global need for forests to stabilize heat-trapping gases and protect biotic potential, and calls for "immediate cessation of deforestation and an increase on the order of 10% or more in the forested area of the earth." Richard A. Houghton looks at global deforestation and the impact on global warming. Darrell A. Posey points to the value of indigenous knowledge for new models of sustained natural resource use. Herman E. Daly notes the evolution of the human economy from an era in which manmade capital was the limiting factor in economic development to an era in which increasingly scarce natural capital has taken its place. Robert Repetto points to the role of governments in forest usage. Jagmohan Maini and Ola Ullsten describe approaches to sustainable development of global forests and propose "Project Green Globe" to establish realistic targets for global forest cover. Ramakrishna proposes an "International Commission on the Conservation and Use of World Forests."

The introduction states that the continuous global destruction of forests is a major factor in the warming of the earth. There is a possibility that this warming may soon reach a point where "additional release of heat-trapping gasses that it induces from accelerated decay of organic matter in forests and soils will exceed the extent of control possible through curtailing use of fossil fuels and improving forest management. The warming will feed on itself by destroying forests more rapidly than they can be regenerated. The effect will be a rapid and progressive reduction in the capacity of the earth to support life, including people. There is no proof of this potential catastrophe. There will be no proof until far too late to deflect the process." **(global warming to feed on itself?)**

109
The Struggle for Land and the Fate of the Forests. Edited by Marcus Colchester (Coordinator, World Rainforest Movement) and Larry Lohmann (Assistant Editor, *The Ecologist*). London and Atlantic Highlands NJ: Zed Books, July 1993/389p/$49.95;$17.50pb.

A 1980 FAO study showed global deforestation at some 114,000 sq. km./yr. A study by Norman Myers estimated 142,000 sq. km. of tropical forest destroyed in 1989, and a further 200,000 sq. km. seriously degraded. Interim results from a new FAO study place the global rate at some 170,000 sq. km./yr. A 1990 study by the World Resources Institute suggests that the figure could be as much as 204,000 sq. km./yr. In Central Africa, the Caribbean, and Southeast Asia, rates of forest loss have surged ahead of this dismaying global average increase. The data and definitions are imprecise and subject to different interpretations, but the overall trend is clear. There is even less agreement as to why this is happening. Many international agencies argue that deforestation results from overpopulation and underdevelopment. But many NGOs lay the blame on transnational corporations, development agencies, and excessive consumption in the North. This study shows that both explanations are partially correct: both wealth and poverty underlie deforestation, as well as the inequitable structures that link the two—i.e., deforestation is an expression of social injustice. Poor people, denied land and livelihood, are being forced into the forests in ever increasing numbers for sheer survival. Those in power hide these unpalatable truths behind a "developmentspeak." Presents case studies of land and forest issues in Guatemala, Brazil, Zaire, Thailand, Indonesia, and the Philippines. There are no quick fixes; long-term solutions require agrarian reform: grassroots movements of forest-dwellers and peasants by which these peoples secure rights to their lands, and outsiders becoming accountable to those whom they would help.

(wealth, poverty, and deforestation)

110
Not By Timber Alone: Economics and Ecology for Sustaining Tropical Forests. Theodore Panayotou and Peter S. Ashton (both Harvard Institute for International Development). Washington: Island Press, Nov 1992/282p/$40.00;$22.00pb.

The world's tropical forests are fast disappearing. Current debate over tropical forests pits economic growth directly against conservation. Exploiting forests for timber is considered the growth-oriented option; production of non-timber goods and services is seen as environmentally sound but unprofitable. Multiple-use management, however, maximizes both growth and conservation (indeed, under certain conditions, the value of non-timber products and services can surpass the value of timber). The most serious problem at present is undervaluation of timber by governments of tropical countries, combined with overvaluation of the net benefits of forest conversion and the common failure to account for non-timber forest products and services. Such products include fuelwood, charcoal, pulp, chips, poles, and non-wood goods such as game, fruit and nuts, cords and fibers, and latexes. Forest services include climate regulation, soil and water conservation, recreation, protection against natural disasters, and preservation of biodiversity. This book seeks to identify these non-timber products and services, determine the extent to which full accounting in a multiple-use management framework can help ensure sustainability, and identify gaps in knowledge. Although knowledge of how tropical forests function is extremely limited, "we do know enough to determine that current policies and practices are promoting destructive and unsustainable uses."

Concludes with an 11-point agenda for reclassifying forestlands, awarding concessions, reforming taxation to eliminate incentives for destructive logging, investing in protection, and promoting private forest investments. **(multiple-use management of tropical forests)**

111

Development or Destruction: The Conversion of Tropical Forest to Pasture in Latin America. Edited by Theodore E. Downing (U of Arizona) *et al.* Boulder CO: Westview Press, Dec 1992/405p/$49.50pb.

A driving force behind the destruction of Latin American tropical forests is the expansion of cattle production into seemingly underutilized areas—a conversion that extracts very high ecological, economic, and human costs. These 28 papers from a 1988 workshop in Oaxaca discuss forest uses, impacts from conversion, production alternatives, perspectives of developers and donors, and community perspectives. Concludes with the "Oaxaca recommendations" for maintaining tropical forest and promoting human and biological diversity: 1) limit physical access to the forest by discouraging new roads; 2) increase the cost of converting forest to pasture; 3) promote diversity by creating a large range of reserves that incorporate sustainable human activities, along with a mix of national parks and forests; 4) intensify cattle production in non-tropical forest regions; 5) promote production and consumption of alternative sources of protein; 6) develop products from tropical forest areas and non-agrarian activities such as tourism and crafts; 7) empower local participation in decisions over natural resources by encouraging and informing local organizations. [ALSO SEE: **Surviving the Cut: Natural Forest Management in the Humid Tropics** by Nels Johnson and Bruce Cabarle (World Resources Institute, Feb 1993/71p/$10.00), on promoting sustainable forestry that is grounded in local communities.] **(saving tropical forests)**

112

Who Will Save the Forests? Knowledge, Power and Environmental Destruction. Edited by Tariq Banuri and Frédérique Apffel Marglin (Smith College). A Publication of UNU/WIDER. London and Atlantic Highlands NJ: Zed Books, July 1993/195p/$55.00;$17.50pb.

The deterioration of forest resources and the environmental crisis in general must be viewed from the angle of the politics of knowledge and a critical approach to the dominant, modern system of knowledge. Societies that treat alternative conceptions of reality as legitimate and respectable are relatively more successful in conserving the environment and empowering marginalized communities. In societies where the alternatives are denegrated, the result is gradual environmental degradation and often local poverty. Modern scientific forestry has three problems: it legitimizes colonization and exploitation, the "scientific" label is often used to justify short-term commercial practices, and direct state control compounds the problem further because of corruption and inefficiency. The managerial solution advocated by mainstream writers may minimize the worst excesses of laissez-faire exploitation, but it undermines the larger social basis of conservation. Chapters compare forestry practices in three regions: 1) among tribes in rural India (with a discussion of the famous Chipko or "tree-hugger" movement that resisted forest destruction); 2) the Forest Karelia region of Finland (where villagers have resisted short-term commercial exploitation in favor of sustained-yield forestry); 3) the corporate/capitalist Maine forests in the US (where the dominant system of knowledge in forestry stands as a warning rather than a model to the world). Concludes that "the modern system of knowledge is harmful for human society in the long run." [**NOTE**: Marglin is co-editor of **Dominating Knowledge: Development, Culture, and Resistance** (Oxford/WIDER, 1990; **FS Annual 1992** #11011), which provides background for this provocative study.] **(forestry in India, Finland, and Maine)**

113

Saving the Forests: What Will It Take? Alan Thein Durning. Worldwatch Paper 117. Washington: Worldwatch Institute, Dec 1993/51p/$5.00. (Brief version in **State of the World 1994**, 22-40; #014.)

Forest conservation efforts have proliferated worldwide in recent years, in response to sweeping destruction of forest ecosystems. For a decade, well-meaning global initiatives, such as the Tropical Forestry Action Plan and the UN Statement of Forest Principles, were launched with fanfare and high hopes. But each effort has proved disappointing, and rates of forest loss have continued to rise. All told, two-thirds of the planet's original natural forests have been cleared, fragmented, or converted to uniform stands of commercial timber. Past attempts to halt deforestation foundered because they set their sights on the proximate causes of deforestation. Ranchers, loggers, developers, etc. are merely the teeth of a saw—the leading edge of a larger force; the saw is a money economy blind to its ecological roots. Saving the forests requires three fundamental changes: 1) *pricing* of wood and other forest products that tells the ecological truth (reflecting soil erosion, decimated fisheries, and other environmental costs); in turn, this would encourage more efficient uses of wood and sustainable forest management; 2) a *property rights* system that allies the interests of forest people with the health of forest ecosystems (this could involve private ownership, collective management by communities, or state control); it would replace present systems where day-to-day users and managers have little incentive to care; 3) decentralizing *power* over forests away from the few who profit; unless the grip of big timber interests and related resource extractors is broken, "all bets for forest conservation are off." **(stopping deforestation)**

114

Defining Sustainable Forestry. Edited by Gregory H. Aplet (The Wilderness Society) *et al.* Washington: Island Press, Nov 1993/328p/$49.95;$24.95pb.

We are now in a period of unprecedented public concern over the alteration and destruction of forests and the accelerating rate of loss of plant and animal species native to forest ecosystems. Consequently, "the practice of forestry in the US is currently undergoing the most profound and rapid change since its establishment a century ago." The evolution from sustained-yield management of a relatively small number of commercial tree species to protection and sustainable management of forest ecosystems is changing some of the key premises of forest management. Reflecting and supporting this trend: 1) the National Research Council issued a 1990 report, **Forestry Research: A Mandate for Change**, calling for a basic redefinition of forest science, broadening research from the agricultural model of improving production to studying healthy forest ecosystems; 2) the Forest Service and the Bureau of Land Management announced in June 1992 new policies adopting an ecosystem management approach for over 200 million acres of Federal forest; 3) the US adopted Agenda 21 and the Forest Principles at the 1992 Earth Summit.

These essays, from a 1992 workshop sponsored by The Wilderness Society, American Forests, and the World Resources Institute, discuss new directions for the US national forest system, institutional constraints on sustainable resource use, ecosystem management in the Northwest and Southeast US, defining an economics of sustainable forestry, and sustainable communities. The four editors conclude that sustainable forestry must be ecologically sound, economically viable, socially desirable, and politically acceptable. A forest is not a perennial row crop in a tree farm; it is something broader. "The future of sustainable forestry lies in finding the level of extraction that also perpetuates the natural dynamics of the forest." **(transition to sustainable forestry)**

115

Clearcut: The Tragedy of Industrial Forestry. Edited by Bill Devall (Prof of Environmental Studies, Humboldt State U, Arcata CA). San Francisco CA: Sierra Club Books/ Earth Island Press, March 1994/291p(12x13")/$50.00.

"The forests of North America are suffering from a massive, human-caused, catastrophic event—an epidemic of massive clearcutting." Some 170 pages of striking color photographs, nearly all of them taken in the past five years, amply demonstrate the devastation in 9 Canadian provinces and 25 US states. Essays describe the need for sustainable forestry (also known as community forestry, ecosystem-based forestry, holistic forest practice, and the practice of right livelihood in the forest), the moral and spiritual dimensions of forests, economic and ecological flaws of clearcutting, myths of the benign industrial clearcut (by Mitch Lansky; see **FS Annual 1994** #12372), why tree farms are not forests (single-species tree plantations of an exotic species can destroy a watershed and wildlife habitat), Canada as "the Brazil of the North" (in the Amazon region, one acre of forest is lost every 9 seconds to logging and forest clearing; in Canada, one acre is lost every 12 seconds), wilderness recovery, the Eco-forestry Institute (see **FS Annual 1994** #12373), the eco-forester's oath, and Chris Maser on attributes of "The 21st Century Forester." [**NOTE**: The essays could grace any anthology, but are difficult to access because the volume is atypically outsized.] **(sustainable forestry needed)**

116

Breaking the Logjam: Obstacles to Forest Policy Reform in Indonesia and the United States. Charles Victor Barber (WRI), Nels C. Johnson (WRI), and Emmy Hafild (Indonesian Environmental Forum). Washington: World Resources Institute, March 1994/120p/$10.00pb.

In its 19th century push westward, the US converted forests into farms, cities, and railroad ties, even as Brazil is now fueling its steel industry with rainforest charcoal and Indonesia is exporting plywood to stoke its rapid economic growth. Once let alone, damaged forests can recover. But in many parts of the world, forests are now under a virtual death sentence. At least 15 million hectares of tropical forest are vanishing each year. In the temperate zones, forests are being fragmented by commercial logging, injured by pollution, converted to tree plantations or farmland, or falling victim to urban/ suburban sprawl. Although Indonesia and the US differ in many ways. both have forestry policies established in an earlier era that are out of step with evolving perceptions of forests' true worth. Focusing on both countries underscores the emerging scientific consensus that forests are in trouble everywhere, and the emerging political consensus that the North cannot deplore tropical deforestation while turning a blind eye to its own troubled woodlands. Overcoming obstacles to more sustainable forest management in both countries will require grappling with three sets of issues: forest property rights regimes, distribution of the costs and benefits of forest exploitation, and the political process for setting forest policies. All forested countries can learn from the experience of these two. A positive vision of good forest management would rest on four pillars: 1) ecological integrity and sustainability; 2) sustainable and equitable human uses of forest products and services; 3) integrated management at the right regional scale; 4) equitable and informed participation by all stakeholders. **(sustainable forestry: US/Indonesia)**

117

Wild Forests: Conservation Biology and Public Policy. William S. Alverson (Botany/U of Wisconsin-Madison), Walter Kuhlmann (Madison WI), and Donald M. Waller (Botany/UW-M). Washington: Island Press, Aug 1994/300p/$49.95;$29.95pb.

North American forests have suffered drastic reduction since European settlement. Logging has removed more than 99% of the great virgin forests from the East, and still proceeds apace in the far West. Overall, 85-95% of America's primary forests had disappeared by the late 1980s. Chapters discuss forest biodiversity, ecological mechanisms and biotic resources, the evolution of forest management, multiple use on national forest lands, and alternative management models. Concludes with a proposed diversity policy: 1) preserve and regenerate large unfragmented blocks of late successional forest; 2) elevate science as a guiding force in forest management and broaden the set of disciplines used; 3) redefine wilderness to restore and preserve ecological integrity and wildness; 4) reinvent multiple use to include dominant-use zoning (all uses cannot coexist at some point on the same acreage); 5) expand legal protection for biodiversity and ecosystems; 6) manage with an understanding of the limits of our knowledge. Central to any vision of 21st century forestry is the need to maintain the biotic richness and complexity of native forests. **(forestry reconsidered)**

118

The Future of the Northern Forest. Edited by Christopher McGrory Klyza and Stephen C. Trombulak (both Middlebury College). Hanover NH: Middlebury College Press/Univ Press of New England, Dec 1994/258p/ $39.95;$19.95pb.

The Northern Forest of Maine, New Hampshire, Vermont, and New York stretches from the Atlantic Ocean to Lake Ontario, encompassing more than 25 million acres. It includes the White Mountains, the Green Mountains, and the Adirondacks. In contrast to forest regions in the western US, humans live and work throughout the area, and much of the land—over 80%—is privately owned. The chief political issue involves the interests and concerns of the forest products industry, environmentalists, and land owners in both a healthy economy and a healthy environment. Success in the

Northern Forest could provide a model for sustainable human and natural communities. These essays from a 1992 conference at Middlebury College include a natural history of the Northern Forest, the Abenaki view, the Northern Forest economy, ethical tensions, the Northern Forest Lands Study process (resulting in the Northern Forest Lands Council), New York's 6-million acre Adirondack Park (the largest area in the US subject to a comprehensive land-use planning and regulating program; the unique set of public-private relationships in this century-old green-line area could serve as a model for the Northern Forest), ten steps to craft a sustainable Northern Forest (eco-efficiency in existing industries, protecting core areas, managing private land for sustainability, sustaining local communities, etc.), and a multi-use working forest. The editors conclude that "the process of changing our views and working for the common need in good faith is often not pleasant, but it is better than the alternative."

(US Northern Forest issues)

10. Land Use

119

The State of the World's Mountains: A Global Report. Edited by Peter B. Stone (Director of Information, Brundtland Commission). London & Atlantic Highlands NJ: Zed Books, June 1992/391p/$60.00;$25.00pb.

A report on "the mountain problematique" as background for the UNCED Earth Summit, prepared by Mountain Agenda (c/o Institute of Geography, U of Berne, Hallerstrasse 12, CH-3012 Berne), with support of the Swiss and German governments. Mountain areas are viewed as the backbone of civilization: they are water towers for human consumption, weather makers for large parts of the world, a source of biological diversity, and privileged places for spiritual and physical recreation. Their major problems include the growing contradiction between marginalization and integration (industrial and political centers are on the plains, but easier access enables integration of remote mountain areas), threats and hopes from agriculture and forestry (the frontier of forest destruction is inexorably moving up the mountains, driven by rapid population increase), new economic opportunities with harmful side effects (mining, large-scale construction, tourism), the impact of climate change (global warming will have a pronounced effect on mountain ecosystems, which can serve as sensitive indicators of the slightest change). Three major steps toward sustainable mountain development: 1) study and support the efforts of mountain inhabitants to reverse present trends and institute new patterns of management (many examples can be found in the hills of Nepal); 2) create more global awareness of the many dangers looming over the mountains, and build an effective global constituency; 3) build a cooperative network of national, regional, and international institutions concerned with sustainable mountain development. Two brief chapters discuss mountain protected areas and climate change in the mountains, with the bulk of the report on specific areas: African mountains and highlands, the Alps, the Himalaya (the most populous and fragile of the world's mountains), the Andes, mountains of the former Soviet Union, and the Appalachians of North America. Briefer attention is given to the eastern Canadian Arctic, Hawaii, Sierra Nevada, Scottish highlands, Pyrenees, the Japanese alps, and six other areas. [**NOTE**: Nothing on the US or Canadian Rockies; otherwise, a thorough and authoritative report. Includes photos and charts.]

(mountain problematique)

120

High Priorities: Conserving Mountain Ecosystems and Cultures. Derek Denniston (Research Associate, Worldwatch Institute). Worldwatch Paper 123. Washington: Worldwatch Institute, Feb 1995/80p/$5.00.

Mountains and high plateaus make up one-fifth of the world's landscape and are home to at least 570 million people (most of them in several thousand tribes or ethnic groups, outside the dominant cultures of the plains). An additional 2 billion people depend on mountains for much of their food, hydroelectricity, timber, and mineral resources. And fully half of the world's people depend on mountain watersheds for fresh water. The expanding world economy is threatening human and natural mountain communities by creating scarcities of arable land, increasing intensive resource extraction, and generating a boom in tourism and recreation. "All over the world, mountain peoples face increasing cultural assimilation, debilitating poverty, and political disempowerment." Mountains have also become critical proving grounds for sustainable development, requiring a deeper understanding of the interdependence between human and natural communities. To elevate the status of mountain peoples and conserve their ecosystems, policy reform is needed in six areas: 1) promoting efforts to secure land tenure or control over local resources (perhaps the most critical reform); 2) reducing the impacts of livestock, timber, hydropower, and minerals production in mountains (e.g., by making more efficient use of trees and minerals, eliminating government subsidies to destructive operations, full-cost pricing of virgin materials and resources); 3) creating regional networks of conservation areas (e.g. corridors of protected areas are being planned for the entire length of the Andes); 4) improving knowledge about mountains through integrated research, public education, and social/environmental monitoring; 5) establishing institutions and cooperative agreements for each major range (e.g., the 1992 Alpine Convention provides a good model); 6) integrating mountains into the policies and projects of development agencies.

(mountain ecosystems and cultures)

121

A New Century for Natural Resources Management. Edited by Richard L. Knight (Colorado State U) and Sarah F. Bates (Utah office, Grand Canyon Trust). Washington: Island Press, March 1995/398p/$47.50;$27.00pb.

The natural resource disciplines (forestry science, range science, fisheries and wildlife management, watershed science, recreation management) originated in the 20th century. At century's end, the old tenets of natural resources management are in crisis, and each of these disciplines finds itself in upheaval. These approaches were designed around the utilitarian conservation philosophy of the 19th century, which assumed needs and conditions different from those that exist today. These essays seek new guidelines for a new century: ***1) The Beginning of Natural Resources Management***: the Federal land management agencies, traditional approaches, traditional education, economics, ethics of nature as property; ***2) The Beginning of a Shift***: a world that takes its environment seriously (by David W. Orr), redefining multiple use, changing values and new perspectives in the Forest Service, the growing power of NGOs, natural resources management by litigation, broadening the economic paradigm toward natural resources, reinterpreting the natural-resource dependent community; ***3) The Future***: the shifting paradigm in ecology (from "the balance of nature" to "the flux of nature"), new approaches to conservation biology (working with ecological processes rather than single species), new and broader directions for education, managing natural resources as social value, ecological economics as a new transdisciplinary science (by Robert Costanza), global environmental ethics, and the shift to a "sustainable resource management future." [**NOTE**: A nice anthology depicting important shifts in thinking.] **(natural resources: new thinking)**

122
Wildlife and Recreationists: Coexistence Through Management and Research. Edited by Richard L. Knight (Colorado State U) and Kevin J. Gutzwiller (Baylor U). Washington: Island Press, March 1995/372p/$49.95; $27.50pb.

As we near 2000, a paradigm shift in land management is nearing. Historically, human activities on public and private lands were extractive (logging, grazing, mining, hunting). Today, our society speaks nearly as one in objecting to primarily utilitarian uses of our once-vast natural resources. With the shift from rural to urban societies, there has been a decades-long increase in outdoor recreational activity and ecotourism, with no decline in sight. But recreation and wildlife are not compatible without some form of overall management. "The principal natural resource management issue for the 1990s and beyond will revolve around conflicts between outdoor recreationists and wildlife—and between different types of outdoor recreationists." Natural-resource agencies are changing their focus from largely extractive uses of land to increasing recreation opportunities; "the next step in agency evolution will be toward land health and stewardship." Essays discuss trends in outdoor recreation (with projections of land- and water-based activities through 2040), human dimensions of wildlife management, wildlife responses to recreationists, impacts of nature tourism, case studies, and Aldo Leopold's land ethic applied to frame recreation as re-creation. **(recreation and wildlife)**

123
The Capacity for Wonder: Preserving National Parks. William R. Lowry (Asst Prof of Political Science, Washington U, St. Louis). Washington: The Brookings Institution, June 1994/280p/$28.95.

The national parks of the US and Canada have always stretched the author's capacity to wonder. His sense of urgency came later, with a growing understanding of politics: "of all the threats facing the parks, the most important is the behavior of government officials." The national park services of the US and Canada are compared. In the US, conflicts over goals have become increasingly contentious, and politicians have moved many decisions away from field managers. The Canadian Parks Service has a growing consensus on agency goals and gives responsibilities to field managers, who increasingly emphasize long-term preservation. Proposals for the parks: 1) make preservation the dominant and explicit goal; 2) reorganize public lands to make the preservation effort more focused (rather than management by four different agencies, all lands should be under NPS); 3) raise entrance and user fees in both nations and funnel them directly to the parks; 4) reject proposals to privatize the parks; 5) involve the public and relevant interest groups in planning efforts; 6) build consensus for preservation through increased public education. **(national parks)**

124
National Parks for the 21st Century: The Vail Agenda. Report and Recommendations to The Director of the National Park Service. Post Mills VT: Chelsea Green Publishing Co (distributor), Jan 1994/142p/$14.95pb.

The National Park Service's 75th Anniversary Symposium, held in Vail CO in October 1991, brought together nearly 700 experts and interested parties from inside and outside the Service to consider the future of the national park system. Today there are 359 parks, representing 16 categories of units within the system, ranging from national battlefields to scenic rivers. From 1980 to 1989, the total number of visits grew 35%, and is now over 258 million annually. The NPS 21st Century Task Force anticipates significant changes in visitation due to the aging and redistribution of the population: from Eastern to Western and Southern parks (demand for recreation areas in the South may exceed supply), from summer to other seasons, and from back country to front country areas. Trends in work and education imply that many visitors will have less time (and thus make shorter visits) and more demand for information and unique experiences. Economic trends suggest that parks as tools of economic development will have strong political support.

Six strategic objectives are proposed: 1) resource stewardship and protection from internal and external impairment; 2) access and enjoyment for the nation's diverse public; 3) education and interpretation, though a wide range of on-site and off-site strategies; 4) proactive leadership in local, national, and international park affairs, actively pursuing the mission of the National Park System and helping others in managing their park resources and values (to this end, NPS should establish an Office of Strategic Planning); 5) a sustained and integrated program of natural, cultural, and social science resource management and research; 6) creating and maintaining a highly professional organization and work force. **(objectives for US national parks)**

125
Ethical Land Use: Principles of Policy and Planning. Timothy Beatley (Chair, Dept of Urban and Environmental Planning, U of Virginia). Baltimore MD: Johns Hopkins U Press, April 1994/302p/$55.00;$17.95pb.

Public policy decisions are inherently ethical decisions as well. This is understood in many areas, but not in land-use policy and planning. Yet the social allocation of land to different uses and activities is fundamentally a problem of ethics, because of many social and environmental impacts. Thousands of planning commissions, city councils, and other citizen boards make difficult land-use decisions on a regular basis, but these decisions are often discussed only in economic, technical, legal, or political terms. Key elements of an ethical land-use policy are proposed: ***1) Maximal Public Benefit***: promoting the greatest quantity of social benefits or welfare, other things being equal; ***2) Distributive Justice***: avoiding actions that serve to lessen the social and economic conditions of the poor; ***3) Preventing Harms***: minimizing or preventing the imposition of harms on people and the environment; ***4) Land-Use Rights***: protecting minimum social and environmental rights due every individual; ***5) Environmental Duties***: obligations to protect and conserve the natural environment; the responsibilities of stewardship, or minimizing the human "footprint"; ***6) Obligations to Future Generations***: protecting landscapes and resources that may enrich the lives of future residents; ***7) Life-Style Choices***: although communities have a moral authority to promote community character, these efforts must be tempered by the need to respect individual rights and life-styles; ***8) Risk-Taking***: requiring full disclosure of all relevant risks and hazards; ***9) Expectations and***

Promise-Keeping: public authorities should keep the promises they make and do everything possible to clarify these promises; ***10) The Privilege of Landownership***: ethical land use views use and development of land as a privilege, not an inviolable right; ***11) Interjurisdictional Obligations***: requiring one jurisdiction to consult with and coordinate with other jurisdictions, and to consider impacts of its land use on communities and citizens beyond its borders; ***12) Fair and Equitable Political Process***: enabling all interested and affected parties to participate. **(land use: ethical principles)**

126

Policy for Land: Law and Ethics. Lynton Keith Caldwell (Prof of Public Affairs, Indiana U) and Kristin Shrader-Frechette (Distinguished Research Prof of Philosophy, U of South Florida). Lanham MD: Rowman & Littlefield Publishers (4720 Boston Way), 1993/332p/$24.95pb.

Exploitation of land for short-term private, political, or economic advantage is common and still regarded by some Americans as a birthright. Disregard for ethical, aesthetic, ecological, and socioeconomic values in the treatment of land remains widespread. The authors believe that "the time is approaching when socially responsible Americans will seriously seek consensus on a new and sustainable land ethics capable of being translated into law and policy." Ending of the frontier and confrontation of the many costs of misused land has opened the way to reconsidering America's cornucopian philosophy. Some of the themes discussed: 1) general political strategies needed for land-use planning (an estimate of probable impacts could become an important instrument of consensus building); 2) an overview of land-related ethical problems (there are at least four land ethics: land-reform ethics, land-use ethics, land-community ethics, and land-rights ethics); these ethics stand little chance of being put into practice so long as the political deck of cards is stacked in favor of short-term economic advantage over all other considerations; 3) a new conceptual basis for land law and policy based on redefining ownership (so that it applies not to land itself but rights to occupy or to use land), and redefining legal rights of occupancy; 4) land use policy as an international issue (people must stop thinking of national sovereignty over their territories as absolute and exclusive); 5) the ecosystem as a criterion: "land policies must be founded on comprehensive, scientific, survival-oriented, long-term, holistic values." **(new land ethics)**

127

Land Conservation Through Public/Private Partnerships. Edited by Eve Endicott (Lincoln Institute of Land Policy; former VP, The Nature Conservancy). Foreword by Sen. John H. Chafee (R-RI). Washington: Island Press, Aug 1993/364p/$45.00;$22.95pb.

The face of US land conservation has been subtly changing over the past two decades. Well before the rhetoric of public/private partnerships became commonplace, government agencies began working with private conservation organizations to preserve natural areas, prime farmland, and community gardens. In the 1970s, such partnerships were the exception. In the 1980s, this modest, piecemeal cooperative effort exploded into a nationwide movement involving every major Federal land-owning agency, every state, hundreds of localities, and many of the 1,000 or so private land protection organizations. "Now it is almost the exception if significant land acquisition is accomplished without a partner from the private sector and one from the public sector, if not more than one from each side." This book provides a detailed look at the partnership movement, showing the broad array of circumstances under which public and private sectors can achieve more and better land conservation together than either one can alone. Chapters discuss The Nature Conservancy and its partners, The American Farmland Trust, The Trust for Public Land (focusing on neighborhood parks and urban community gardens), public/private partnerships in and around national parks, National Park Service partnerships beyond the national parks, state partnerships in Rhode Island and New York, Vermont's Housing and Conservation Trust Fund, local land trusts (there are now nearly 1,000 land trusts in the US, more than double the number of ten years ago), financing arrangements, and state land conservation programs. **(growth of land conservation partnerships)**

128

Visions Upon the Land: Man and Nature on the Western Range. Karl Hess Jr (Cato Institute; Senior Associate, FREE). Foreword by John A. Baden (President, Foundation for Research on Economics and the Environment). Washington: Island Press, Oct 1992/278p/$22.00.

Landscape visions are the mental constructs through which people encounter and react with the natural world. They are the link between people and their environments. Over the span of a century, three visions have brought a degree of uniformity to the diverse landscapes of America's western range: 1) the agrarian vision, infused with Jeffersonian principles, did not enshrine democracy or sustain thriving farms; 2) the progressive conservationist vision did not mend broken dreams or rehabilitate public lands; 3) the environmental vision and its righteous management has had a dismal record of conservation and stewardship. This last vision concludes what the visions of Jefferson and Pinchot began: "the making of the environmental welfare state." Government has made the bold leap from being a neutral arbiter and facilitator to becoming a guardian of visions that presume to know what is right and best for people and nature on the western range. Ideological blinders are steering the environmental welfare state on an unwavering and arrogant course.

What is needed to heal overgrazed lands and enrich western landscapes is a *market of landscape visions*—an atmosphere of political and environmental tolerance in which individuals and groups are free to pursue their dreams without the benefit of political force or the largesse of Federal subsidy. Such a market will bring visions down to earth, making them answerable to the land, to wildlife, and to the human communities that bear many of the costs and benefits of environmental abuse and benevolence. It would buffer the environment against ill-advised visions, and enlist all Americans in the destiny of their western lands, in contrast to "a handful of trained experts and impassioned special interests (who) exercise unchallenged dominion over one-third of the nation's land." Hess proposes that, over a 20-year period, the Federal government be required to divest itself of the surface rights to all lands managed by the Forest Service and BLM—the most sweeping land reform in American history. Every American would receive 100 shares in these public lands, which they could sell to others or apply to land purchase. An infinite array of decentralized "little commons" would spring forth—a commons that people could be part of and take pride in. "The western range would at long last be... opened to the promise of many minds," creating a "virtuous republic of independent, caring, and responsible stewards." [**NOTE**: Eloquent and intriguing. Hess Jr is the son of a widely-known libertarian thinker.]

(market of landscape visions proposed)

129
Crossing the Next Meridian: Land, Water, and the Future of the West. Charles F. Wilkinson (U of Colorado). Washington: Island Press, Nov 1992/376p/$25.00; $17.00pb.

The American West has been ruled by "the lords of yesterday": 1) the Hardrock Mining Law of 1872 that dedicates more than half of all public lands to mining as the preferred use; 2) the practice of allowing free and unregulated grazing of cows and sheep on public domain (grazing fees are set at a fraction of their market value, and poor grazing practices have devastated western rangeland and rivers that receive resulting eroded soil); 3) the turn-of-the-century policy of treating logging as the dominant use of the national forests (despite changed perceptions of the forests and clear evidence that government timber sales are unprofitable); 4) the dams and other development practices of the Pacific Northwest that have crippled the runs of Pacific salmon and steelhead; 5) the prior appropriation doctrine for water, under which diverters of water obtain vested property rights that cannot be taken away. The costs to the lands and waters of the West are now coming due, with a growing number of rivers, aquifers, forests, rangelands, and farms in decline.

In perhaps the greatest book ever written about the West, **Beyond the Hundredth Meridian**, Wallace Stegner showed how the mountain men, hardrock miners, and other early explorers were the first era in the settlement of the region west of the Dakotas by non-Indians. The second era, of which Stegner wrote, involved the populating of the lands beyond the barrier of the dry line. We are now in a new era where the task is to move beyond settlement and to achieve resource sustainability, economic stability, and social justice. To do so, we must cross a new meridian, not of geography, but of intellectual, social, and government commitment. The lords of yesterday arose in an age when the controlling idea was unfettered development (with this extreme laissez-faire approach multiplied several times by subsidies). The new unifying principle must be the idea of sustainability, which must involve community planning and personal responsibility.

(a new era of sustainability in US West)

130
Mining the Earth. John E. Young (Research Associate, Worldwatch Institute). Worldwatch Paper 109. Washington: Worldwatch Institute, July 1992/53p/$5.00pb.

From 1750 to 1900, the world's overall use of minerals increased tenfold while population doubled. Since 1900, it has jumped at least 13-fold again. Of the non-fuel minerals, stone, sand and gravel are produced most widely and in the largest quantities. Other nonmetals (phosphates, lime, soda ash, clays) are more valuable and travel further from mine to user. The most valuable minerals are metals, of which the most important is iron. About 30 times as much iron is produced as aluminum, the next most common metal. But scouring the planet for its minerals has damaged large areas of land. Scarcity of deposits does not appear likely to constrain production of most important minerals in the foreseeable future. But can the world afford the human and ecological price? Answering this question depends on a true accounting of the costs of taking minerals from the earth. Today's low prices reflect only the immediate economics of extraction and distribution; they fail to consider the costs of denuded forests, eroded land, dammed or polluted rivers, acid rain from smelters, and uprooting or decimation of indigenous peoples. Although much of this damage is in poor nations, the rich nations bear a special responsibility. Mine cleanups should be funded by taxes on virgin mineral consumption. The best strategy to minimize new damage is not merely to make mineral extraction cleaner, but to reduce the need for virgin materials by promoting recycling and making products more durable and repairable. [**NOTE**: This overdue focus on the neglected environmental costs of mining parallels the shift in thinking about fossil fuels, from scarcity to environmental costs of usage.]

(environmental impacts of mining)

131
Mining and the Environment: International Perspectives on Public Policy. Edited by Roderick G. Eggert (Colorado School of Mines; Editor, *Resources Policy*). Washington: Resources for the Future, June 1994/172p/$25.00pb.

Writing in 1556 in the first scientific text on mining, Georgius Agricola noted how the fields were devastated by mining operations, the woods and groves felled, the brooks and streams polluted, and the beasts and birds exterminated. For centuries, this was considered as part of the price for mineral production. Even in the 1960s and 1970s, most early environmental legislation did not consider mining. In the 1990s, though, mining has become the focus of greater environmental scrutiny, and most major mining companies now have senior managers in charge of environmental affairs—"something unheard of a decade ago." Mining is significant: according to UNEP, such activities (including oil and gas production) account for about three-quarters of annual solid wastes generated in Canada, and one-third in the US. These five lectures consider: 1) the issue of land access for mineral exploration, and cost-benefit analysis as a necessary tool for focusing public debate about land use; 2) the polluter-pays principle as the starting point for cleaning up mine wastes (but the principle is inappropriate for dealing with past mining, because many sites have changed ownership several times); 3) environmental policy in Chile (legislation is expected by the end of the 1990s to govern abandoned mines, tailings, and water and soil quality); 4) the growing EC influence on environmental matters in Europe, likely to bring tougher requirements for mining; 5) the need for new and better methods of mining and mineral processing, especially in developing countries, where much environmental degradation results from obsolete technology and out-of-date work practices. **(reducing mining impact)**

11. Pollution

132
Toxics Watch 1995. INFORM, Inc. NY: INFORM, Inc, March 1995/c800p/$125.00pb.

The first in a biennial series on the full range of toxics in the US environment, with analysis of conditions and trends, current and future influences on such trends, and progress in source reduction. Chapter topics: ***1) Sources of Environmental Toxics***: the role of product use as a source of contamination is significant and in some cases predominant; indoor air pollution is more dangerous to many people than outdoor air pollution; ***2) Toxics in Commerce***: there are more than 72,000 chemicals in commerce in the US (excluding foods, drugs, cosmetics, and pesticides), but almost none of them have been fully characterized for their ability to cause environmental and health effects; new chemicals enter commerce through manufacture or import at the rate of 200-1000 per year; "no clear picture of US chemical production or use can be had"; ***3) Toxics in Waste***: generation of toxic waste increased between the last two reporting years of data available for this report: the EPA's Toxics Release Inventory shows an increase of 3% between 1991 and 1992 in production-related waste; the EPA's Biennnial Reporting System showed an increase of 9% in toxic waste between 1989 and 1991; ***4) Toxics in Waste by Industry***: relatively few industries and facilities dominate toxic waste generation; some facilities, companies, and smaller industries are achieving significant reductions, but they were more than offset by lack of progress by others; "in the aggregate no net reduction occurred in waste generation of TRI chemicals from 1991 to 1992, nor is any projected in the next two years"; ***5) Geography and Waste***: Texas, Mississippi, and Louisiana generate the largest amounts of TRI chemicals; ***6) Toxics in the Environment***: despite hundreds and perhaps thousands of databases on toxic chemicals, fewer than a score are both national in scope and broad in coverage (and few of these are current); ***7) Environmental Justice***: the movement seeking *adequate* protection for all emerged in the late 1960s and has steadily evolved and broadened; ***8) State Databases***: data collected by New Jersey and Massachusetts permit a much clearer picture of toxic chemical use than does TRI; ***9) Tackling Industrial Toxic Waste Through Voluntary Action***: EPA's 33/50 program launched in 1991 encourages voluntary reduction (33% by 1992 and 50% by 1995); it is too early to assess the program's impact, but evaluation procedures are discussed. [**NOTE**: Extensive detail and sophisticated analysis.] **(first US toxics report)**

133

Toxics A to Z: A Guide to Everyday Pollution Hazards. John Harte (Energy and Resources Group, U of California-Berkeley), Cheryl Holdren, Richard Schneider, and Christine Shirley. Berkeley: U of California Press, Dec 1991/479p/$75.00;$29.95pb.

Introductory sections discuss thinking about risk, causes of death and disease, the language of toxics, basic medical aspects of the toxics problem (testing, classification, entry and exit, nutrition and susceptibility), major sources of toxic exposure (air, water, food, consumer products), toxics and the environment (ecological effects of toxics, global environmental hazards), toxic metals, petrochemicals, pesticides, radiation, and managing toxics (the waste crisis, regulation, the right to know about toxics in the workplace and the community). About half of the book is devoted to information (alternative names, properties, exposure and distribution, health effects, protection and prevention, regulatory status) on about 110 commonly encountered toxics such as acetic acid, aluminum, asbestos, benzene, caffeine (the most popular drug in the US), CFCs, DDT, dioxin (one of the most toxic chemicals ever made), extremely low frequency electromagnetic fields, fluoride, food colors, lead, mercury, nicotine, noise, paraquat, radon, saccharin, sulfites, ultrasound, vinyl chloride, warfarin, xylene, and zinc. Appendices provide a glossary, list of commonly used abbreviations, sources for home testing equipment, eight national hotlines for further information, and a list of 10 effective ways to reduce toxics in your environment: don't smoke, read labels and follow directions carefully, conserve energy, reduce and recycle, substitute where possible, test for radon, test for home water quality, don't ignore workplace safety guidelines, support sound environmental legislation, and share your knowledge with others. [**NOTE**: Very well-conceived and quite readable.] **(guide to toxics)**

134

Reducing Toxics: A New Approach to Policy and Industrial Decisionmaking. Edited by Robert Gottlieb (UCLA Pollution Prevention Educ. and Research Center). Washington: Island Press, April 1995/447p/$35.00.

Citizens in the US and throughout the world have begun to forcefully insist that the range of environmental hazards present in everyday life needs to be reduced in all its manifest forms. This argument is particularly linked to generation, use, and disposal of toxic and hazardous materials. For more than 25 years, various government policies and industry responses have been developed, but they "have failed to significantly address the problem of toxics generation and use beyond the question of management and disposal." Pollution control has remained the dominant framework, but new pressures seek to shift the focus to minimizing or reducing wastes. Despite growing interest, "transition from control to prevention continues to remain an elusive objective." Chapters discuss policy tools in pollution control, the media-specific approach to toxics policies, occupational risk, consumer product hazards, new approaches to toxics (production design, right-to-know, definition debates), green marketing (subject to abuse), green labeling (often limited to a single aspect of product use), the US chemical industry, pollution prevention for the emerging electric vehicle industry, the aerosols packaging industry, 3M's voluntary "Pollution Prevention Pays" program (despite widespread praise, 3M remains a major polluter), and workplace involvement as supportive of pollution prevention initiatives. Concludes that "pollution prevention still lacks coherence in terms of its definition, institutional implementation, and day-to-day application." **(pollution prevention difficulties)**

135

Unequal Protection: Environmental Justice and Communities of Color. Edited by Robert D. Bullard (Prof of Sociology, U of California-Riverside). San Francisco: Sierra Club Books, May 1994/392p/$25.00.

America's environmental laws, regulations, and policies have not been applied fairly across all segments of the population. Pollution in urban ghettos and barrios, rural poverty pockets, and Native American reservations "presents potential threats to public health that individuals with affluence or political clout are unwilling to accept." The dominant environmental protection paradigm institutionalizes unequal enforcement, trades human health for profit, places the burden of proof on the "victims", promotes risky technologies such as incinerators, subsidizes ecological destruction, delays cleanup actions, and fails to develop pollution prevention as an overarching strategy. The mood has shifted, however, so that affected populations are launching assaults on polluting industries and decisionmakers who view them as expendable. These essays offer several case studies (Indian Creek AL, Warren County NC, Texarkana, Dallas, Louisiana's "Cancer Alley", the Hopi and Navajo reservations, Los Angeles), and discuss building networks, PUEBLO (People United for a Better Oakland) fighting against lead poisoning, grassroots activism by women of color, the First National People of Color Environmental Summit (held in Washington in October 1991), and recommendations to the EPA. [ALSO SEE: Bullard's **Dumping in Dixie**, 2nd edition (Westview, July 1994/199p/$50.00;$19.95pb), and **Toxic Struggles: The Theory and Practice of Environmental Justice**, edited by Richard Hofrichter (New Society Publishers, June 1993/260p/$39.95; $16.95pb).] **(environmental justice)**

136

40 Chernobyls Waiting to Happen, Maurice Strong (Secretary General, UNCED), *The New York Times* (Op-Ed), Sunday, 22 March 1992, E15.

All of the 60 nuclear reactors in the former Soviet Union and in Central and Eastern Europe have safety problems that have been building up for many years. The International Atomic Energy Agency warns that 26 have "serious" safety deficiencies, and 14 have "considerable" ones. Adding to this condition, the morale of managers and workers at these reactors has been destroyed by a lack of clear government policy and support, uncertainty about the future, a lack of incentives and suitable training, and shortages of funds, spare parts, and supplies. "All this has led to a serious deterioration in operating and maintenance performances in plants that already are accident prone because of design and construction flaws." The governments are aware of the problems but faced with a Hobson's choice: shutting the plants would deprive the countries of energy that is critical to weakened economies; alternative sources of energy require unaffordable investment or unavailable foreign exchange; replacing or rebuilding the plants takes time and money. "The only alternative is a large-scale rescue operation by the international community, which could not come at a worse time." But this potential crisis cannot be ignored. "The costs of acting now to prevent or contain a new environmental disaster will be far less than the costs of letting it happen." [**NOTE**: A classic warning of a highly undesirable future that we ought to heed, but probably won't.]
(**deteriorating nuclear plants in ex-Soviet bloc**)

137
Preparing for Nuclear Power Plant Accidents. Edited by Dominic Golding, Jeanne X. Kasperson, and Roger E. Kasperson (all Clark U). Boulder CO: Westview Press, Jan 1995/537p/$47.00pb.

Since the accident at Three Mile Island in 1979, vast improvements have been made to upgrade emergency planning around nuclear power reactors, particularly in overall regulatory guidance, communications systems, and a regime of exercises and drills. Nevertheless, notable problems still remain in existing structures and programs: 1) organizational roles and responsibilities are sharply divided and lack flexibility, thus encouraging a checklist mentality that favors regulatory compliance over safety; 2) an optimal geography arrangement for emergency planning does not now exist (there should be a graded emergency response at a variety of distances to cope with a spectrum of accidents); 3) plans are too often conceived as evacuation exercises applying only to the 10-mile emergency planning zone, but downwind evacuation to 15 miles and sheltering out to 25 miles could prove necessary, with localized control on milk consumption where radiation hot spots occur; 4) the apparent mindset of many officials is that worst-case accidents are so unlikely as to be incredible, and thus they fail to plan seriously for large accidents; medical planning is still hopelessly inadequate to deal with a worst-case accident, and plans for ingestion-exposure controls are rudimentary or nonexistent; 5) emergency planners do not take account of public fears and how the public might respond in an emergency.
(**nuclear accident planning faulted**)

138
Air and Water Pollution Regulation: Accomplishments and Economic Consequences. Martin Freedman (Prof of Accounting, SUNY-Binghamton) and Bikki Jaggi (Prof of Accounting, Rutgers U). Westport CT: Quorum Books/Greenwood, July 1993/266p/$65.00.

After 20 years of Federal air and water pollution regulations and billions of dollars spent, what has been accomplished? The authors examine two industries: electric utilities (the largest industrial air polluter in the US) and the pulp and paper industry (a major contributor to water pollution). Chapters discuss the air and water pollution laws, pollution by plants and companies, economic performance of the two industries, and measuring pollution performance. Concludes that "we have no doubt that from a societal perspective, the benefits of environmental legislation outweigh the costs." But in the absence of Federal enforcement, the effect of the pollution legislation has never been tested. Even with weak enforcement, it has been possible to reduce pollution. Though industries were forced to spend billions of dollars on pollution abatement, they were able to survive and in many cases prosper. But due to different philosophies of management and different regulatory constraints, there was a vast difference in emissions of plants within the same industry and even within the same firm. This is not fair. An effective and fair system of environmental regulation requires business/government/environmentalist cooperation in the standard-setting process, experimentation and creative solutions to achieve the standards, and uniform enforcement of the laws. Legislation that imposes a single solution to the problem (whether it be best available technology, emissions trading, taxing emissions, or fines) creates an adversarial relationship between the regulators and the regulated. (**fair and effective regulation needed**)

139
Clearing the Air in Los Angeles, James M. Lents and William J. Kelly (both South Coast Air Quality Management District, Los Angeles), *Scientific American*, 269:4, Oct 1993, 32-39.

Southern California's air quality remains the worst in the US, despite dramatic improvements in recent decades. From 1955 to 1992, the peak level of ozone in Los Angeles declined from 680 to 300 parts per billion. This improvement took place while population grew from 4.8 to 14 million, and the number of motor vehicles rose from 2.3 to 10.6 million. Yet air pollution in the region still reaches unhealthy levels on half the days each year. Los Angeles must reduce pollution even further to meet Federal and state health standards. By doing so, the region's citizens will gain $9.4 billion annually in health benefits (by preventing premature death, and eliminating restricted activity and discomfort). Los Angeles has 50 years of experience in studying the local atmosphere and experimenting with various policies, and could serve as exemplar to other US cities. Present plans seek to attack pollution in three stages: **Tier I** involves 135 measures using existing technologies, to be adopted by 1996 (e.g. low-polluting burners and catalysts on electric power plants, programs encouraging carpools); **Tier II** will rely on technologies that have just entered the commercial market (e.g. alternative fuels for autos); **Tier III** requires technologies that have not been fully developed but are likely to be available in the next decade or so (fuel cells, new paint coatings). The Air Quality Management District has found that it could not simply play the role of pollution cop, but also had to provide free technical assistance and a loan program for small businesses to purchase pollution-control equipment. (**Los Angeles as air pollution leader**)

140
Redesigning the American Lawn: A Search for Environmental Harmony. F. Herbert Bormann, Diana Balmori, and Gordon T. Geballe (all Yale School of Forestry and Environmental Studies). New Haven CT: Yale U Press, June 1993/166p/$19.95.

Americans love their lawns with a passion rarely seen in other countries. Lawns arose primarily in Western civilization, and mainly in France and England. In 1830, the Englishman Edward Budding invented the lawn mower, enabling modest householders to keep their lawns tidily cut without scythes or sheep. The mowed lawn appeared in the mid-19th century along with the creation of the American suburb. By 1870, detached housing surrounded by a yard had emerged as the suburban style of choice. Town centers in New England were transformed in the 19th century from bare stamped earth to lawned and treed parks, now called "greens." Today, healthy, green grass has become a symbol of success and plenty. A 1991-1992 survey indicates that 58 million American households (62% of the total) are engaged in lawn care. These lawns cover an estimated 20 million acres, with an average size of one-third of an acre. They range from the highly manicured "estate lawns" or the Industrial Lawn (all grass, no weeds or pests, continuously green, closely mowed) to the minimum-care Freedom Lawn. It is believed that lawns can add up to 15% of a home's worth.

But the American lawn as a human-modified ecosystem has environmental costs: 1) use of fossil fuels in lawn-care equipment and in producing and transporting fertilizers and pesticides (one hour of using a power lawn mower emits pollutants equivalent to driving 350 miles); 2) excess fertilizers and pesticides wash off our lawns and run into wells, streams, and lakes (homeowners use up to ten times more chemical pesticides per acre than do farmers); 3) lawn irrigation can exacerbate regional water-supply problems (up to 30% of urban water on the East Coast is used for lawn irrigation; in the West it claims 60% of urban water use); 4) lawn wastes are major contributors to the national solid waste problem (yard waste is the second largest component of the waste stream, and three-quarters of yard waste is grass clippings in large polyurethane bags); 5) replacement of millions of acres of naturally occurring ecosystems by the American lawn displaces native habitat and plays an important role in the decline of plant and wildlife species. Alternatives involve selecting the right grass, reducing or eliminating inputs (energy, fertilizer, pesticides, herbicides, water), selecting an ecologically-minded lawn care company (many companies lead us "down the road to what might be called the lawn care equivalent of drug addiction"), switching to a no-care Freedom Lawn, reducing the amount of lawn (more shrubs, trees, patios, or garden space), or outright lawn replacement (e.g., Japanese gardens use crushed stone or raked sand). In sum, "we need not cease to love the lawn... we can adapt it to our time." [**NOTE**: Brings the environmental problematique to our front doorstep. ALSO SEE: **The Lawn: A History of an American Obsession** by cultural historian Virginia Scott Jenkins (Smithsonian Institution Press, May 1994/237p/$14.95pb), who points to lawns as symbols of domestic values and good management, as well as control of the environment. Unlike Bormann *et al.*, Jenkins also mentions the role of golf courses.] **(environmental impact of lawns)**

141

Polluting for Pleasure. Andre Mele (Kingston NY). NY: W.W. Norton & Co, July 1993/224p/$22.95.

"Pleasure boating produces as much hydrocarbon pollution as all the cars, trucks, and buses—all the road vehicles—in America." Mele, a yacht designer, points out that more than 12 million powerboats are registered in the US. If every boat were used 20 times per season, there would be more than 240 million boat trips for joyriding, water-skiing, fishing, etc. The typical boat with an outboard motor produces 68 horsepower and consumes 20 gallons of gasoline and 3.5 pints of lubricating oil during a 3-4 hour afternoon of operation. "The total agglomerate oil and hydrocarbon pollution caused by pleasure boating is 420 million gallons per year—the equivalent of as many as 40 *Exxon Valdez* disasters every year." The average pleasure boat unleashes 70 times more hydrocarbons per gallon of fuel consumed on the environment than the average car. One hour of pleasure boat operation is the pollution equivalent of more than 700 miles in a car. Today, 30% of all Americans participate in recreational boating every year. Alternatives include clean energy, rethinking the propeller (where 30-70% of energy is lost), low-resistance boats, emission limits for pleasure boats similar to those for road vehicles, annual vessel emissions inspections, efficiency ratings for boats, etc. [**NOTE**: Stunning examination of a major but widely overlooked source of pollution, similar to the discussion of lawns (above). The first chapter discusses various calculations and assumptions.] **(powerboat pollution)**

142

Indoor Air Pollution: A Health Perspective. Edited by Jonathan M. Samet (Prof of Medicine, U of New Mexico) and John D. Spengler (Prof of Public Health, Harvard U). Baltimore MD: Johns Hopkins U Press, July 1991/407p/$65.00;$24.95pb.

A comprehensive review of the rapidly enlarging evidence on indoor air quality and health, with chapters on sources and concentrations of indoor air pollution, dynamics of air entry and air flow into buildings, assessing indoor air quality, understanding personal exposures, control strategies for air quality and thermal comfort, legal aspects, and the health effects of ambient tobacco smoke, nitrogen dioxide and carbon monoxide from combustion processes, wood smoke, formaldehyde, volatile organic compounds (from building materials, adhesives, household products), indoor air as an agent for infectious disease, biological agents and allergic diseases, radon, and building-related illnesses (research on "sick-building syndrome" has been hampered by lack of definitions).

The editors note that as outdoor air has become cleaner, the importance of indoor air has become increasingly evident. Most indoor living and working environments have air pollutant sources from people, pets, heating, cooking tobacco, soil gases, structural components and furnishings, consumer products, and a plethora of biologic sources. Thus "it should not be surprising to find indoor environments that are substantially more polluted than the nearby outdoor air." High concentrations of indoor pollution may be associated with discomfort, irritation, illness, and even death. Indoor air pollution can be quite high in remote villages in poor countries, where concentrations of organic and particulate matter can exceed by factors of 10 to 100 those in homes with modern cooking and heating. But modern buildings have their own problems. America's Clean Air Act provides a regulatory framework for pollutant levels in outdoor air, but we lack a comprehensive legal and regulatory framework for indoor air pollution. [**NOTE**: Readily attains the stated purpose of a "state-of-the-art" review.] **(indoor air pollution overview)**

143

Preservation of Near-Earth Space for Future Generations. Edited by John A. Simpson (Dept of Physics, U of Chicago). Cambridge UK: Cambridge U Press, June 1994/247p(8x12")/$79.95.

Proceedings of a June 1992 symposium on space debris. **Donald J. Kessler** (NASA-Houston) argues that about 200kg of meteoroid mass is within 2000km of the earth's surface, whereas the 7,000 man-made orbiting objects (the number might increase to 8,000 by 1995) within the same 2000km has a total mass of about 3 million kg. There is a "near certainty" that two of these objects will collide in the relatively near future, and thus a need to make immediate changes in operational practices. **Molly K. Macauley** (Resources for the Future) estimates the probability of a geostationary communications satellite being ruined by debris as 1 in 1000 during the expected life of the spacecraft, but notes that, due to proliferation of debris, the probability will increase 400-fold by 2000. She goes on to discuss the costs and benefits of a sustainable space environment. Other contributors consider the space debris problem as assessed in Europe and Japan, minimization and mitigation techniques, insurance aspects, legal issues, expected development of international minimization standards, UN initiatives, and elements of a multilateral treaty. **(growing space debris problem)**

12. Toxic and Solid Waste

144

World of Waste: Dilemmas of Industrial Development. K. A. Gourlay. London & Atlantic Highlands NJ: Zed Books, June 1992/247p/$55.00;$17.50pb.

Waste is what we do not want or what we fail to use. We live in a world dominated by waste, a World of Waste, most of it undesirable. Chapters discuss; 1) dilemmas of describing waste; 2) waste makers and their wastes: domestic and municipal waste, industry, mining and smelting, agriculture and fish farming, the nuclear sector, burners of fossil fuel, CFCs; 3) what happens to waste: into and onto the earth, into the water, into fire by combustion, into the air; 4) eliminating waste: responses from science and industry, the toxic waste trade, the ozone treaty, policies on greenhouse gases, and ideologies of development. Concludes that a concerted push from environmentalists, and enlightened politicians may yet wake up the world, but "my personal inclination is to doubt it." Greed, shortsightedness, and stupidity are far too widespread. [**NOTE**: Very good overview of where waste comes from and where it goes. Very poor and overly (?) gloomy conclusion on what can be done about reducing—not "eliminating"—the problem.] **(waste overview)**

145

Getting at the Source: Strategies for Reducing Municipal Solid Waste. World Wildlife Fund. Washington: Island Press, June 1992/142p(8x11")/$15.00pb.

Final report of the WWF Strategies for Source Reduction Steering Committee, which addresses the question of "How can the design and use of products be altered to reduce the amount and toxicity of municipal solid waste?" Proposals include: 1) a national definition for source reduction that suggests multiple actors and multiple activities; 2) wide adoption of a three-part Life-Cycle Assessment model consisting of: an inventory of materials, energy used, and environmental releases at all stages in the life of a product or process; an analysis of potential environmental effects; an analysis of needed changes; 3) setting a national policy goal that serves to inspire—rather than dictate—implementation plans by others and quantifiable goals for specific sectors of the economy; 4) a framework for evaluating source reduction opportunities; 5) an annual national awards program for outstanding achievement in municipal solid waste source reduction (recipients could include manufacturers, public interest groups, state and local governments, and educational institutions); 6) programs for labeling consumer products as "environmentally preferred" and "standard setting"; 7) various types of research (developing methods to measure the amount of waste generated and the extent of source reduction accomplished, improving social/psychological knowledge of waste generation patterns); 8) various strategies that can be implemented now (backyard composting, charging households by the quantity of garbage, using nontoxic paper and inks). [**NOTE**: An outline of growing complexity in future waste consciousness.] **(solid waste reduction strategies)**

146

Germany, Garbage, and the Green Dot: Challenging the Throwaway Society. Bette K. Fishbein (Sr. Fellow, INFORM). NY: INFORM, April 1994/261p/$25.00pb + $3.00 p&h.

Germany has initiated the world's most ambitious national solid waste policy, aimed at promoting source reduction, reuse, and recycling. Its 1991 Packaging Ordinance extends industry's responsibility for its packages to the end of their life cycles, including the cost of recycling. Germany is also considering legislation to extend industry responsibility to products such as automobiles, electronic equipment, newspapers, and batteries. A trademark green dot (*der Grüne Punkt*) is now placed on 90% of packages, signifying that they will be recycled, and most German retailers will not carry packages without the dot. The German system is described in great detail, as well as some lessons learned: the lack of adequate recycling capacity for collected materials, and many companies failing to pay their green dot fees on time or in full.

Concludes with questions for US solid waste policymakers. *1) How Should Responsibility for Waste Be Allocated?* In Germany, the "polluter pays" principle shifts responsibility to industry, and has already helped change the way companies think. France and Canada have opted for shared responsibility between the private and public sector. *2) What Should Be the Consumer's Role?* Making people pay for their trash by the bag or bin is common in Germany, and has been adopted in the US by more than 1,000 communities as of late 1993. *3) How Can Markets Be Created for Secondary Materials?* The lack of markets is the biggest problem facing recycling programs worldwide. *4) What Is Recycling?* The lack of standard terms often makes data comparison impossible, and published recycling rates are misleading because they are commonly based on materials collected rather than those actually recycled. *5) What Recycling Rates Can Realistically Be Achieved?* No consensus exists as to what is practical; any material can be recycled at high rates if cost is not a concern. *6) How Much Should Society Spend on Recycling?* Some argue that Germany's system mandates recycling at any cost; on the other hand, massive recycling could lead to economies of scale and development of new technologies. *7) Should Refillable Beverage Containers be Promoted?* Germany maintained its system, and its refill rate for beer and soft drinks is 80%; the US sold 84% of beer and soft drinks in refillable containers in 1964, but this has now dropped to 5%; benefits from expanding refilling most be weighed against the costs of reviving the US refill system. [**NOTE**: Excellent questions, intelligently discussed.] **(solid waste and recycling policy)**

147

Salvaging the Land of Plenty: Garbage and the American Dream. Jennifer Seymour Whitaker (Director, Committees on Foreign Relations, Council on Foreign Relations, NYC). NY: William Morrow, May 1994/336p/ $25.00.

America's infatuation with commerce and progress has brought more people more goods than any other society in history, but this plenty threatens to make our world unlivable. Trash and sewage, industrial wastes, incinerator and landfill emissions, agricultural chemicals, mining debris, and nuclear stockpiling are devastating our quality of life. We are being forced to face a new definition of pollution that is not just a matter of filth but of space. We generate more waste than any other people in the world. Until about 50 years ago, recycling and source reduction were integral to the lives of most Americans. When cities began collecting garbage around the turn of the century, many municipal officials saw it as a resource. Before WWI, more than 70% of US cities ran source-separation programs. From the 1920s on, the spread of incineration and dump trucks changed the economics of waste disposal. In the late 1960s and early 1970s, the notion of recycling reappeared, reinforced in the 1990s by serious constraints on waste disposal. But short-term expectations for recycling have been inflated (even Japan recycles only 30-35% of its household garbage), and most new recycling technologies are not yet at the stage of commercial development (these technologies get only minuscule amounts of research assistance). Germany is leading the way in getting industries to reduce wastes and design for recycling [see above].

Policy proposals: 1) rethink cost-benefit analysis (costs are frequently overstated); 2) use life-cycle analysis to increase awareness of impacts and use resources more effectively; 3) use government policy to reshape institutional responses (government must take a strong lead to shift the direction of incentives; this is well-understood in all other advanced nations); 4) substitute taxes on materials for many of the levies on income, sales, and property ("the most effective way to curb waste"); 5) rules specifying outcomes to be achieved are usually more efficient than those prescribing technologies to achieve them; 6) experiment with deposit-refund programs, especially for wastes that retain their original identity during use (car batteries, motor oil). We must adapt our idea of progress: "we must begin to judge our well-being by the quality and quantity of what we *use* rather than what we *own*." How America responds to its environmental crisis will also greatly affect what happens in the rest of the world. If aspiring developing countries can build their economic revolutions on conservationist technologies geared to 21st century materials efficiency, we will all be much better off. [**NOTE**: Thoughtful and well-documented.] **(rethinking waste)**

148

Making Less Garbage: A Planning Guide for Communities. Bette K. Fishbein and Caroline Gelb (both INFORM Solid Waste Program). NY: INFORM Inc, Dec 1992/179p(8x11")/$30.00pb + $3.00p&h (212/689-4040).

The average US resident produces 50% more waste than 30 years ago (4.0 pounds per person per day), and the EPA projects this to grow to 4.9 pounds by 2010. The US faces a solid waste crisis: we are less willing and able to bear the costs of managing ever-increasing amounts of garbage. Source reduction tops the list of solid waste management options adopted by the EPA (followed by recycling, incineration, and landfilling). Reducing the amount and/or toxicity of waste entering the waste stream is still in its infancy, and the potential is enormous. INFORM looked around the US for successful source reduction efforts, organizing strategies into 8 categories: 1) government source reduction programs through changing procurement policies or modifying operations (four state efforts are profiled); 2) institutional source reduction programs (correctional facilities, hospitals, schools); 3) government programs to stimulate reduction (technical assistance, waste audits, materials assessments, promoting backyard composting, leaving grass clippings on the lawn, awards and contests, reuse programs); 4) business source reduction programs (double-sided copying at AT&T, reducing waste in airline food services, reducing hotel waste); 5) educating the public; 6) economic incentives and disincentives (variable waste disposal fees generally based on volume, taxes or fees to include the cost of waste disposal in the price of a product, deposit/refund systems); 7) regulatory measures (required source reduction plans, environment-friendly labeling, disposal bans, reducing packaging waste); 8) reducing toxic materials. Concludes with a 13-page "Source Reduction Planning Checklist." **(garbage source reduction guide)**

149

Taking Out the Trash: A No-Nonsense Guide to Recycling. Jennifer Carless. Washington: Island Press, June 1992/249p/$35.00;$16.00pb.

Chapters discuss benefits of recycling, methods, common and uncommon recyclables, what individuals can do, how to set up a recycling program (at school, work, and in communities), what's being done (Federal legislation, state and local programs, industry efforts, National Park programs), and the $25 billion waste management industry (dominated by two major companies, Waste Management Inc. and Browning-Ferris Industries, both big recyclers). But over the past three decades, the US recycling rate has only increased from 7% of the waste stream to today's 13%. Recycling rates for some major products (paper, glass, aluminum) are gradually increasing, but plastics still lag behind. Barriers include: **1) Financial Factors**: tax advantages for users of virgin and nonrenewable resources often discourage recycling; **2) Markets**: current markets for recycled markets are weak and relatively unstable; setting recycling goals of 25-50% are futile without higher goals for purchasing recycled materials (government procurement of products made from recycled materials is essential); **3) Technology**: the US has no sophisticated R&D for recycling, and lacks recycling facilities with state-of-the-art equipment; **4) Public Habits and Attitudes**: many people still have well-entrenched buy/use/dispose habits, and are skeptical about claims of degradability and recyclability—often rightly so; **5) Contamination**: a small amount of noncompatible substance can ruin a whole batch of recyclable material at various stages in the recycling chain. Appendices list 35 non-profit organizations, 51 state recycling offices, 20 sources of recycled products, 17 relevant periodicals, and a bibliography of 160 items. **(recycling overview and guide)**

150

Recycling Solid Waste: The First Choice for Private and Public Sector Management. Thomas E. Duston (Associate Prof of Economics, Keene State College, NH). Westport CT: Quorum Books/Greenwood Press, April 1993/204p/$55.00.

"Recycling has come of age: it can be the first choice for solid waste management in the vast majority of waste disposal situations in the US." It can also be the first choice (the least costly method) for the more specialized solid waste where people live, work, and play: office buildings, shopping malls, warehouses, schools, colleges, and recreation areas. Chapters discuss the anatomy of a solid waste system, stages of a well-planned recycling program, how to measure gains from recycling (savings from costs avoided, sale of recovered materials), general questions that people ask about recycling, evaluating a recycling program, the microeconomics of recycling, and integrated waste management. Appendices include New Hampshire guidelines for composting yard waste, vendor lists of various recycling technologies, a sample recycling brochure, etc. [**NOTE**: Duston served as chairman of the committee that initiated a recycling program in Chesterfield NH, reducing the town solid waste budget by 35%.] **(solid waste recycling)**

151

In Defense of Garbage. Judd H. Alexander (former Executive VP, American Can Co and former Chairman, Keep America Beautiful Inc). Westport CT: Praeger, March 1993/239p/$22.95.

"Of all the great challenges faced by the US, the garbage problem is the easiest to solve." The garbage crisis is not a physical crisis or a financial crisis; it is a political and informational problem. We need new disposal capacity, rather than untested, blank check programs for source reduction and recycling. A modern, efficient, safe, user-friendly waste disposal system will be far less costly and far more responsive to the needs of people. **1)** Plan urban landfills for a 5-year instead of a 30-year life. Nearby property values would most likely go up, not down, when the site selection is announced. If neighbors knew that in five years the landfill would close and be replaced by a park or other desirable open space, property values would rise, and much NIMBY resistance would melt away. **2)** Pay rural communities to take the waste of distant cities—the money is often sufficient to finance a large share of the town budget. **3)** Call garbage "national solid waste" instead of "municipal solid waste," and establish a National Solid Waste Reserve with ten disposal sites scattered around the US on government land, especially military bases due to close. This Reserve could take care of a third of the municipal waste generated in the US in the next 100 years. These mega-landfills in the midst of vast tracts of land would have sufficient volume to justify sophisticated materials recovery systems. Access could be limited to rail, which is the cheapest and most energy-efficient shipping system. **(national garbage disposal proposed)**

152

Whose Backyard, Whose Risk: Fear and Fairness in Toxic and Nuclear Waste Siting. Michael B. Gerrard (Columbia U School of Law). Cambridge MA: MIT Press, Dec 1994/335p/$39.95.

A practicing environmental lawyer argues that, although the modern era in environmental law is nearly 25 years old, "we have nothing approaching a coherent national system of waste disposal." Instead, there are at least two dozen completely different kinds of waste streams, each the subject of its own scheme of regulation or chaos. Waste is being trucked from every state to every other state, or sitting in often leaky containers until some other state will take it. Siting laws are based on: 1) a fundamental conceptual error, framing the problem as finding the best locations to dispose of hazardous waste/radioactive waste, rather than finding the system that maximizes social welfare (taking full account of social and environmental costs while achieving fairness); 2) factual mistakes (e.g., illegal dumping has little to do with disposal capacity); and 3) policy blunders (separate regulation of different kinds of waste streams, insisting on technically perfect sites, imposing a site on an unwilling community). A new alternative is proposed, based on local control, state responsibility, and national allocation of burdens among the 50 states, based on how much waste they each generate. The states would ask communities to volunteer to host facilities and offer financial inducements, and the Federal government would make available contaminated military sites. Social welfare would be maximized by reducing waste disposal to a minimum through price incentives, and eliminating hidden subsidies for waste generation. **(toxic and nuclear waste siting)**

153

Race and the Incidence of Environmental Hazards: A Time for Discourse. Edited by Bunyan Bryant and Paul Mohai (both School of Natural Resources and Environment, U of Michigan). Boulder CO: Westview Press, Oct 1992/251p/$33.00pb.

"A resurgence of the civil rights movement may be taking place as minority and oppressed communities across the nation begin to redefine their struggle in terms of a safe and clean environment being a basic right for all, regardless of race or color." These papers from a January 1990 conference in Ann Arbor focus on how minority communities are disproportionately exposed to environmental hazards. Topics include toxic waste in black communities, how the environmental movement can attract the support of minorities, the environmental record of the Congressional Black Caucus (it has consistently scored higher than other groups), a model of environmental discrimination, environmental blackmail in minority communities, minority anglers and toxic fish consumption, the effects of occupational injury and illness on the health status of black Americans, hazardous waste incineration in minority communities (a "least cost" option for incinerators), the effects of uranium production on Navajo communities, the unequivocal evidence for environmental racism, pesticide exposure of US farm workers, pesticide exports to the Third World, and dumping of toxic waste in African countries. Final chapter reprints the 17 Principles of Environmental Justice agreed on at the October 1991 First National People of Color Environmental Leadership Summit in Washington, e.g.: 1) environmental justice affirms the sacredness of Mother Earth and the right to be free from ecological destruction; 2) public policy should be based on justice for all peoples; 3) environmental justice mandates the right to ethical, balanced, and responsible uses of land and renewable resources in the interest of a sustainable planet; 4) policies should seek to clean up and rebuild cities and rural areas in balance with nature. **(civil rights and environmental justice)**

154

Nuclear Waste: The Problem That Won't Go Away. Nicholas Lenssen (Research Associate, Worldwatch Inst). Worldwatch Paper 106. Washington: Worldwatch Institute, Dec 1991/62p/$5.00. (Brief version in *World Watch*, 5:2, March-April 1992, 10-17.)

"The waste problem has emerged as the central nuclear issue of the next millennium—and beyond—and the one that may determine the final verdict on the nuclear age." The world's 413 commercial nuclear reactors, producing about 5% of the world's energy, created some 9,500 tons of irradiated fuel in 1990, bringing the total accumulation of used fuel to 84,000 tons—twice as much as in 1985 (the US accounts for a quarter of this). Within eight years, the global figure could pass 190,000 tons. In the Cold War years, governments downplayed problems of waste while promoting production of nuclear weapons and creation of civilian nuclear industries. Most governments had decided by the 1960s that burying nuclear waste deep in the earth is the best means of protecting the public. But like mirages, safe and permanent methods of isolating radioactive materials seem to recede from reach as societies begin to examine them closely. It is now clear that geologic burial cannot guarantee that these materials will remain sealed off from the biosphere. Nor have past efforts worked as planned: old burial and storage sites have proved leaky, and nuclear weapons facilities have become major public health hazards. No government has yet come up with a course of action acceptable either to those dedicated to an expanding nuclear industry or to those determined to stop production of more nuclear waste. A stalemate has formed in nearly every country, and attempts to deal with the ever-mounting waste threat continue to be pushed further into the future. **(nuclear waste storage)**

155

A $200 Billion Scandal (Special Report), Douglas Pasternak, *U.S. News & World Report*, 14 Dec 1992, 34-47.

Over the past four decades, disposal of nuclear waste at America's 17 principal and 100 secondary weapons factories was a problem whose solution was endlessly delayed. In the course of maintaining its Cold War military edge, the US contaminated billions of gallons of water and millions of tons of soil. Cleaning it up would be the largest public works project in US history. The Department of Energy, which is responsible for the cleanup effort, initially estimated the cost at $35-64 billion over the next 30 years. In the past four years, the DOE estimate has risen; "many of those involved now concede that the final bill could go much higher than $200 billion." Adding to the daunting cost, a review of the effort to date indicates deep problems that call into question the competency of the cleanup effort and the likelihood of even a moderately successful outcome. The outgoing head of the cleanup effort acknowledges that waste and inefficiency will account for up to 40 cents of every dollar spent, due to uncontrolled costs, lax standards, and overly close ties between DOE and its plant contractors.

(nuclear waste cleanup problems)

156

Siting Low-Level Radioactive Waste Disposal Facilities: The Public Policy Dilemma. Mary R. English (Energy, Environment, and Resources Center, U of Tennessee). Westport CT: Quorum Books (Greenwood Press), June 1992/278p/$49.95.

Low-level radioactive waste excludes high-level waste: spent fuel from nuclear power reactors and waste from reprocessing spent fuel (such wastes have long-lived levels of radioactivity and heat, requiring special shielding and isolation for thousands of years). LLW includes depleted uranium slag, spent resins and filters, solidified sludge and liquids, and contaminated trash, tools, hardware, and plant components. In 1990, 1.1 million cubic feet of LLW was sent to disposal sites. Of this waste, nuclear power plants accounted for 56% by volume and 79% by radioactivity, the industrial sector accounted for 31% and 19% respectively, government accounted for 6% and 2%, and academic and medical institutions accounted for 6% and 0.2%. Many nuclear reactors will be shut down between 2010 and 2030, producing large quantities of LLW, despite recent efforts at waste minimization. LLW is in some respects a harbinger of things to come for hazardous and solid waste disposal.

In 1980, at the urging of South Carolina, Washington, and Nevada—the only states with active facilities for LLW disposal—Congress enacted the Low-Level Radioactive Waste Policy Act (LLWPA), requiring each state to ensure disposal capacity for LLW generated within the state. The act encouraged formation of interstate compacts for regional disposal facilities, and specified that compacts could close their borders to outsiders by 1986. When it became apparent that new disposal facilities would not be available for some time, the deadline was extended to 1993. Only California and Nebraska anticipate having facilities open by then. Worse than "getting to 1993," the big problem is that, in 1996, storage or disposal becomes the problem of the states and not the generators. English describes issues of authority, trust, risk, justice, and legitimacy, concluding that LLWPA is only the beginning of a long and complicated endeavor, and that commitment to an open and just process will help to create legitimacy for all parties. [**NOTE**: Lays out the issues in great detail but does not offer prescriptions for this knotty problem.]

(low-level radioactive waste disposal)

157

Public Reactions to Nuclear Waste: Citizens' Views of Repository Siting. Edited by Riley E. Dunlap (Prof of Sociology, Washington State U), Michael E. Kraft (Prof of Environmental Studies, U of Wisconsin-Green Bay), and Eugene A. Rosa (Prof of Sociology, Washington State U). Durham NC: Duke U P, Nov 1993/332p/$49.95;$24.95pb.

How Americans view nuclear waste is critically important to waste disposal policy and management. This book assembles some of the best recent work on the topic by psychologists, sociologists, political scientists, and economists. Radioactive waste is perceived to be a particularly dangerous hazard, and the public has little trust in the Department of Energy. The high level of perceived risk and low level of trust are more important in predicting opposition than is public knowledge about waste disposal. Some successful technological fix might mollify public concerns somewhat, but not nearly enough to ensure easy acceptance of a waste facility. The nuclear subgovernment still holds an unshakable belief that increased public understanding—the knowledge fix—will translate into support for nuclear technologies. The nuclear waste problem has been defined as a scientific and technical problem, creating a process of self-fulfilling defeat. Conventional fixes (technological, judicial, knowledge) overlook the most important fix of all: a fix through participatory democracy. Involving the public does not guarantee success, but not involving it virtually guarantees failure. **(nuclear waste problem ill-defined)**

158

Burying Uncertainty: Risk and the Case Against Geological Disposal of Nuclear Waste. Kristin S. Shrader-Frechette (Distinguished Research Prof of Philosophy, U of South Florida). Berkeley CA: U of California Press, Dec 1993/360p/$40.00;$15.00pb.

Opposes present policy to bury "radwastes" as profoundly misguided on both scientific and ethical grounds, with special emphasis on the proposed facility at Yucca Mountain, Nevada. Questions the arguments of professional risk assessors: "one of the greatest abuses of quantitative risk assessment has been to cloak the value judgments behind a veneer of technical precision." In virtually all cases of estimating the magnitude of the risks of geological disposal, scientists and engineers have erred through an overconfidence bias. In nearly all evaluations of the acceptability of the radwaste risks, DOE assessors employed subjective and questionable value judgments. We should postpone the question of geological disposal for at least a century; until then, we should use monitored, retrievable, above-ground storage. [ALSO SEE: **Risk and Rationality** by Shrader-Frechette (#439), arguing that democratic procedural reforms in risk management are needed.] **(nuclear waste: expert views questioned)**

159
Toxic Debts and the Superfund Dilemma. Harold C. Barnett (Prof of Economics, U of Rhode Island). Chapel Hill NC: U of North Carolina Press, March 1994/334p/ $49.95;$19.95pb.

The Superfund was created in 1980 to clean up the toxic legacy of the industrial and chemical revolutions. Over a decade later, the consensus is that the program has failed: too much has been spent and too little accomplished. Since 1980, EPA has identified over 34,000 potential Superfund sites, and placed 1,236 on the National Priorities List. The states have also identified tens of thousands of sites needing attention, many of which will not be addressed through Superfund. EPA has made progress toward permanent cleanup at over 300 sites, but only certified 33 sites as "clean." Superfund has failed due to conflict over who will pay the toxic debt, estimated to exceed $100 billion. Other problems involve the failure to resolve conflict over how clean is clean, insufficient EPA staff resources relative to its mandate, and the cleanup and enforcement burden on the states that often exceeds fiscal capacity. Resolution depends on forging a broad consensus regarding an acceptable tradeoff between efficiency and equity, and building this trade-off into a program that rewards cooperation and penalizes recalcitrance.

(toxic waste cleanup failure)

160
Superfund: The Political Economy of Environmental Risk. John A. Hird (Asst Prof of Political Science, U of Mass-Amherst). Baltimore MD: Johns Hopkins U Press, July 1994/315p/$48.50;$16.95pb.

The language of environmental policy was until recently dominated by words like "conservation" and "stewardship." Although such issues are still a central element, the terms of the environmental policy debate are increasingly driven by public concerns about health risks, chiefly from chemical products and wastes. Nowhere has the fear of chemical hazards been more pronounced than with disposal of hazardous wastes. Opinion polls show that Americans routinely rate hazardous waste sites as the most important environmental problem. But experts rate health risks from hazardous waste sites well down on the list of environmental and public health priorities. In spite of these disagreements, the Superfund program was initiated in 1980 to treat abandoned hazardous waste sites. It was hurriedly pieced together with little concern for such issues as fairness, acceptable site cleanup levels, and the appropriate level of government to implement the cleanup. The basic controversy that Superfund must resolve is the difference between the public's view of risks and that of the experts. The only reasonable way to bridge this difference is to create institutions that allow the public to be more actively involved in seeking cooperative risk management solutions. "The technocratic view of risk management is not only excessively limited, but it also fails to enjoin the public in debate over environmental priorities." Other proposed reforms: 1) state and local implementation of parts of Superfund would be more appropriate and effective; 2) divide all inactive hazardous waste sites into those closed before 1981, and those closed after 1981 (when firms should have been aware of liability); 3) establish a National Environmental Restoration Fund to finance pre-1981 site cleanups; 4) allow a citizen-based committee to make allocative decisions to remediate environmental hazards so that they can participate meaningfully and recognize opportunity costs of spending clean-up funds perhaps better invested in other problems.

(toxic waste cleanup and risk)

161
Beyond Superfailure: America's Toxics Policy for the 1990s. Daniel Mazmanian (Prof of Public Policy, Claremont Graduate School) and David Morell (President, EPICS International, Oakland CA). Boulder CO: Westview Press, July 1992/278p/$55.00;$16.95pb.

The 1977 Love Canal tragedy startled the American public into an awareness of the mismanagement of hazardous wastes and toxic materials, and has been generally regarded as the starting point for intensifying concern. Congress quickly enacted the Superfund program to clean up hazardous waste dumps, and most states passed their own cleanup laws. But the legislation has not worked very well, and numerous microsuccesses were overshadowed by overall macrofailure. The first toxics decade began from ground zero, without any real appreciation by the general public of the scope or severity of toxics problems. A fundamental rethinking of toxics strategy is required: rather than piling on more and more microsuccesses, the shortest path to successful dispute resolution and policy implementation is a participatory democratic process. Few technical solutions will be implemented without public trust and confidence in experts, business, and government. Some promising policies for the "second toxics decade": 1) Broad Strategies to Direct Decisions of Individual Firms: sunshine disclosures and information systems, bans on selected activities or products, assigning environmental rights and calling upon market mechanisms (privatization), waste-based taxes (a powerful incentive to minimize, substitute, and treat), fee-and-rebate programs to remove incentives for illegal dumping, no-fault cleanup; 2) Positive Incentives: positive-action permitting (to encourage accelerated, simplified permit approvals), positive-action compliance (simplified regulatory procedures), positive-action cleanup; 3) Community Contracts for Siting Facilities: compacts between facility owners and locals as a means of resolving siting controversies; 4) A Public Utility Model for Siting, Permitting, and Management (akin to regulated gas and electric utilities; the best US example is the Gulf Coast Waste Disposal Authority in Houston); 5) Regional Compacts and Management Strategies (as currently unfolding in California); 6) Comprehensive Waste Management Through Auctions (whereby states periodically auction off the right to collect and treat all hazardous wastes in each of about 70 categories); 7) Democratic Discourse: siting efforts based explicitly on the politics of equity, so that the YIMBY response (Yes In Many Backyards) replaces NIMBY as the hallmark of siting in the 1990s. Concludes that there is no single best answer; the use of several complementary approaches is best. [**NOTE**: Sophisticated overview of a complex and controversial issue. ALSO SEE: **Borrowed Earth, Borrowed Time: Healing America's Chemical Wounds** (Plenum, 1991/ 298p/$27.50) by Glenn E. Schweitzer, the first Director of the EPA Office of Toxic Substances).]

(hazardous waste/toxics policy)

162
Cleaning Up the Mess: Implementation Strategies in Superfund. Thomas W. Church and Robert T. Nakamura (both College of Public Affairs, SUNY-Albany). Washington: The Brookings Institution, March 1993/209p/ $14.95pb.

Environmental pollution is a classic example of market failure. Toxic cleanup is a technically uncertain enterprise, with extraordinarily high costs, uncertain benefits, and controversial financing. Superfund "almost certainly is the most expensive environmental program in the history of the US, or of any other nation." With the 1990 reauthorization, the cumulative commitment of tax dollars stands at $15.2 billion. State and local funds, along with mandated contributions by liable industrial parties, will probably total many times the Federal expenditure. One estimate originally believed to be excessively pessimistic set the total cost of the Superfund at $100 billion. A 1991 estimate now suggests that the plausible upper bound of total societal expenditures necessary to bring about a rigorous cleanup of all US hazardous waste is $1.7 trillion. The authors describe the operation of Superfund, with an overall picture of lengthy delays, high costs, and conflict within the EPA, between the EPA and other agencies, and between government and potentially responsible parties. Concludes with "ruminations on reform." [**NOTE**: **Beyond Superfailure** (above) has a broader vision of toxics policy reform.] **(toxic waste cleanup)**

II. TOWARD SUSTAINABILITY

1. Overviews

163

The Global Partnership for Environment and Development: A Guide to Agenda 21. UN Conference on Environment and Development. Geneva: UNCED, April 1992/116p/$9.95pb. (Available from UN Publications, 2 UN Plaza, NYC 10017; #E.92.I.15.)

Sustainable development is an imperative, in both environmental and economic terms. The difficult but feasible transition will require a major shift in priorities for governments and people, "involving the full integration of the environmental dimension into economic policies and decision-making in every sphere of activity and a major redeployment of human and financial resources at national and international levels. This global partnership is essential to set the world community onto a new course for a more sustainable, secure and equitable future" (Maurice F. Strong in Foreword).

This summary guide to Agenda 21, the UNCED action plan for the 1990s and beyond, is organized around seven themes: **1) The Prospering World**: revitalizing growth with sustainability; **2) The Just World**: sustainable living that combats poverty, changes consumption patterns, and meets basic health needs; **3) The Habitable World**: developing sustainable human settlements with safe water supplies, sound management of solid wastes, and appropriate air pollution control; **4) The Fertile World**: efficient use of land and water, sustainable agriculture and forest development, managing fragile ecosystems, conserving biodiversity, sound management of biotechnology; **5) The Shared World**: protection of the atmosphere and oceans, sustainable use of living marine resources; **6) The Clean World**: managing toxic chemicals and hazardous and radioactive wastes; **7) The People's World**: enhancing participation and responsibility through education and training, and by strengthening the role of major groups (women, children and youth, indigenous people, NGOs, farmers, local authorities, trade unions, business and industry, and the sci/tech community). Essential means to attain these objectives include better information for decision-making, national capacity building, international legal instruments, sci/tech for sustainability, and improved international institutions. **(Agenda 21 summary)**

*164

AGENDA 21: The Earth Summit Strategy To Save Our Planet. Edited by Daniel Sitarz. Introduction by Sen. Paul Simon (D-IL). Boulder CO: EarthPress (800/748-1175), March 1993/321p/$24.95. (Quantity discounts available; also distributed by National Book Network.)

An abridgement by a lawyer/environmentalist of the final official text of AGENDA 21, as adopted by all 172 nations (including the US) attending the UN Conference on Environment and Development (UNCED) in June 1992. The first paragraph of the final unedited version of the Preamble reads as follows: *Humanity stands at a defining moment in history. We are confronted with a perpetuation of disparities between and within nations, a worsening of poverty, hunger, ill health and illiteracy, and the continuing deterioration of the ecosystems on which we depend for our well-being. However, integration of environment and development concerns and greater attention to them will lead to the fulfillment of basic needs, im-*

proved living standards for all, better protected and managed ecosystems and a safer, more prosperous future. No nation can achieve this on its own; but together we can—in a global partnership for sustainable development.

The initial chapter, "Understanding AGENDA 21," describes the strategy as a blueprint for action in all areas of human activity, with 40 separate sections of concern and 120 action programs. Subsequent chapters are devoted to seven key themes: the quality of life on earth (controlling population growth, eradicating poverty, changing consumption patterns, improving health), efficient use of natural resources (land, fresh water, energy, agriculture, forests, deserts, mountain areas, biodiversity, biotechnology), protection of our global commons (atmosphere, oceans), managing human settlements, managing chemicals and waste, sustainable economic growth (trade policies, decision-making, transfer of environmentally sound technology), and implementation (roles of women, youth, children, indigenous people, private organizations, farmers, local authorities, workers and unions, business and industry, the sci/tech community, and education). "The secretariat of the Earth Summit has estimated the average annual costs (1993-2000) of implementing all of the activities in AGENDA 21 in developing countries to be over $540 billion." For the far-ranging programs of AGENDA 21 to be successful, "a concern for the environment must begin to be integrated into every human action and every personal decision." Humanity must begin the difficult task of taking responsibility for its actions. [ALSO SEE: the full official text, **AGENDA 21: The UN Programme of Action From Rio** (UN document E.92-38352 in 3 volumes).] **(AGENDA 21 abridgement)**

165

The Earth Summit's Agenda for Change: A Plain Language Version of Agenda 21 and the Other Rio Agreements. Michael Keating. Geneva, Switzerland: Centre for Our Common Future (52, rue des Paquis), April 1993/70p/$10.00pb (20 or more copies, $7.50 each). (Also in French, German, Italian, Russian, and Spanish.)

A brief and popularized version of AGENDA 21, along with shortened versions of 1) the Rio Declaration on Environment and Development (stating that environmental protection should be an integral part of the development process, etc), 2) the Statement of Principles on Forests (committing nations to take part in "the greening of the world," etc), 3) the UN Framework Convention on Climate Change (requiring nations to provide information on greenhouse gases they release and how much is absorbed by their sinks, etc), and 4) the Convention on Biological Diversity (requiring nations to develop national strategies for sustainable use of biodiversity, restore degraded ecosystems, etc). Following the Preamble, AGENDA 21 is condensed here into four sections: **1) Social and Economic Dimensions**: international cooperation, combating poverty, changing consumption, controlling population, protecting and promoting health, sustaining human settlements, making decisions for sustainable development; **2) Conservation/Management of Resources**: protecting the atmosphere, managing land, deforestation, desertification and drought, mountain development, rural development, biodiversity, managing the oceans and fresh water, safer use of toxic chemicals; **3) Strengthening the Role of Major Groups**: women, children and youth, indigenous people, NGOs, local authorities, farmers, workers and trade unions, business and industry, scientists; **4) Means of Implementation**: financing sustainable development, technology transfer, science, education and public awareness, creating capacity, organizing for sustainable development, international law, and information for decision-making. [**NOTE**: This 70-page version has a different arrangement than the 321-page "abridgement" (above). Both propose a guiding vision for many people worldwide.] **(AGENDA 21 plain language version)**

166

The Earth Summit: A Planetary Reckoning. Adam Rogers (Editor, *Earth News*, Los Angeles). Foreword by Noel Brown (UNEP). Afterword by David Suzuki (UBC). Los Angeles: Global View Press, Feb 1993/351p/$16.95pb.

A journalist's report of the UNCED Earth Summit held in June 1992, with chapters on the road to Rio (beginning with the Stockholm conference of 1972), the Rio experience, the unofficial Global Forum (attracting 7,892 NGOs from 167 countries), voices of children and indigenous people, summaries of and comments on the major speeches (by George Bush, Brian Mulroney, Fidel Castro, Gro Harlem Brundtland, and others), the five pillars of UNCED (the Conventions on Climate Change and Biological Diversity, the Rio Declaration, the Statement of Forest Principles, Agenda 21), the missing agenda (the lack of attention paid to transnational corporations, enforcement of the agreements, financing, timetables, overpopulation, the military, nuclear power, and the meat industry), the alternative treaties (highlights of 43 alternatives, such as the Earth Charter, Treaty for NGO cooperation, the NGO Code of Conduct, the Treaty on Alternative Economic Models, etc.), and the road from Rio ("not a single highway but hundreds of avenues leading to the 21st century"). A key indicator of what happens after Rio is how effectively the UN implements UNCED agreements via the new Commission on Sustainable Development. Also discussed is the new Costa Rica-based Earth Council, which will evaluate the progress of governments and corporations in achieving UNCED objectives. In the Afterword, David Suzuki laments that the Earth Summit "failed to meet even the most pessimistic of expectations" and offers his Declaration of Interdependence to promote a reconnection with the family of life. [**NOTE**: A readable and seemingly objective report with astute comments.] **(report on 1992 Earth Summit)**

167

Appraising the Earth Summit: How Should We Judge UNCED's Success? Peter M. Haas (U Mass-Amherst), Marc A. Levy (Princeton U), and Edward A. Parson (Harvard U), *Environment*, 34:8, October 1992, 6ff.

With 20 years of hindsight, we have some idea of the outcomes of the 1972 Stockholm conference on the environment. It served to boost concern among governments and other social actors, to spawn a number of institutions (notably UNEP), and to set in motion or accelerate negotiating processes that resulted in numerous environmental measures. The 1992 UNCED Earth Summit could play an even more influential role in the next 20 years. The media attention, level of participation, and involvement of NGOs were all much greater in Rio than in Stockholm. More than 100 heads of state attended UNCED, while only 2 went to Stockholm. More than 1,400 NGOs were accredited by UNCED, whereas only 134 NGOs were at Stockholm. Substantively, UNCED shows more promise as well. **Agenda 21** (#164) reflects a far more sophisticated appreciation of the ecological links that must be addressed, than did the 21-page Stockholm Action Plan. The agreements signed in Rio created several new international institutions, notably the Sustainable Development Commission. This new UN body, composed of representatives from the 54 member states of the UN Economic and Social Council, will monitor implementation of Agenda 21 and related activities throughout the UN system. Two other important new international institutions inspired by UNCED: the Business Council for Sustainable Development (#248) and the Planet Earth Council, a privately financed NGO to be led by UNCED Secretary-General Maurice Strong and a committee of 28 well-known scientists. The Council may emerge as an analogue to Amnesty International, producing impartial reports on national efforts for sustainable development. The authors also provide ***A Summary of the Major Documents Signed at the Earth Summit and the Global Forum*** (pp 13-15ff), such as Agenda 21, the Convention on Biodiversity, a non-binding declaration on forests, and various products of NGOs such as the Earth Charter. **(widespread UNCED influence likely)**

168

Earth Summit: Conversations with Architects of an Ecologically Sustainable Future. Steve Lerner (Director, Commonweal Research Institute). Published by Commonweal and Friends of the Earth, U.S. Bolinas CA 94924: Common Knowledge Press, August 1991/263p/ $9.95.

Interviews with 19 leading environmentalists and social activists, focusing on the 1992 UNCED meeting—the greatest opportunity of this decade to encourage a global discussion of how to build a framework for a just and sustainable global system. This book seeks to raise the level of public dialogue. ***1) The Big Picture***. **Jim MacNeill** (Institute for Research on Public Policy, Ottawa) estimates a grace period of 20-30 years in which to complete transitions in population, agriculture, energy, industry, and equity; but he fears pressures on leaders at UNCED to negotiate an empty framework convention, with no goals, no targets, no funds, and no means of compliance; if such an "empty" global warming convention is signed, it will remove pressure to do something substantive; **Jessica Mathews** (VP, World Resources Institute) argues that sometimes a treaty that is meant to be a fake can in fact move the process along, and that there is an enormous horizon of potential from reinventing technology on nature's example; **Herman Daly** (World Bank Environment Program) points to several rules of thumb that move toward sustainability, but worries that we are still in a phase of "huge denial". ***2) Building and Protecting Sustainable Communities***: essays on building sustainable municipalities in Norway and Czechoslovakia, the Regional Network of Environmental Experts in southern Africa, the Social Ecological Network in Uruguay, the Western Canada Wilderness Committee, and the Haribon Foundation in the Philippines which convened the Green Forum coalition of NGOs. ***3) Reforming the Institutions***: pressuring the World Bank to make its loans promote sustainable development (including more public accountability and freedom of access to information), reforming the IMF and the World Bank, the desirability of taxing pollution instead of payrolls. ***4) Organizing, Lobbying, Networking, Negotiating***: the U.S. Citizens Network on UNCED, the Asian NGO Coalition for Agrarian Reform and Rural Development (including over 2000 groups), the Greenpeace effort for a visionary response, and the origins of UNCED out of the Brundtland Commission. **Michael Lerner** (President, Commonweal) summarizes major themes: whether a meaningful convention on global warming will be signed, whether incremental movement toward sustainable development is sufficient, working at the grass-roots level, ways in which activists hope to use the UNCED process, and the pivotal degree to which the Reagan and Bush Administrations have blocked progress on critical environmental issues. [**NOTE**: Sophisticated conversations at the frontiers of global environmental politics.] **(UNCED and a sustainable future)**

169

Beyond the Earth Summit: Conversations with Advocates of Sustainable Development. Steve Lerner (Research Director, Commonweal). Bolinas CA 94924: Common Knowledge Press/Commonweal (PO Box 316), May 1992/294p/$12.50pb.

Interviews with various leaders during the months leading up to the UNCED meeting in Rio. ***1) Overviewers***: **Lester R. Brown** (President, Worldwatch Institute) sees the conference as the official event marking the transition to a new era dominated by the need for ecological sustainability; **Sen. Al Gore** (D-TN) thinks that "UNCED is the designated time and place for the emergence of a new world ethic"; **Maurice Strong** (Secretary-General, UNCED) discusses "launching an era of international environmental negotiations"; **James Gustave Speth** (President, WRI) views UNCED as potentially the most important multi-party negotiation since the ones that set up the UN and the World Bank. ***2) Activists on the Front Line***: **Wangari Maathai** (Coordinator of the Green Belt Movement in Kenya), on making tree planting a vehicle to reach women; **Martin Rocholl** (European Youth Forest Action), on encouraging European groups to do environmental actions together; **Vandana Shiva** on the dispersed Chipko Movement in India; **Regina Barba Pirez** on promoting eco-tourism in Mexico; **Uchita de Zoysa** (Coordinator, Public Campaign on Environment and Development, Sri Lanka) on the alternative Citizen's Report to UNCED (competing with the National Report of the Sri Lankan Government). ***3) Defenders of Biological Diversity***: **Thomas E. Lovejoy** (Smithsonian Institution); **Darrell Addison Posey** (Ethnobiological Institute of the Amazon); **Jason W. Clay** (Cultural Survival); **Oren R. Lyons** (Onondaga Nation, NY) on sustainable lifestyles of indigenous people. ***4) South vs. North***: **Martin Khor** (Third World Network, Malaysia), **Amb. Robert J. Ryan Jr** on the US position at UNCED; ***5) Reformers of International Organizations***: **David Runnals** (IRPP, Ottawa) on reforming the UN to coordinate sustainable development activities; **Hilary F. French** (Worldwatch Institute) on environmental treaties and amending the GATT; **Scott A. Hajost** (Environmental Defense Fund) on lobbying at international environmental negotiations; **William R. Pace** (Director, Center for the Development of International Law) on the 1989 Hague Declaration on the Environment ("an extraordinary declaration" calling for a major new global environmental authority to protect the earth's atmosphere, and for strengthening the enforcement mechanisms of the World Court; it was opposed by the UK and US, "so it just sank"); **Stephen Collett** (Friends World Committee) on UNCED as the start of a process to redefine the UN vision; **Harris R. Gleckman** (UN) on the code of conduct being developed by the Centre on Transnational Corporations. **Michael Lerner** (President, Commonweal) concludes that there is widespread agreement that UNCED is the beginning of a process of achieving global bargains on sustainable development, that success of this process will depend on strengthened international institutions, and that NGOs will be crucial to this process.

(sustainable development leaders interviewed)

*170

Caring for the Earth: A Strategy for Sustainable Living. IUCN/UNEP/WWF (edited by David A. Munro and Martin W. Holdgate). Gland, Switzerland: IUCN-The World Conservation Union, Oct 1991/228p/$19.95. (Distributed in US by Island Press.)

World Conservation Strategy: Living Resource Conservation for Sustainable Development was published in 1980 by the International Union for Conservation of Nature, the UN Environment Programme, and the World Wildlife Fund, resulting in conservation strategies in over 50 countries. This report re-states current thinking about conservation and development, seeking to secure widespread commitment to a new ethic for sustainable living. "Living sustainably must be a guiding principle for all the world's people." The change to living sustainably and caring for the Earth will be a major one for most people. To live within the limits of the Earth, population growth must stop everywhere, and the rich must stabilize (and in some cases reduce) their consumption of resources; ways exist to do this without reducing the real quality of life. Legal, social, economic, and technical measures aimed at sustainability must be integrated in planning and action at all levels, particularly in national governments.

Chapters are organized in three parts: **1) Principles for Sustainable Living**: respect and care for the community of life, development as improving the quality of human life, conserve the Earth's vitality and diversity, minimize depletion of non-renewable resources, keep within Earth's carrying capacity, enable communities to care for their own environments, provide a national framework for integrating development and conservation, create a global alliance to help poor countries develop sustainably, environmental education as an integral part of formal education at all levels; **2) Additional Actions for Sustainable Living**: develop explicit national energy strategies for the next 30 years, build concern for the Earth into the whole apparatus of business and industry, adopt high environmental performance standards backed up by economic incentives, develop efficient national and global systems for waste management, develop a sustainable urban transport policy, create a national strategy for agricultural sustainability, strengthen community management of forests, improve information for sustainable water management; **3) Implementation and Follow-Up**: promote the Strategy for Sustainable Living through broad-based publicity, reorder budget priorities, monitor and evaluate the Strategy, and finance new investment through Earthcare bonds, reduced military outlays, and more corporate sponsorship. In all, 132 actions are recommended, and targets are recommended for 1995, 2000, and 2010. [**NOTE**: Free 24-page summary available from Thaddeus C. Trzyna (CIPA, Sacramento; Chairman of the IUCN Commission on Environmental Strategy and Planning), 916/442-2472.] **(ethic for sustainable living)**

171

The 2050 Project (quarterly newsletter), World Resources Inst., 1:1, Winter 1994. (Free from WRI, 1709 New York Ave NW, Washington DC 20006; 202/662-2517.)

In January 1993, WRI, The Brookings Institution, and the Santa Fe Institute jointly launched a four-year collaborative project to explore the question, *How can we achieve sustainable existence on this planet by the middle of the next century?* The year 2050 is far enough in the future for critical population, energy, and global warming trends to become manifest and for large-scale social and economic transitions to occur; yet, it is within the life span of today's children. The Project will: 1) examine alternative scenarios of sustainability and explore desirable future conditions, the actions needed to reach them, and the potential for unexpected developments that could profoundly alter future conditions; 2) define the traits of a sustainable world in 2050 and determine what must be done to achieve these; 3) develop regionally-specific policy proposals and action plans for the next decade, along with regional policy workshops; 4) provide strategic guidance to policymakers worldwide, and stimulate thinking in the general public. The Project, directed by Rob Coppock, intends to be "a landmark study" and to release TV programs and tapes, publications for general readership, and computer programs. **(2050 project underway)**

172

Saving Our Planet: Challenges and Hopes. Mostafa K. Tolba (Exec. Director, UN Environment Programme, Nairobi). London & NY: Chapman & Hall (29 W 35th St), Aug 1992/287p/$25.00pb. (Also in French and Spanish.)

Outlines trends for the 10 major environmental issues of the past two decades: urban air pollution (lower in most developed world cities but markedly worse in LDCs), rapid advances in understanding stratospheric ozone depletion, increased understanding of causes and possible effects of climate change, increased marine pollution from land-based sources (about 60% of world population lives within 100 km. of coastal areas), assuring quantity and quality of freshwater, growing desertification and degradation of arid lands, deforestation and wetland destruction, loss of biological diversity, reducing the risk of human-caused disasters, and control of hazardous wastes and toxic chemicals. Chapters also discuss sustainable development (in agriculture, industry, energy, transport, and tourism), human well-being (population growth, settlements, health, peace and security), and public perceptions and attitudes. Some priorities for the next two decades: **1) Regulation**: by 1995, a global agreement on reforestation targets for each decade of the 21st century, a global plan to combat marine pollution from land-based sources, agreement to ban all exports of hazardous wastes to LDCs, an international code of conduct for technology transfer; **2) Assessment**: by 2000, comprehensive assessments of urban air quality, freshwater resources, land and soil degradation, and impacts of new technologies; **3) Environmental Management**: by 2000, a 30% reduction from 1990 levels in hazardous waste generated, environmental and natural resource accounting in the national accounts of every country, natural resource imports and exports included in international trade statistics, an end to net global deforestation; **4) Environment and Economics**: by 1995, estimates of the global costs of failing to deal with various problems, and of resources needed for transferring information and technology to LDCs. [**NOTE**: Valuable and concise overview.]
(global trends and priorities)

173

Saving the Planet: How to Shape an Environmentally Sustainable Global Economy. Lester R. Brown, Christopher Flavin, and Sandra Postel (Worldwatch Institute). NY: W. W. Norton & Co, Nov 1991/224p/ $19.95;$8.95pb.

The first in a new Worldwatch Environmental Alert Series, edited by Linda Starke. Chapters describe the June 1992 UNCED conference in Brazil (and the hoped-for Earth Charter as a Magna Carta for the planet), and compare the general inability to comprehend the scale of the ongoing degradation of the planet to passengers on the *Titanic* who could not imagine the reality of a sinking ship. The still widely-held belief that the global economy can continue along the path it has been following stems in part from a narrow economic view of the world. "The Shape of a Sustainable Economy" is described in six brief chapters devoted to the energy efficiency revolution, building a solar economy, reusing and recycling materials, protecting the biological base for food and raw materials (croplands, grasslands, forests, and fisheries), adequately feeding a world population 50% larger than today's 5.4 billion people (the authors believe that human numbers can go no higher than 8 billion; the World Bank now projects at least 12 billion people late in the 21st century), and stabilizing world population. Chapters are also devoted to five "Instruments of Change:" 1) abandoning "growth" as an overriding goal and seeking "sustainable progress" instead; 2) better indicators of human welfare that recalculate and go beyond GNP; 3) reshaping government incentives that work against sustainability (underpriced irrigation, below-cost timber sales, subsidized road-building, subsidies for pesticides); 4) green taxes (eight potential taxes are described, e.g.: carbon content of fossil fuels, hazardous wastes generated, pesticide sales, SO_2 emissions, etc.); 5) ecologically sound banking (a new generation of World Bank reforms).

Concludes that the goal of a sustainable society depends on redirecting the engine of economic development. "Unless nations address the twin problems of growing Third World poverty and increasing international inequity, global economic and environmental decline are certain to accelerate." And, as national governments begin to take action, new measures of international leadership (rather than size of military or GNP) will likely be developed. The UN will have to assume a more prominent role; most of its environment programs are underfunded or staffed by people with little training.

("sustainable progress" goal needed)

174

Save the Earth. Edited by Jonathon Porritt (London UK). Foreword by HRH The Prince of Wales. Introduction by Robert Redford. Atlanta GA: Turner Publishing Inc (One CNN Center; distributed by Andrews & McMeel), Sept 1991/208p(9x11")/$29.95.

A glossy coffee-table companion to the "Save the Earth" TV series on TBS SuperStation in early 1992. Includes essays by Norman Myers on the disappearing rainforest, Lester Brown on erosion of croplands, Peter Raven on biodiversity, Stephen Schneider on global warming, and Michael Walsh on the dangers of air pollution. Other topics: grasslands, threats to the world's mountain ranges, population growth, urbanization in poor countries, resources in rural villages, honoring the wisdom of tribal peoples, water shortages, failing fisheries, intense pressures on shorelines and estuaries, the abuse of rivers and lakes, draining of wetlands, and accepting responsibility. Chapters are punctuated with 240 color photographs and brief statements by such luminaries as Gro Harlem Brundtland, David Suzuki, James Lovelock, Karan Singh, Maurice Strong, Jesse Jackson, Walter Cronkite, Margaret Atwood, Paul McCartney, Mostafa Tolba, Carl Sagan, Petra Kelly, Paul and Anne Ehrlich, Desmond Tutu, and Sen. Al Gore.

[**NOTE**: A British import, similar to the two Gaia atlases by Norman Myers (**FS Annual 1992** #10925), which have more ideas, and **Imperiled Planet** by Edward Goldsmith *et al* (**FS Annual 1992** #11089), which has even better photos.]

(global environmental problems overview)

175

If You Love This Planet: A Plan to Heal the Earth. Helen Caldicott, M.D. (founder, International Physicians to Save the Environment). NY: W. W. Norton, March 1992/231p/$19.95;$10.95pb.

Australian author of **Nuclear Madness** and **Missile Envy** describes symptoms of our planet's "disease processes": ozone depletion, the greenhouse effect, deforestation, acid rain, toxic pollution, species extinction, human overpopulation, First World greed and Third World debt, the manufacture of consent, pathology induced by the corporate ethic, concentrated press ownership, and the low standards of corporate-controlled journalism. Her prescription is "Love, learn, live, and legislate." Unless we connect directly with the earth, we won't know how to save it. Prescriptions include a carbon tax to reduce global greenhouse gas emissions, altered lifestyles so that we do no more harm, deep love for our fellow creatures, compulsory voting as in Australia, proportional representation for women, and independent green candidates running in every congressional district who "will cause a sea change in the thinking and behavior" of the major parties. [**NOTE**: A spirited but superficial polemic, followed by a notably skimpy "plan."]

(prescription for planetary disease)

176

Healing the Planet: Strategies for Resolving the Environmental Crisis. Paul R. Ehrlich and Anne H. Ehrlich (both Biological Sciences, Stanford U). Reading MA: Addison-Wesley, Oct 1991/366p/$22.95.

A companion volume to **The Population Explosion** (Simon & Schuster, 1990; **FS Annual 1991** #10302) warning that the human enterprise is rapidly outstripping the capacity of Earth to nurture civilization. Carrying capacity of the planet for human beings is threatened virtually everywhere. Yet our society's leaders do not get the message. A root cause of environmental deterioration is human population, as summarized by the I=PxAxT equation (Impact = Population size x per-capita Affluence x Technologies that supply consumption). The 1990s must be a decade of environmental politics, or humanity will have blown its last good chance. Chapters discuss our life-support systems, energy and the environment, meeting future energy needs, global warming, the ozone shield, air and water pollution, pressures on land resources, making agriculture sustainable, and risk-benefit decisions.

Concludes that people cannot escape the I=PxAxT equation. Humanity must take control of its long-term future and mitigate its total impact through all three factors in the equation: controlling population growth, reduced consumption by the rich, and more environment-benign technologies. "We believe that economists, perhaps even more than ecologists, hold the key to the human future." But mainstream economics does not address itself to the critical question of our day: how to sustain the human enterprise. A new ecological-economic paradigm could be constructed that unites nature's housekeeping and society's housekeeping, with first priority given to keep nature's house in order. The President's Council of Economic Advisors should be subsumed under a new "Council of Ecological and Economic Advisors." Other proposals to move toward a sustainable society and escape the "human predicament": 1) an educational system enabling citizens to grasp the big picture ("failure to produce citizens equipped to deal with the modern world is a disgrace"); 2) policies to shrink population in LDCs ("stop at two" programs backed with incentives and penalties); 3) a global effort to arrange controls on CO_2 and other emissions; 4) fundamental changes in the attitudes of people toward nature and toward each other; 5) substantial increases in US gas taxes and "feebates" (a rebate or fee assessed at the time of an automobile purchase); 6) US military conversion, with part of today's military force serving in a National Service Corps; 7) global institutions providing a forum for discussing how to modify the market system to deal properly with externalities and how to adjust the terms of international trade. A final chapter describes tasks for natural scientists, social scientists, engineers, journalists, business executives, attorneys, teachers, and others; anyone can help to improve the human prospect. **(escaping the human predicament)**

177

Earth in the Balance: Ecology and the Human Spirit. Senator Al Gore (D-Tenn). Boston: Houghton Mifflin, Jan 1992/407p/$22.95.

"Our ecological system is crumpling as it suffers a powerful collision with the hard surfaces of a civilization speeding toward it out of control." The damage is remarkably sudden and extensive, but most of us act as if we don't perceive a collision at all (partly because the crunching and shattering take place over a longer time span that we associate with a violent collision, say of automobiles). "The potential for true catastrophe lies in the future." The downslope that pulls us toward it is becoming steeper, and sooner or later our momentum will take us beyond the point of no return. We need a positive feedback loop to accelerate the pace of changes now so urgently needed.Chapters discuss the history of climate and civilization, the political struggle to control atmospheric pollution at the global level, threats to the earth's water system, de-forestation, the threat of genetic erosion to the global food system, the floodtide of garbage, the deep crisis of our political system that dumps huge mountains of debt and pollution on future generations, forces working against stewardship, the blindness of classical economics, the crisis in education coinciding with the flood of information, our dysfunctional civilization (we seem increasingly eager to lose ourselves in technology, media, and consumption rituals; the price we pay is the loss of our spiritual lives), and the common environmental threads of world religions.

Concludes that "we must make the rescue of the environment the central organizing principle for civilization." This means securing widespread agreement that it *should* be the organizing principle, and embarking on an all-out effort to use every policy and program to preserve and nurture our ecological system. Toward this end, a Global Marshall Plan is proposed, which, like the post-WWII plan for European recovery, focuses on strategic goals and emphasizes actions likely to remove bottlenecks. The five strategic goals that must inform our efforts: **1) Stabilizing World Population**: ensure that birth control devices are ubiquitously available, fund carefully targeted functional literacy programs for sustainable agriculture; **2) Developing and Sharing Appropriate Technologies**: a worldwide Strategic Environment Initiative is needed (similar to SDI) to develop a series of breakthroughs for environmentally appropriate technologies in agriculture, forestry, energy, building technology, waste reduction, and recycling; **3) A New Global Eco-Nomics**: redefine definitions of GNP and productivity, eliminate government use of inappropriate discount rates and subsidies, full disclosure of corporate responsibility for environmental damage, more and better information on environmental impacts, an Environmental Security Trust Fund with payments based on the amount of CO_2 put into the atmosphere, a Virgin Materials Fee imposed on products at the point of manufacture or importation based on the quantity of nonrenewable virgin materials, tree planting programs as part of community workfare programs; **4) A New Generation of Treaties and Agreements**: the US must take a leadership role, with the Montreal Protocol on phasing out CFCs as a prototype; **5) A Cooperative Plan for Educating the World's Citizens**: a comprehensive program for monitoring changes now under way, a massive effort to disseminate information on local and regional threats, establishing environmental training centers throughout the world. It is also urged that the Plan should have as a general and integrating goal the establishment of social and political conditions most conducive to emergence of sustainable societies: social justice, human rights, adequate nutrition and health care, high literacy rates, political freedom and participation. [**NOTE**: Remarkably sophisticated and well-informed thinking by a world-class political leader for the 21st century. ALSO SEE: **World on Fire: Saving an Endangered Earth** by former US Senate Majority Leader George J. Mitchell of Maine (Scribner's, Jan 1991; **FS Annual 1992** #11106), which has similar ideas, but no focus on technology and education. **Environmental Gore** (#378) provides a book-length critique.]

(Global Marshall Plan for environment)

178

The Rapid Growth of the Human Population: Sustainable Economic Growth, Al Gore (Vice President of the US), *Vital Speeches of the Day*, 60:24, 1 Oct 1994, 741-745.

An August 1994 presentation at the National Press Club on the post-Cold War "opportunity and responsibility to address long-neglected future-oriented concerns that will determine what kind of world we leave to our children and grandchildren." Three factors have produced a radical change in the relationship between human civilization and the ecological system of the earth: 1) the habit of denying responsibility for the long-term consequences of our actions; 2) the unprecedented and accelerating revolution in science and technology; 3) the rapid growth of human population, which often contributes to degradation of natural resources. Population trends add to the problem of low wages and poverty, challenge societies' ability to invest in infrastructure and human capital, and often put strains on social stability. "Realizing the vision of sustainable development... must be a centerpiece of international cooperation in the post-Cold War period." The most effective way to reduce population growth and abortion (there are more than 25 million abortions each year worldwide) is through a comprehensive global strategy that makes family planning information and services as widely available as possible, that promotes sustainable economic development, that increases child survival and literacy, that fosters women's health, that strengthens families, and that focuses on education and empowerment of women. "We have learned that family planning by itself rarely succeeds, and that economic development by itself does not automatically reduce population growth."

(realizing sustainable development)

*179

Earth Politics. Ernst Ulrich von Weizsäcker (President, Wuppertal Institute for Climate, Energy and Enviroment, Wuppertal, Germany). Foreword by Ricardo Diez-Hochleitner (President, Club of Rome). London & Atlantic Highlands NJ: Zed Books, Jan 1994/234p/ $60.00;$19.95pb.

To make sustainable development a reality, we need a new set of values, a new culture, and a new set of incentives to make millions of actors in the North and the South act differently. The days of the Economic Century are numbered; we are now entering a Century of the Environment, where the hallmark of the realist must be regard for the environment. "In the Century of the Environment it is the ecological imperative which will become the dominant determinant for law and administration, for city planning and agriculture, for the arts and for religion, for technology and indeed for the economy." The political task associated with this impending transformation is Earth Politics. Chapters are in four parts: ***1) Setting the Framework***: the rape of nature in the economic century, classical environmental policy as pollution control, EC environmental policy, emerging global visions (Club of Rome, UNCED); ***2) Five Key Issues***: truth-telling prices for energy and materials, ecological transport policy, breaking the vicious circle in agriculture, a new model of wealth and progress for the Third World, endangered biodiversity; ***3) Towards Coherent Solutions***: prices that tell the truth, limits to command and control, ecological tax reform [see #236 for book co-authored by von Weizsäcker], economy-friendly environmental policy, cooperation with business instead of harassment, urban environmental policy (requiring cities to pay full-cost prices for their pollution); ***4) New Models of Wealth***: environmental technology (characterized by cleanliness, productivity, information intensity, error friendliness, and suitability for the informal economy), redefining productivity, responsible science, thinking less about the formal economy, a new role for adult education, and further development of human rights to include rights of future generations. [**NOTE**: A rich package of original, leading-edge ideas. The author is a Club of Rome member, former President, U of Kassel, and Prof of Interdisciplinary Biology, Essen U.] **(politics for the Century of the Environment)**

*180

Our Country, The Planet: Forging a Partnership for Survival. Shridath Ramphal (President, World Conservation Union-IUCN). Foreword by Seymour Topping (President, American Society of Newspaper Editors). Washington: Island Press, May 1992/291p/$15.00pb.

Former foreign minister of Guyana and secretary-general of the 50-nation Commonwealth (1975-1990), and a member of all the major international commissions that reported in the 1980s (including the Brandt and the Brundtland Commissions), prepared this personal statement on the agenda of the Earth Summit at the request of Maurice Strong. Chapters focus on: **1) A Fragile World**: "the crisis of survival we face arises from the propensity for destruction, including self-destruction, our species has displayed;" **2) Air and Water**: an overall picture of unsustainable demands on the planet's natural resources; **3) Earth and Fire**: on desertification, deforestation, impacts of burning fossil fuels; **4) The Profligate Rich**: the 25% of the human species in the OECD countries account for 70% of world fossil fuel consumption, 90% of automobiles, 85% of chemical products, and 85% of military expenditures; **5) The Powerless Poor**: the great majority of the poor subsist in the Earth's less productive, more fragile, and most disaster-prone locations; **6) Population Pressures**: world population has grown 42% in just 20 years, and will increase by a further 3 billion by 2025; **7) A Feudal World**: a feudal system holds people in permanent subordination and most of today's developing countries still suffer from the days of colonialism; sustainable development—that which is guided by environmental considerations—requires greatly enlarged flows of aid; "the integration of development and environment is the major issue confronting human society ... the rich must improve the quality of their development to relieve the stress they now place on the planet, while the poor must develop on a basis of growth that respects the Earth's capacity to sustain life. There is no other way;" **8) Ethics of Survival**: "We urgently need a new universal ethic and a reordering of global priorities;" such an ethic would teach that all life is sacred and promote a sense of human identity transcending national loyalties.

Concludes with three scenarios: **1) Muddling Through**: continuing the present pattern of inadequate ad hoc responses to developments as they become critical; by 2000 there is slow transition to renewable energy, further global warming, an expanded and deepened ozone hole, continued land degradation and deforestation, etc.; **2) An Ordered World**: economic control increasingly wielded by IMF and the World Bank; a glaring failure in tackling global environmental issues due to imposing external values and methods on human communities that often have more understanding about how to manage resources than outsiders; **3) Enlightened Change**: the path of shared responsibility for our common future; a significant degree of multilateral commitment to environment and development; ascendancy of democratic values worldwide; an enlarged role for internationalism and the UN; a middle path on lifestyle issues ("between ascetic self-denial and complacent self-indulgence"); a multilateral fund to finance sustainable development financed by an international tax. [**NOTE**: An eloquent globalist/Third World view, referring to seemingly all of the recent and important writing on global environmental affairs.]

(Third World view of global environment)

181

Beyond Interdependence: The Meshing of the World's Economy and the Earth's Ecology. Jim MacNeill (Institute for Research on Public Policy, Ottawa), Pieter Winsemius (Director, McKinsey & Co, Amsterdam), and Taizo Yakushiji (Graduate School of Policy Science, Saitama U, Japan). Introduction by Maurice Strong (Secy-Genl, UNCED). A Trilateral Commission Book. NY: Oxford U Press, Aug 1991/159p/$24.95;7.95pb.

Since 1900, the world's population has multiplied more than three times, and its economy has grown twentyfold. Consumption of fossil fuels has grown by a factor of 30, and industrial production by a factor of 50. "If current forms of development were employed, a further five- to tenfold boost in economic activity would be required over the next 50 years to meet the needs and aspirations of 10 billion people." Since WWII, governments have been preoccupied with economic interdependence. "But the world has now moved beyond economic interdependence to ecological interdependence—and even beyond that to an intermeshing of the two ... This is the new reality of the century, with profound implications for the shape of our institutions of governance, national and international. It raises fundamental questions about how economic and political decisions are made, and their implications for sustainability." The maxim of sustainable development is not "limits to growth," but "the growth of limits." We must learn to recognize, and live within, limits of physical impact beyond which degradation of ecosystems are inevitable. "The environment and the economy must be integrated in our major institutions of decision-making: government, industry, and the home. This is perhaps the most important condition for sustainable development."

The June 1992 Earth Summit in Rio seeks to launch a global transition to sustainable development. The summit will address a range of reform proposals, many of which are discussed in this volume: 1) integrating environment into economic decision-making; 2) correcting perverse interventions in the market (OECD's 24 member countries spend over $250 billion a year on agricultural subsidies that encourage overproduction and depletion of basic farm capital); 3) reforming economic accounting systems ("the single most important change in public-sector incentives"); 4) comprehensive indicators on sustainable development; 5) broadening the concept of national security (an environmental interpretation is simply an extension of the economic interpretation); 6) global bargains (helping certain countries eliminate the use of CFCs to protect the ozone layer is one of the most promising and timely strategies); 7) reducing greenhouse gas emissions (an increasing number of studies show substantial potential reductions through energy efficiency and other measures that could return a profit or at least break even); 8) reforming our great international institutions; 9) new global structures such as a World Environment and Development Forum that bring signatories together at least once a year to evaluate progress and decide on necessary actions. Concludes that the 1992 UNCED summit could place the world community on a new course toward a sustainable future. "The Earth Summit will likely be the last chance for the world, in this century at least, to seriously address and arrest the accelerating environmental threats to economic development, national security, and human survival." [**NOTE**: Important background on UNCED. MacNeill was the main author of the seminal Brundtland Commission report on **Our Common Future** (Oxford, 1987; **FS Annual 1987** #8090), and former director of environment for OECD.] **(UNCED: background issues)**

*182

Beyond the Limits: Confronting Global Collapse, Envisioning a Sustainable Future. Donella H. Meadows (Environmental Studies Program, Dartmouth College), Dennis L. Meadows (Prof of Management Studies, U of New Hampshire), and Jörgen Randers (Prof of Policy Analysis, Norwegian School of Management). Post Mills VT: Chelsea Green Publishing Co (PO Box 130; 802/333-9073), April 1992/277p/$19.95.

In 1972, under the auspices of the Club of Rome, the three authors (along with William W. Behrens III) published **The Limits to Growth**, concluding that the physical limits to human use of materials and energy were somewhere decades ahead. The conclusion of this 20th anniversary update: in spite of improved technologies, greater awareness, and stronger environmental policies, many resource and pollution flows have grown beyond their sustainable limits. "The human world is beyond its limits. The present way of doing things is unsustainable. The future, to be viable at all, must be one of drawing back, easing down, healing." Without significant reductions in material and energy flows, there will be in the coming decades an uncontrolled decline in per capita food output, energy use, and industrial production. Chapters discuss overshoot (if a correction is not made, a collapse of some sort is certain "within the lifetimes of many who are alive today"), exponential growth of population and capital as the driving force (which simply cannot and will not go on for long), sources and sinks as limits, dynamics of growth in a finite world (the World3 model, also used 20 years ago, indicates four possible behavior modes: continuous growth of carrying capacity and population/economy, sigmoid growth where population/economy limits itself, overshoot and oscillation, and overshoot and collapse), the Montreal Protocol on ozone as an example of learning about overshoot and backing off from a clear limit, and why technology and markets alone cannot avoid overshoot (an answer to the most common criticisms of putative World3 model oversights).

Concludes that the human world can respond in three ways to signals that resource use and pollution emissions have grown beyond sustainable limits: 1) deliberately disguise, deny, or confuse the signals (build higher smokestacks, dump toxics in someone else's territory, control prices trying to rise in response to scarcity); 2) alleviate the pressures with technical or economic fixes without changing underlying causes (search for more resources, use resources more efficiently, recycling); 3) acknowledge that the human socioeconomic system as currently structured is unmanageably headed for collapse, and change the structure by shifting perceived costs and benefits, increasing the time horizon, and seeing the social whole. Thirteen scenarios are sketched, including that of a sustainable society of 7.7 billion people in the early 21st century, with enough food, consumer products, and services to support everyone in comfort—provided that appropriate policies are begun in 1995. Similar to the Agricultural Revolution and the Industrial Revolution, the Sustainability Revolution could lead to enormous gains and some losses. Everyone can contribute to this necessary revolution, "already well underway." Five tools to facilitate the transition: visioning what is really wanted, networking to inform and sustain us, truthtelling to overcome verbal traps in public discourse, learning, and loving.

[**NOTE**: A more urgent message than the 1972 **Limits**, with a more sophisticated vision of "sustainable society" replacing the easily misunderstood 1972 call for "the equilibrium state." The authors do not favor zero growth, but "forms of growth that are equitable, affordable, and sustainable." To support this important analysis, 70 figures are employed. But every attempt is made to keep the argument non-technical. Incidentally, the 1972 book was not the first to highlight the limits. **Resources and the American Dream: Including a Theory of the Limit of Growth**, by Samuel H. Ordway Jr (Ronald Press, 1953), criticized the "Cornucopian Faith" and advocated a "Balanced Civilization." In the same year, Fairfield Osborn published **The Limits of the Earth** (Little, Brown, 1953). Five years earlier, in **Road to Survival** (William Sloane Associates, 1948), William Vogt warned of a falling carrying capacity over most of the earth. The earliest intimations of these problems are widely attributed to George Perkins Marsh's **Man and Nature** (1864). Will enough of us ever learn?]

(overshoot of earth's limits)

183

World Development Report 1992: Development and the Environment. The World Bank. NY: Oxford U Press, May 1992/308p/$32.95; $16.95pb.

The 15th in an annual series, and the first to focus on the need to integrate environmental considerations into development policymaking. "The value of the environment has been underestimated for too long, resulting in damage to human health, reduced productivity, and the undermining of future development prospects. The Report argues that continued, and even accelerated, economic and human development *is* sustainable and can be consistent with *improving* environmental conditions, but that this will require major policy, program, and institutional shifts." A two-fold strategy is required: 1) aggressively exploiting positive "win-win" links between efficient income growth and the environment (e.g., by removing distortionary policies that encourage overuse of natural resources, expanded emphasis on population programs, more female education, better agricultural extension, more local participation in development programs); 2) strong policies and institutions to encourage corporations, governments, households, and farmers to adopt less damaging forms of behavior. Costs of protecting and improving the environment may require raised investment rates in LDCs by 2-3% of GDP by 2000, but these costs are modest in comparison with their benefits and with the potential gains from economic growth. The final chapter describes total costs for water and sanitation, efficient electric power, lowering emissions from urban vehicle traffic, industrial emissions, sustainable agriculture, family planning, and female education ("improving education for girls may be the most important long-term environmental policy in the developing world"). Concludes that "Good environmental policies are good economic policies and vice versa. Efficient growth need not be an enemy of the environment, and the best policies for environmental protection will help, not hurt, economic development." Also provides 33 tables of "World Development Indicators" for some 180 nations and territories.

(environment and efficient growth)

184

For Earth's Sake: A Report from the Commission on Developing Countries and Global Change. Alvaro Soto (Executive Secretary) *et al.* Ottawa K1G 3H9: International Development Research Centre (PO Box 8500), 1992/145p/CA$14.95pb.

The Commission was established with support from IDRC and the Swedish Agency for Research Cooperation with Developing Countries. Its seven members include Alvaro Soto (Latin American Environmental Network, Columbia), Anil Agarwal (Director, Center for Science and Environment, New Delhi), Julia Cabias (Ecology Laboratory, Mexico), Martin Khor Kok Peng (Third World Network, Malaysia), Thandika Makandawire (Council for Development of Economic and Social Research, Senegal), Adolfo Mascarenhas (Director of Graduate Studies, U of Dar-es-Salaam, Tanzania), and Erna Witoelar (founder of Indonesia Environmental Forum). The Commission is based on three key propositions: 1) global environmental problems have potentially catastrophic implications for many developing countries; 2) Third World perspectives must be integrated into the international agenda on global environmental change; 3) social dimensions of these issues must be understood and resolved.

The environment worldwide is in crisis at the same time that most parts of the Third World are facing a severe economic and social crisis. In many cases, the socio-economic crisis is the result of development styles that destroy both human potential and the environment. The two crises are the result of unsustainable systems of production and consumption in the North, inappropriate development models in the South, and a fundamentally inequitable world order. To date, Northern concerns have directed the global environmental debate; they reflect a definition of sustainability in which the physical environment is the primary focus and long-term intergenerational issues are key. But the major message from virtually all quarters of the South is that social concerns, economic issues, and *intra*generational equity are the keys to resolving the environment/development crisis. "It is critical to any meaningful approach to sustainable development that environmental issues be integrated with issues of equity, social justice, human rights, and development." Whereas Northern concerns focus primarily on long-term impacts of climate change and ozone depletion, the most immediate and pressing environmental problems in the South relate to depletion and degradation of the biomass base, on which the majority of the population continues to be directly dependent. An additional priority is pollution and resulting health impacts associated with inadequately controlled industrial development and misapplication of chemical technologies. The North tends to frame many of its concerns in the context of the "global commons," but extension of this concept to nationally-based resources threatens Southern sovereignty. Although the economic debt of the South has received abundant attention, the environmental debt of the North has been greatly underplayed. The North bears primary responsibility for many global environmental problems, but the South is likely to experience greater hardships as a result of these problems. "The issue of how the burden of adjustment will be distributed is critical to any meaningful global environmental negotiations." Concludes with a proposed research agenda on eight interacting themes: resource depletion, pollution, natural disasters, poverty and basic needs, economic development models, politics and participation, appropriate technologies, and processes of cultural change (e.g., the population issue should be seen as one of women's rights and women's development). [ALSO SEE: The South Commission report (Oxford, 1990; **FS Annual 1992** #10992).]

(North/South differences on environment)

185

Impoverishment and Sustainable Development: A Systems Approach. Gilberto Gallopin. Winnipeg: International Institute for Sustainable Development (204/958-7700), March 1994/79p(8x11" spiral)/C$20.00.

While environmental destruction is happening almost everywhere, the extreme forms of poverty are concentrated among the rural people of developing countries. Here the environment is being degraded not only to support economic growth, but for the bare survival of ever-rising numbers of poor. No lasting progress is possible unless the plight of the environment is treated as one with the plight of the people. And, with rapidly increasing economic and ecological interdependence, there can be no separate solutions for the South and the North. Poverty eradication can be viewed as an end in itself, as a way of removing a major obstacle to sustainable development, or as a means of moving towards sustainable development. A systems perspective broadens the scope of the poverty issue from a state (poverty) to a dynamic process (impoverishment), broadens the boundaries of the meaningful to include the entire socio-ecological system at local and global levels, highlights fundamental systemic properties, and emphasizes horizontal and vertical linkages. The generally unquestioned "developed country scenario" of poverty alleviation will prove to be either an impossible mirage or an unstable situation. For this reason, strategies for eradicating poverty must include changes in the values and lifestyles of the whole society. The quest for eradicating poverty through sustainable development should combine top-down approaches with bottom-up initiatives, broaden space and time horizons, and integrate economic, social, cultural, political, and ecological factors.

(poverty and sustainable development)

186

Visions of a Sustainable World. *Engineering & Science*, 55:3, Spring 1992/60p/$2.00 from California Inst of Technology, Pasadena CA 91125 (818/356-3630).

Highlights of an October 1991 symposium at Caltech, dedicated to Harrison Brown (1917-1986), Caltech professor, author of **The Challenge of Man's Future** (Viking, 1954), and co-author of **The Next Hundred Years** (Viking, 1957). **Murray Gell-Mann** (Prof of Theoretical Physics and Nobel laureate) focuses on the key concept of achieving "sustainable quality" and discusses transitions over the next few decades in human population, technology, economic accounting, inter- and intragenerational equity, and ideology. **Harlan Cleveland** (President, World Academy of Art and Science) considers managing human behavior in the global commons (space, the atmosphere, oceans, Antarctica) and the need for a UN Trusteeship Commission to act as trustee for our four great surrounding environments. **Carl Djerassi** (Stanford U) outlines prospects in contraception and proposes a rallying cry of "Make abortion unnecessary." **Gregory Benford** (UC-Irvine) imagines the 21st century as "a biological century" informed by living technology and the demographic explosion. Various panels discuss a sustainable global commons (with special emphasis on biodiversity), development for a sustainable future (requiring a much stronger commitment to equity and social justice than at present), the technological transition to a hydrogen economy, the economic transition to recognizing total costs, the transition to global governance, and cultural and ideological transitions.

(Caltech sustainability symposium)

187

Visions of Sustainability (Special Issue). Edited by Doug McKenzie-Mohr and Michael Marien. *Futures*: 26:2, March 1994/152p/£31.80 single copy.

Brings together the latest thoughts of some leading thinkers on sustainability issues to illustrate the complexities involved in a profound and multifold transition, now just under way. All 13 essays are written by members of HOLIS: The Society for a Sustainable Future; 11 of them are based on presentations in three sessions at the World Future Society's Seventh General Assembly, held in Washington in June 1993. ***1) Key Elements of Sustainability***. **Lester W. Milbrath** (SUNY-Buffalo) aims at the inadequacies of the "dominant social paradigm" and the societal imperatives of the "new environmental paradigm." **Hazel Henderson** (St. Augustine FL) explains the broader set of indicators that would be used by any society concerned with sustainability. **Virginia Abernethy** (Vanderbilt U) warns that US population is growing faster than any other industrial country, sending the wrong signals to the rest of the world. **Robert Goodland, Herman Daly**, and **John Kellenberg** (all World Bank) consider the startling message of the I=PAT formula (Impacts = Population x Affluence x Technology), and list imperatives for both high- and low-income nations. **Robert L. Olson** (IAF) also considers I=PAT and how it would grow under three scenarios of Continued Growth, Technology Transformation, and Social Transformation. **Kenneth A. Dahlberg** (Western Michigan U) elaborates on the necessary transition from industrial-era agriculture to regenerative food and fiber systems. **Mary E. Clark** (Cottage Grove, OR) points to the necessity of addressing basic human needs such as genuinely shared meaning. **Sally Lerner** (U of Waterloo) focuses on the basic task of any sustainable society to seriously consider work, employment, and income distribution.

2) Moving Towards Sustainability. **Dennis Pirages** (U of Maryland) argues that sustainability is an evolving process, requiring something on the order of a "sociocultural genome project." **Walter H. Corson** (Global Tomorrow Coalition) furnishes lengthy outlines of needed changes, barriers to change, general strategies, specific agents of change, and how to monitor progress through indicators of sustainability. **Doug McKenzie-Mohr** (St. Thomas U, Fredericton, New Brunswick) offers a psychologist's view of social marketing for sustainability applied to promoting residential energy conservation. **Duane Elgin** (Mill Valley CA) provides an integrated vision of our evolutionary journey toward a sustainable species-civilization, viewing us in the midst of seven major stages of development. **Michael Marien** (*Future Survey*) explores two key barriers to sustainability—infoglut and other competing problems—and sketches three scenarios of False Success, Mini-Success/Regress, and Evident Regress. **(sustainability: HOLIS essays)**

188

Development Betrayed: The End of Progress and a Coevolutionary Revisioning of the Future. Richard B. Norgaard (Prof of Energy and Resources, U of California-Berkeley). London and NY: Routledge, April 1994/280p/$59.95;$18.95pb.

Modernity promised control over nature through science, material abundance through superior technology, effective government through rational social organization, and peace and justice through a higher individual morality and superior collective culture. During the second half of this century, yesterday's promise, only partly fulfilled, fostered material madness, accentuated inequality, imprisoned modernity's devotees in bureaucratic gridlock, accelerated resource depletion and environmental degradation, and pitted people against people. Accelerating progress through planned development has been the recent product of modernity, yet development has gone seriously awry. Modernism and development have betrayed progress. Chapters discuss the challenge of sustainability, the challenge of defining sustainable development, change as a coevolutionary process, lessons of coevolutionary environmental history (experimentation should be undertaken cautiously on a small scale; experiments that entail very long time commitments should be avoided; systems will more likely survive with diversity), the imbalance between technological and cultural progress, philosophical roots of the betrayal, two maladaptive determinisms (environmental determinism of the "limits to growth" school vs. the cultural determinism of Julian Simon and others), the coevolution of environment and human culture, principles of coevolutionary cosmology, the tyranny of liberal individualism (the excessive emphasis on exchange relations and the individual), democratizing knowledge and progress, and a coevolving patchwork of discursive communities. In sum, "the challenge of sustainable development is to derive a viable image of the future, to change the vocabulary of our political discourse... so that a patchwork of coevolutions of social organization, knowledge, technology, values, and ecosystems is sustainable." [**NOTE**: Heavyweight thinking, with 58 pages of bibliographic essays.]

(coevolving patchwork for a sustainable future)

189

Ensuring Prosperity: Implementing Sustainable Development. The Report of the Future Environmental Directions for Alberta Task Force. Edmonton, T5K 2M4: Alberta Envir'l. Protection Info. Centre (9920 108th St; 403/422-2079), March 1995/112p(8x11" spiral)/free.

The Alberta Round Table on Environment and Economy began its consultation process in 1990 to formulate a vision of sustainable development for the province. The Future Environmental Directions for Alberta project was initiated in 1993 to identify and recommend an action agenda. Its "most ambitious input mechanism" was a two-stage, modified Delphi survey conducted in Spring 1994, with 380 respondents to one or both questionnaires. The resulting five Task Force priorities: 1) make sustainable development a fundamental value in the way we govern and conduct our everyday lives (minimize or avoid pollutants, apply the environmental precautionary approach, strengthen sustainable development education in schools and workplaces); 2) implement sustainable resource management practices to protect air, water, and land (use government policy and local coordinated resource planning, implement systematic natural resource accounting, make sustainable tourism a reality); 3) maintain the "Alberta Advantage" (become a leader in environmental management and communicate the progress, develop environmental technology to meet future needs); 4) develop innovative, cost-effective ways of doing business to support sustainable development (implement full-cost accounting, pursue objectives-based alternatives to environmental regulation, review fees/taxes/subsidies to determine their effect on sustainability; 5) ensure greater environmental empowerment and accountability for all sectors of Alberta (promote public and private sector reporting on sustainable development, improve access to environmental information, encourage individual environmental responsibility). [**NOTE**: Many good ideas, handsomely packaged.]

(sustainable development for Alberta)

190

Choosing a Sustainable Future: The Report of the National Commission on the Environment. Chaired by Russell E. Train (Chairman, WWF). Washington: Island Press, Jan 1993/180p/$25.00;$15.00pb.

The 19-member Commission (including Peter A.A. Berle, Madeleine Kunin, Alice M. Rivlin, William D. Ruckelshaus, and James Gustave Speth) was convened by the World Wildlife Fund. It found that the natural processes supporting life on Earth are increasingly at risk, and the US must choose between two very different futures. If America continues down its current path, environmental quality will continue to deteriorate and eventually the economy will decline as well. If leadership is based on a new ethic of environmentally responsible behavior, it is far more likely that the coming years will bring a higher quality of life. Key recommendations, with a chapter devoted to each: **1) The Goal of Sustainable Development**: "sustainable development should be the primary goal of environmental and economic policy;" **2) Technology**: policies should encourage development and adoption of technologies compatible with sustainable development (design for the environment must take its place alongside cost, safety, and health as a guiding criterion for development); **3) Economics**: eliminate price-distorting subsidies, tax environmentally harmful activities, revise the measurement of economic activity to incorporate resource and environmental factors; **4) Literacy**: "schools, families, religious institutions, the media, businesses, and government must cooperate in fostering an environmentally literate citizenry"; **5) Governance**: environmental considerations must become integral to all governmental policies; **6) America's New Global Role**: the US has a vital interest in leading efforts to protect the global environment, moderate world population growth, improve the standard of living in LDCs, and promote the effective implementation of Agenda 21; "the urgent need to put the world on the road to sustainability provides a common purpose that can and must unite the global community;" **7) Energy**: the most critical technologies for sustainable development are energy technologies; highest priority should go to energy efficiency; there must be a deep-seated change in how our country produces and uses energy, including "a progressive shift away from fossil fuels as quickly as possible"; **8) Pollution**: efforts to halt pollution should become more integrated and holistic; pollution prevention should take priority over pollution control; **9) Habitats**: environmentally sensitive management of public and private land.

(sustainable development as primary US goal)

191

A Sustainable World: Defining and Measuring Sustainable Development. Edited by Thaddeus C. Trzyna (Chairman, IUCN Commission on Environmental Strategy and Planning; President, CIPA). Published for IUCN-The World Conservation Union. Sacramento: Intl. Center for the Environment and Public Policy/California Institute of Public Affairs, July 1995/272p/$28.00.

"Sustainability" and "sustainable development" have become guiding principles of environmental policy and international development. But many questions remain about what these concepts mean, how to implement them, and how to measure progress. **Caring for the Earth** (IUCN, 1991; #170) defines sustainability as "a process or state that can be maintained indefinitely" and sustainable development as "improving the quality of human life while living within the carrying capacity of supporting ecosystems." The 14 essays consider ***1) Definitions***: the need to develop a protocol for assessing sustainability, what we need to know to achieve sustainability (by Stephen Viederman), sustainable authentic development (by Denis Goulet), 12 "Isms" that threaten the environment (economism, technocratism, ecologism, worldism, NGOism, etc); ***2) Measuring Progress***: monitoring for sustainability, a possible annual report on global progress toward environmentally sustainable development (now being considered by the World Bank), the European Sustainability Index Project (coordinated by the International Institute for the Urban Environment in Delft), a new approach to assessing sustainable well-being; ***3) Indicators of Sustainability***: limits to measuring complex natural systems, environmental indicators for Latin America and the Caribbean, reforming national economic indicators, poverty alleviation measures in Zimbabwe, indicators of social environmental learning (by Matthias Finger and James Kilcoyne), and linking sustainability levels to performance goals at national and subnational levels (by Walter H. Corson). [**NOTE**: Important thinking to advance the important notion of sustainability.]

(sustainability measures)

192

What Works: An Annotated Bibliography of Case Studies of Sustainable Development. D. Scott Slocombe (Prof of Planning, Wilfrid Laurier U), *et al.* Foreword by Thaddeus C. Trzyna. Published for the IUCN Commission on Environmental Strategy and Planning. Sacramento CA: International Center for the Environment and Public Policy (916/442-2472), Nov 1993/55p (8x11")/$20.00pb (plus $2.00 p+h).

Abstracts 146 case studies of projects designed to promote sustainable development, especially interventions that address problems in a systematic way. The geographic index lists about 90 countries and 9 world regions, with special emphasis (more than 10 items) on

Africa, Kenya, India, Nepal, Philippines, and Costa Rica. Important main themes in the subject index (more than 10 items) include agriculture, agroforestry, community participation, forestry, protected areas, soil conservation, traditional knowledge, water resources, and the role of women. A second edition is due in 1995. In the Foreword, Trzyna (Chair of the IUCN Commission) notes that this project identifies a need for more and better case studies, effective ways of communicating them, and trying to understand what works.

(sustainable development case studies)

193

Sustainable Development Tool Kit. Donald R. Lesh (President, GTC) and Diane G. Lowrie (VP, GTC). Washington: Global Tomorrow Coalition, June 1995/350p (8x11" looseleaf)/$85.00 individuals; $165.00 orgs.

The GTC, founded in 1981, is a non-profit alliance of organizations and individuals acting to assure a more sustainable America. The Tool Kit is in three parts: ***1) Understanding Sustainable Development***: chapters on sustainability as "a new path of progress," the history of sustainable development, considering full costs, demanding good information, stewardship for future generations; ***2) Sustainable Development in Action***: policy framework for sustainable development, principles of sustainability, criteria for effective management, building consensus (discussing the Canadian Round Tables on Environment and Economy and the round table approach in the US), private sector responses (full-cost accounting, codes of conduct, sustainability awards, corporate reporting), public sector responses (national, state, and local actions), indicators of sustainability, the Earth Summit and Agenda 21; *3) Resources*: national information directory, international information directory, a 26-page briefly-annotated bibliography (by Walter H. Corson), glossary, GTC members and business alliance partners, information on GTC (including a Tool Kit evaluation and update form). [**NOTE**: A wealth of information in an imposing 12x12" three-ring binder, enabling updates and removal of some 100 charts for training and presentation. Also includes an 18-minute GTC video. ALSO SEE: the IISD **EarthEnterprise Tool Kit** (#249), now in a compact 7x9" edition, and the 530-page **Environment Industry Yearbook 1995** (Environment Press, 26 Brock St, Bath BA1 2LN, UK; 0225-330312), listing 2,650 organizations.] **(GTC Tool Kit)**

2. Philosophy and Ethics

194

Awakening Earth: Exploring the Evolution of Human Culture and Consciousness. Duane Elgin (Mill Valley CA). NY: William Morrow, Oct 1993/390p/$23.00.

Author of **Voluntary Simplicity** (#199) warns that Earth's biosphere is being severely wounded, even crippled, by humanity. Yet through humanity, the Earth is awakening as a conscious global organism. "Pushed by the harsh reality of an injured Earth, the human family is being challenged to realize a new level of identity, responsibility, and purpose." To be sustainable, a civilization must maintain the integrity of its physical, social, and spiritual foundations. To seek only to survive is not sufficient for long-term sustainability—we need a shared and compelling sense of purpose and potential for living together as a world civilization. Two views of evolution are dominant today: *materialism* (where matter is considered the primary reality) and *transcendentalism* (which emphasizes rising above the material world). But these old approaches no longer work in isolation from each other. A third, *co-evolutionary view* is emerging that integrates East and West, viewing reality as comprised equally of matter and consciousness in a mutually supportive spiral of development that can produce a sustainable, planetary civilization.

Our highest potential as a species is our ability to achieve full self-reflective consciousness. Developing our capacity for reflective knowing, both personally and socially, is a paramount evolutionary challenge. To fully coevolve our capacity, humanity must work through major stages of development. Chapters are devoted to: 1) contracted consciousness and the archaic era, beginning roughly 35,000 years ago; 2) surface consciousness and the era of awakening hunter-gatherers; 3) depth consciousness and the era of agrarian-based civilizations (includes most of recorded human history; basic arts of civilization developed, although most individual lived as impoverished peasants); 4) dynamic consciousness and the scientific-industrial era (an era of moral relativism, intellectual absolutism, nation-state egotism, and technological giantism, that separated humans from nature, the divine Life-force, and other persons; 5) reflective consciousness and the era of communication and reconciliation (an epoch of profound stress and challenge, as humanity copes with the problems of the scientific-industrial era, and engages in economic, ecological, political, spiritual, racial, ethnic, gender, and generational reconciliation); 6) oceanic consciousness and the bonding and building era (will begin with great confidence in knowing we made it through a perilous transition and now share a common vision of a workable future; a fermenting global culture will emerge, and a system of global governance could be developed); 7) flow consciousness and the surpassing era (the strength of human bonding enables the world to liberate its creativity); 8) integral awareness and the initial maturity of planetary civilization (emergence of a wisdom culture that balances being and becoming, unity and diversity). Humanity could move very fast through the stages of growth essential to realizing maturity, perhaps within another dozen generations, or roughly 500 years (due to exponential growth in communications technology and the push of global crises). But our future is not predetermined: we can fill out these stages of development in whatever way and with whatever timing we choose. And we are only halfway through the stages of growth to initial maturity.

The problem at the moment is that five different perceptual paradigms coexist on the planet: hunter-gatherers (much less than 1%), agriculturalists (roughly half of the world population), industrialists (those in urban settings, soon to be a majority), communicationists (those influenced by the global consciousness of the new communications), and Gaians (planetary citizens with a strong sense of global family and global bonding). By the mid-21st century, a majority of humanity could begin to coalesce within the Gaian perceptual context as a common organizing paradigm. Pitfalls on the path to maturity include inflexibility and delay, lack of vision, severe ecosystem disruption, resource wars, authoritarianism, lack of rich-poor reconciliation, information wars between authoritarian leaders and hackers, biological warfare (the poor terrorist's nuclear weapon), religious fanaticism, or planetary psychosis. Priority needs for building a sustainable and satisfying future: breaking the cultural hypnosis of consumerism, designing ecological ways of living, sharing a compelling vision of the future, developing a conscious democracy, nurturing a reflective perceptual paradigm, and reconciling our polarities. [**NOTE**: Challenging, thoughtful, original, and well-documented.] **(evolution to a sustainable society)**

195

The Universe Story: From the Primordial Flaring Forth to the Ecozoic Era—A Celebration of the Unfolding of the Cosmos. Brian Swimme (San Francisco) and Thomas Berry (Bronx, NY). San Francisco: HarperSanFrancisco, Dec 1992/305p/$22.00.

The Director of the Center for the Story of the Universe at the California Institute of Integral Studies teams with the author of **The Dream of the Earth** (Sierra Club, 1988; **FS Annual 1990** #9591) to provide a comprehensive story of the universe for modern times. Aimed at the general reader, the authors seek to provide a new unity to the educational process and "the comprehensive context of the future" as we enter the Ecozoic era. Chapters describe the primordial flaring forth 15 billion years ago, galaxies and supernovas, the sun, the first life on Earth 4 billion years ago, plants and animals in the Paleozoic era 550 million years ago, human emergence 2.6 million years ago, Neolithic settlements of 12,000 years ago, classical civilizations, the rise of nation-states, and the change of consciousness toward viewing the universe as an irreversible sequence of transformations. The final chapter describes transition from the Cenozoic era of the last 67 million years to the Ecozoic era where the well-being of the planet is a condition for the well-being of any of the component members of the planetary community. The first law of economics must be to preserve the economic viability of the planet; the first commitment of the medical profession must be to preserve the health of the planet; the basic concern of religion must be to preserve the natural world as the primary revelation of the divine. The dominant issue of the immediate future will be the tension between plunderers and preservers, the mechanistic and the organic, and anthropocentric and biocentric norms. The immediate goal of the Ecozoic is to alter the mode of consciousness that has been responsible for devastation of the planet. The overall objective is to assist in establishing a mutually enhancing human presence on Earth, based on an Earth-centered language. [**NOTE**: A very ambitious cosmic integration of past, present, and future, based on contemporary science: is there any better "story of the Earth" for our time?] **(transition to Ecozoic era)**

196

Voices for the Earth: Vital Ideas from America's Best Environmental Books. Daniel D. Chiras (U of Denver; President, Sustainable Futures Society, Evergreen CO). Boulder CO: Johnson Books, April 1995/247p/$16.95pb.

Twelve authors, educators, scholars, and activists were asked to nominate books that they felt offered the clearest analyses of contemporary problems and the best solutions to restructure for sustainability. The nominated authors were then asked to write summaries of their books. They include Anne LaBastille on the extraordinary extinction of the giant grebe in Guatemala, Dave Foreman on threats to wilderness and principles of Earth First!, Donella Meadows *et al.* on the "overshoot" of society beyond its limits and steps to avoid collapse, Donella Meadows on the notion of exponential growth, Garrett Hardin on carrying capacity and living within limits, Daniel Kemmis on restoring community and citizenship with a new sense of place and possibility, Thomas Michael Power on an environmental view of the local economy and the total economy, Stephen Schmidheiny on business and sustainable development, Steve Van Matre on principles of earth education, David W. Orr on ecological literacy for the 21st century, James Swan on nature as teacher/healer and reawakening nature kinship, Daniel D. Chiras on learning to live sustainably on the Earth, and Lester R. Brown *et al.* on the sustainable and efficient economy of 2030 (from **State of the World 1990**). Includes brief biographies of each author. **(best environmental books summarized)**

197

Deep Ecology for the 21st Century. Edited by George Sessions (Philosophy Chair, Sierra College, Rocklin CA). Boston MA: Shambhala, Jan 1995/488p/$20.00pb.

Readings on the philosophy and practice of the Deep Ecology movement, inspired by Aldo Leopold, Rachel Carson, David Brower, Paul Ehrlich, Gary Snyder, and, especially, Norwegian philosopher Arne Naess. The 38 selections include Thomas Berry on the viable human community (which must overcome the deep cultural pathology of anthropocentrism), Fritjof Capra on the new ecological paradigm to replace the mechanistic worldview, Gary Snyder on creating a "culture of the wilderness", Dave Foreman on the new conservation movement based on biodiversity as the fundamental value, George Sessions on ecocentrism, Arne Naess on the eight key ideas of the deep ecology movement and on various questions (the "Eight Points" revisited, equality and rights, self-realization, ecosophy, the place of joy in a world of fact, deep ecology and lifestyle), Warwick Fox on the deep ecology/ecofeminism debate, George Sessions on deep ecology and the New Age movement (commenting on Buckminster Fuller, Teilhard de Chardin, James Lovelock, and Henryk Skolimowski), Jerry Mander on space colonies and Disney's EPCOT Center, George Sessions on global ecosystem protection zoning and ecological restoration, Edward Grumbine on wildness vs. the Wise Use Movement, Donald Worster on the shaky ground of sustainability thinking and "the new permissiveness in ecology" (rather, our first priority must be "the careful and strict preservation" of our evolutionary heritage), Wolfgang Sachs on the "global change" buzzword of international ecology experts, Arne Naess on the three goals of the ideal Green society (ecological sustainability, peace, and social justice), Gary Snyder on the rediscovery of Turtle Island (the old/new name for the continent), and Arne Naess on deep ecology for the 22nd century. [**NOTE**: An authoritative overview of purist thinking.] **(deep ecology essays)**

198

The Way: An Ecological World-view. Edward Goldsmith (Editor, *The Ecologist*). Boston: Shambhala, Spring 1993/442p/$20.00pb. (Published in UK by Rider, 1992.)

A statement of 66 basic principles underlying an ecological world-view. They are all interrelated, forming an all-embracing and self-consistent model of our relationship with the world. A sampling: 1) ecology is a unified organization of knowledge; 2) ecology studies natural systems in their Gaian context; 3) ecology reflects the values of the biosphere; 4) Gaia is alive; 5) "life processes are anticipatory: for natural systems to achieve their goal of maintaining their own and Gaian stability, they must be able to predict changes to which they may have to adapt" (there is every reason to suppose they are capable of doing so, e.g. cacti storing water in their pores, species migrating when winter approaches); "prediction is an essential component of perception, the first step in self-regulation"; 6) living things seek to understand their relationship with their environment; 7) increased complexity leads to greater stability; 8) vernacular man follows The Way (the concept was probably entertained in some way by all traditional societies); "our modern society has quite clearly set out to deviate from The Way," with an overriding goal of replacing the biosphere with the technosphere; 9) progress—the world view of modernism—is anti-evolutionary and anti-

Way; 10) The Great Reinterpretation requires a conversion to the world-view of ecology (principle #66).

(66 principles of ecological world-view)

199

Voluntary Simplicity: Toward a Way of Life That Is Outwardly Simple, Inwardly Rich (Revised Edition). Duane Elgin (Mill Valley CA). Introduction by Ram Dass. NY: William Morrow/Quill, April 1993/240p/$10.00pb.

The pastoral vision of simplicity has much appeal to those in the West for whom life is full of confusion, distraction, and complexity. At the heart of voluntary simplicity, as first articulated by Richard Gregg in 1936, is an emphasis on harmonious and purposeful living. It is not impoverished living, a turning away from economic progress, a barren plainness, or necessarily living in rural areas. Those choosing a simpler life generally tend to work on developing the full spectrum of their potentials, feel an intimate connection with the earth and a compassionate concern for the world's poor, lower their overall level of personal consumption, favor products that are durable and non-polluting, shift their diet toward natural and healthy foods, reduce undue clutter and complexity in their lives, and prefer smaller-scale living and working environments. "The push of necessity and the pull of opportunity" lead to more ecological approaches to living. A *Time*/CNN survey (*Time*, 8 April 1991) found that 69% of those questioned said they would like to "slow down and live a more relaxed life." Another survey in *Fortune* (14 Aug 1989) found that 75% of working Americans between the ages of 25 and 49 would like "to see our country return to a simpler lifestyle, with less emphasis on material success."

Concludes with broad reflections on our season of "systems breakdown," the winter of Western industrial civilization where there is loss of social cohesion around a compelling civilizational purpose. Three scenarios are possible: **1) Overshoot and Collapse**: instead of rising to the challenge of sustainability, a majority of persons respond with denial, feelings of helplessness, blaming others, and escapism; the mass media promote this collapse by aggressively promoting a consumerist consciousness; **2) Dynamic Stagnation**: society is unwilling to make a creative leap forward, and expends all its energy trying to perpetuate the status quo and keep from falling backward; as everyone struggles to hold on, the social order sinks ever deeper into malaise; **3) Reconciliation and Revitalization**: a sustainable future emerges when citizens recognize the absolute necessity of change and seek the most healthy pathway; democratic processes are revitalized as we directly engage our predicament. [**NOTE**: Simply written and more timely than ever. This new 248 page edition with more white space is appropriately shorter than the original 312 page version (Morrow/Quill, 1981; **FS Annual 1981-82** #3744). ALSO SEE: Elgin's **Awakening Earth** (#194).]

(necessary/desirable shift to a simpler life?)

200

Wisdom of the Elders: Honoring Sacred Native Visions of Nature. David Suzuki (Prof of Genetics, U of British Columbia) and Peter Knudtson (Vancouver BC). NY: Bantam Books, Aug 1992/274p/$22.00.

A juxtaposition of thoughts on man and nature by the elders of Western science and of native peoples worldwide, with each chapter providing a brief sketch of modern scientific perspectives as a backdrop to 48 vignettes of Native visions that could provide the foundations for a new, global environmental ethic, e.g.: **1) Recognizing the Kinship of All Life**: the still-unfolding evolutionary story of life is perhaps the single most compelling statement of interconnectedness in Western societies (the Desana of the Amazon see the Sun as Creator of the Universe, the Hopi of the American Southwest think of the beginning goddess, the Koyukon of Alaska view Raven and his heroic Distant Time companions serving as cosmic catalysts); **2) Nature as a Living System**: modern ecology continues the ancient human quest for understanding the web of relationships (the Desana vision of nature as an orderly solar-powered energy circuit, the Waswanipi Cree view of the bonds of human kinship); **3) Ways of Seeing Nature**: many Western scientists still make sense of the natural world by breaking it into fragments (the Chewong of the Malaysian rainforest believe that Tohna the creator will destroy life on earth because humans have neglected their responsibility). Other chapters contrast shaman and scientist on the kinship of humans with beasts and with vegetation, the relationship between humans and land, the rhythms of nature and time as a circle, world renewal and maintaining balance, and the fate of the earth.

In the Foreword, Suzuki [co-author of **It's a Matter of Survival** (Harvard, 1991; **FS Annual 1992** #11095)] challenges the prevailing scientific paradigm and states that "we need a radically different way of relating ourselves to the support systems of the planet." The knowledge and worldview of aboriginal peoples is relevant in a time of imminent global eco-catastrophe. Major problems such as global warming, species extinction, and worldwide pollution will not be solved in the long run by perpetuating the current worldview and applying Band-Aids such as tax levies, greater efficiency, and recycling. Not only does the Newtonian worldview fail to comprehend the complexity of life on earth; we have barely begun to understand its dimensions. "If biodiversity and ecosystem integrity are critical to salvaging some of the skin of life on earth, then every successful fight to protect the land of indigenous peoples is a victory for all of humanity and other living things." [**NOTE**: Interesting, challenging, and handsomely produced.]

(wisdom of native peoples and modern eco-crisis)

201

Our Responsibility to the Seventh Generation: Indigenous Peoples and Sustainable Development. Linda Clarkson (Winnipeg), Vern Morrissette (Winnipeg), and Gabriel Regallet (IISD). Winnipeg, Manitoba: International Institute for Sustainable Development, May 1992/87p(8x11")/Can$20.00.

Indigenous people around the world face a paradoxical situation: policies and institutions have pushed them to the fringes of society and economy, yet the world is coming to understand their feeling that the current development path is unsound. Indigenous people from Canada (including Clarkson and Morrissette) are aware of the symbiotic relationship of humans to the earth. The law passed down from their ancestors crystallizes the sacred responsibility to be caretaker of all that is on Mother Earth and to be responsible to ensure the survival of the seventh generation into the future. Chapters describe this Indigenous perspective, modernization and the processes of impoverishment, why Indigenous peoples have been destroyed by development schemes (the latest threats being large water development projects and toxic waste sites), and the necessity for Indigenous people to create mechanisms of change based on their original teachings. Primary components of an international program should include: 1) protecting the traditional way of life (issues relating to rights and protection should be dealt with on a regular and permanent basis by the UN); 2) documentation, promotion, and protection of traditional knowledge; 3) healing programs to ensure the well-being of current and future generations (strategies to address the suppressed sorrow and repressed pain that are obstacles to healthy development); 4) education for cultural survival (the full decolonization of indigenous education; "traditional knowledge systems must be recognized as equivalent to those of Western societies"); 5) economic self-reliance ("eco-tourism and alternative tourism could provide a livelihood for Indigenous peoples in ways that are culturally and ecologically sustainable"); 6) development of a communication capacity between indigenous people from different regions. [ALSO SEE: **The State of Native America: Genocide, Colonization, and Resistance**, edited by M. Annette Jaimes (Boston: South End Press, 1992/460p/$16.00pb), and **The Gaia Atlas of First Peoples: A Future for the Indigenous World** by Julian Burger (Anchor/Doubleday, 1990; **FS Annual 1992** #10970).]

(sustaining Indigenous people)

202

Why Posterity Matters: Environmental Policies and Future Generations. Avner de-Shalit (Hebrew U of Jerusalem). London & NY: Routledge, March 1995/161p/$49.95; $14.95pb.

For many years the environment has been treated as a free commodity, but it is becoming clear that we cannot continue to exploit it without incurring costs. Policies that improve the welfare of present generations often cause severe environmental risks for future generations. This raises moral questions about the relations between people who live in different times, even different centuries. Our obligations to those who will live on after us are a matter of justice. We must consider these problems of intergenerational distribution. A theory of intergenerational justice is possible, and can serve as the moral basis for environmental policies. Utilitarian, contractarian, and rights-based theories fail to provide adequate justification of our obligations to future generations because of their individualistic biases. A "communitarian" theory is proposed, involving the concept of a transgenerational community extending into the future. This theory of intergenerational justice holds that in each community people should decide on certain policies and refrain from certain acts because they owe it to future generations.

(future generations and environmental policy)

203

Gathering Hope: A Citizens' Call to a Sustainability Ethic for Guiding Public Life. Stanley R. Euston (President, The Sustainability Project). Santa Fe NM: Sustainability Project Civic Dialog Program (4 Gavilan Rd; 505/466-2052), April 1995/20p(8x11")/free.

The great challenge facing American democracy today is finding common values and actions that lead to a socially just, economically equitable, and environmentally sustainable future. But the institutions of American democracy are not responding, and public opinion substitutes for public virtue. A Public Philosophy of Sustainability is proposed in this working paper, as a call to ethical responsibility and the beginnings of a new social compact with the future. Its central values and guidelines for action include: 1) concern for future generations—listening now to the wider claims of the future, both human and nonhuman; 2) respect for ecological integrity; 3) responsible earth-keeping (stewardship in economic and social arrangements; 4) rededication to justice (including full participation in decision-making and a new standard of enough for all without excess or wastefulness); 5) respect for community; 6) rededication to civic democracy and civic dialog. Principles of the sustainability imperative include restoration and maintenance of ecological health, technologies assessed and driven by their contribution to sustainability, a political economy that provides meaningful work, market mechanisms and indicators that reflect full social and environmental costs, an effective citizens movement, and restoring trust in government.

(Public Philosophy of Sustainability)

204

Ecological Prospects: Scientific, Religious, and Aesthetic Perspectives. Edited by Christopher Key Chapple (Associate Prof of Theology, Loyola Marymount U, Los Angeles). Albany NY: State U of New York Press, Jan 1994/236p/$59.50;19.95pb.

The continuing assault on the natural world by industrial processes has raised awareness of the need for corrective action. To be effective, such action must arise from sources as deep as life itself—from a new or renewed understanding of cosmological and ecological processes that sustain all forms of life. These essays discuss Gaian views, the human role in homogenizing the planetary biome, rethinking the wilderness idea, ecological theory and natural resource management, two approaches to ecophilosophy, environmental action choices, neo-Confucianism as an ecological cosmology, emerging options in ecological Christianity, ecofeminist ethics, the religious radicalism of Earth First!, and the land aesthetic. A concluding chapter on an ecological virtue ethic focuses on 11 virtues referred to by the authors: holistic thinking, reverence, eros, compassion, attunement, humility, acceptance of finitude, earthiness, asceticism, creativity, and inclusivity.

(eco-ethics)

205

A Sacred Place to Dwell: Living With Reverence Upon The Earth. Henryk Skolimowski (Prof of Ecological Philosophy, U of Lodz, Poland; U of Michigan). Rockport MA: Element Books (42 Broadway; 508/546-1040), Dec 1993/148p/$13.95pb.

Author of 17 books, including **Eco-Philosophy** and **Eco-Theology**, extends his thinking to propose new concepts of divinity (as diffused throughout all creation), responsibility (as we co-create with the universe/evolution/God), spirituality, God, and ethics. Various religious traditions are beginning to acquire a distinctly green shade or to be transformed into a new, more universal religion based on ecological spirituality. This means more than simply understanding the interconnectedness of life. "Religion in the ecological epoch will need to be based on a sense of deep communion with all beings." Ecological religion is not painting green the edges of past religions, but making sense of the new story of creation.

To understand the meaning of evolution is the beginning of wisdom. We have already seen the first two stages in the discovery of evolution: geological and biological. The next two stages in our discovery of evolution are happening now: the recognition of conceptual evolution (realizing that our knowledge is evolving, and that, as Karl Popper notes, all knowledge is conjectural), and the recognition of theological evolution—that we no longer accept one fixed and immutable God. "We have no choice but to accept the idea that we are God in the making . . . the right reading of evolution is participating in God-making." God is the gigantic light of the future, and spiritual people are those who develop inner mirrors to catch glimpses of this light. We should not negate any previous attributes of God (Hebrew, Christian, Muslim, or Hindu). But we should be sensitive to his/her recent messages. Any complete God, in our day and age, must wish us to engage in a thoroughgoing ecological reconstruction to safeguard God's creation. In 1979, Pope John Paul II proclaimed St. Francis as the Patron Saint of Ecologists. But the main problem still remains, of giving man dominion over every living thing. We should return to the teachings of St Francis in earnest, and the Pope should proclaim that all creatures are sacred. We also need a new theory of social justice (Franciscan Marxism), a new form of socialism (Franciscan Socialism), and a New Gospel (viewing the world as a sanctuary). [**NOTE**: A futurist/philosopher offers a bundle of leading-edge ideas, lucidly and simply presented. Recommended to anyone interested in theological and spiritual dimensions of the future.] **(evolution of God for our times)**

206

Living Philosophy: Eco-Philosophy as a Tree of Life. Henryk Skolimowski (Chair of Ecological Philosophy, Technical U of Lodz, Poland; formerly U of Michigan). London and NY: Arkana/Penguin Books, Feb 1992/255p/ $10.00pb.

Author of **Eco-Philosophy: Designing New Tactics for Living** (Marion Boyars, 1981; **FS Annual 1981-82** #3125) continues his exposition on a new eco-cosmology as foundation for a new world view and civilization, and as a matrix for our action. In contrast to Faustian Man "at the mercy of indulgence," ecological individuals understand the interconnection of all things, feel responsible for future generations, attempt to cultivate frugality, and hold a reverence for life. Eco-ethics is an extension and articulation of reverence for life, valuing "the pursuit of wisdom contrasted with the pursuit of mere information." Chapters describe the need for a new cosmology, creative evolution, eco-philosophy contrasted with the debacle of contemporary philosophy (which is piecemeal, spiritually dead, socially unconcerned, and environmentally oblivious), restoring the unity of knowledge and values, the transition from arrogant humanism to ecological humanism in the age of stewardship, following the teachings of Albert Schweitzer and Mother Teresa, the ecological person and the celebration of life, the evolutionary concept of reverential power, the quest for quality in architecture, and ecological consciousness as the next stage of evolution. **(eco-ethics and eco-cosmology)**

207

Gaiasophy: The Wisdom of the Living Earth: An Approach to Ecology. Kees Zoeteman (Head, Physics and Chemistry Dept, Netherlands Institute for Environmental Health; Coordinator for National Environmental Planning). Hudson NY: Lindisfarne Press (RR4 Box 94 A1), Nov 1991/374p/$16.95pb. (Translation of **Gaiasofie**, Stuttgart, 1989.)

Gaiasophy sees the Earth, Gaia, as a living organism, made up of components which each bear a resemblance to human organs. The industrial area and the urban area are the dominant organs in the landscape. Chapters discuss holism and systematic thinking, spiritualism vs. materialism, the origins of the physical Earth, the biosphere, the Earth as organism, restoring health to urbanized areas, the landscape organism in The Netherlands, organizational levels for the Earth organism, a healthier approach to ecosystems, the alienation of nature and technology, improving the Earth's steering mechanism, and the basic law of environmental policy aimed at sustainable development: "at every level our activities will have to be aimed at harmonization with the natural cycles of raw materials." **(gaiasophy and sustainability)**

208

Gaia 2: Emergence: The New Science of Becoming. Edited by William Irwin Thompson (Founder and Director, The Lindisfarne Association, Cathedral of St. John the Divine, NYC). Hudson NY: Lindisfarne Press (RR4, Box 94 A1), Nov 1991/272p/$14.95pb.

Presentations and discussion from a 1988 conference in Italy. **James E. Lovelock**, who first introduced the Gaia hypothesis of a living Earth in 1972 (and elaborated in two 1973 papers with Lynn Margulis as coauthor), gives a brief account of the development of Gaia theory and argues that it adds to Darwin's vision, offering a new way of looking at the Earth that no longer considers the evolution of the species separately from the evolution of their environment. **Lynn Margulis/Ricardo Guerrero** describe the new arithmetic of life, where One is always Many and Many often make One. **Wes Jackson** favors reverence but not "New Age breathlessness," and views the scaling down of the Cartesian world and emergence of the dialectical or ecological approach which sees our every act involved in creating the world (but to say Earth is an organism—the popular view of Gaia theory—may "turn out to be bad poetry"). **John Todd** and **Nancy Jack Todd** consider biology as a basis for design. **Susan Oyama** explores biology as basis for design and taking more responsibility for the Nature we construct. **William Irwin Thompson** interprets Gaia as essentially a new vision of the little and the large; in essence, the immune system of our planet. "In a Gaian understanding of planetary dynamics, we can begin to use a new language for understanding the dynamics of cultural ecologies within a biosphere." The transition from an ideological to an ecological mode of consciousness requires that we move to another level of awareness beyond Left and Right, and for this level Gaia theory shows how oppositions can be constitutive of the whole in global dynamic behavior. [**NOTE**: Arrogant and abstruse to some; avant-garde augury for others. ALSO SEE: **Gaia: A Way of Knowing: Political Implications of the New Biology**, edited by Thompson (Lindisfarne Press, 1987/ 217p/$10.95pb), from a 1981 conference.] **(Gaian understanding)**

209

Earth Covenant: A Citizens' Treaty for Common Ecological Security (1991, 1p), c/o Global Education Associates (Gerald F. Mische, President), 475 Riverside Drive (Suite 456), New York NY 10115.

A companion to GEA's Project Global 2000, individual signatures to the Covenant were solicited worldwide and presented at the 1992 UNCED "Earth Summit" in Brazil. The Covenant has been translated into 20 languages and promoted in over 80 countries (over 200,000 people signed it in the Philippines alone, including President Aquino). Its basic principles and commitments include: 1) all life forms are sacred, and we will work to preserve natural life processes and prevent ecological degradation; 2) each human being has the right to a healthful environment, and the duty to work for these rights for present and future generations; 3) to be sustainable, economic development must preserve Earth's life-support systems; 4) protection and enhancement of life on Earth requires adequate systems of governance (local, national, regional, international) that are empowering, participatory, and based on openness of information. *The Earth Covenant News* is published three times a year for Earth Covenant Network members ($10 membership). **("Earth Covenant" citizen's treaty)**

210

Spirit and Nature: Why the Environment Is a Religious Issue. Edited by Steven C. Rockefeller and John

C. Elder (both Middlebury College, VT). Boston: Beacon Press, June 1992/226p/$30.00;$16.00pb.

Essays from a Fall 1990 symposium at Middlebury College on religion, ethics, and the environmental crisis. Speakers included Audrey Shenandoah (Clan Mother of the Onondaga nation, who relates the thanksgiving prayer acknowledging that all elements of creation sustain life, and stressing the tradition that "we are responsible for seven generations into the future"), Ismar Schorsch on a Jewish perspective for learning to live with less, Sallie McFague (a liberal Christian theologian on the convergence of ecological and feminist modes of thought, and the need to deconstruct models and metaphors that are oppressive and dangerous to our planet: "the planetary agenda is the most serious, most awesome fact facing us"), J. Ronald Engel on the American tradition of spiritual democracy and the failure of "we the people" to take moral responsibility for our world (the ecological crisis is a crisis of citizenship: a failure to preserve the common good and share the wealth of the earth), Seyyed Hossein Nasr on the Islamic love of nature as manifesting "signs of God," the Dalai Lama on basic Buddhist precepts that urge "loving-kindness for the entire world" and the need for a practical ethic of caring for the earth as our home, and Robert Prescott Allen on the IUCN's updated World Conservation Strategy [see #170] and the need for a world ethic of sustainability. Includes participants' dialogue.

In the introduction, the editors conclude that "The global environmental crisis, which threatens not only the future of human civilization but all life on earth, is fundamentally a moral and religious problem." The major theme of human history in recent centuries has been a struggle to overcome oppression and a quest for social and individual freedom. But the dangers that come with greater freedom are becoming apparent in humanity's relations with the biosphere. "Humanity may destroy the possibilities for life on earth unless the freedom and power that we have acquired are channeled in new creative directions by a spiritual awareness and moral commitment that transcend nationalism, racism, sexism, religious sectarianism, anthropocentrism, and the dualism between human culture and nature. This is the great issue for the 21st century."

(global crisis as moral/religious problem)

211

Earthkeeping in the '90s: Stewardship of Creation (Revised Edition). Edited by Loren Wilkinson (Calvin College). Grand Rapids MI: William B. Eerdmans Publishing Co, Oct 1991/391p/$19.95pb.

First published in 1980, much of what appeared in that edition remains. "Christian" has been dropped from the earlier subtitle, "Christian Stewardship of Natural Resources," because stewardship is not only a Christian task, but belongs to all humans. Chapters discuss the state of the planet, the medieval view of nature, the scientific revolution, economics, valuing creation, appropriate technology, and Biblical principles of earthkeeping. Concludes with "A Steward's Hymn" and 20 "Guideposts" to direct any use we make of the Earth, including working toward a just human sharing of creation, responsibility for both human and nonhuman creatures, using creation in a sustainable way, knowing as fully as possible where the things that sustain us come from and how they are produced, evaluating our activities on their consequences for "the household of life," reforming those societal structures that damage the ecosphere, and maintaining a clear distinction between price and value.

(guideposts for stewardship)

212

Loving Nature: Ecological Integrity and Christian Responsibility. James A. Nash (Exec Director, Churches' Center for Theology and Public Policy, Washington). Nashville TN: Abingdon Press, Fall 1991/256p/$16.96pb.

The ecological crisis is definitely dangerous. The scale and scope of the problems are severe. "Thus the ecological crisis warrants a prime place on the Christian moral and political agenda, ranking with the quest for economic justice and international peace." The crisis is more than biophysical; it is a philosophical-ethical challenge for every religion and culture, since distorted dogmas of dominion come in other religious and secular versions unrelated to Christianity. If ecological ethics is given its due, significant redefinitions of moral responsibilities and relationships will be necessary in all branches of Christian ethics. Christianity does bear part of the burden of guilt for our ecological crisis, and "ongoing repentance is warranted." Nash seeks to develop an ecological ethic grounded in Christian theology—Christian love in an ecological context—and to show some of the major implications for public policy. An ecologically sound and morally responsible public policy will protect the interests of future generations ("future generations can be said to have *anticipatory rights*, and every present generation, therefore, has *anticipatory obligations*"), continually resolve the economics/ecology dilemma, include public regulations that are sufficient to match social and economic needs, protect nonhuman species, and promote international cooperation.

(ecological ethics and Christianity)

213

Eco-Socialism: From Deep Ecology to Social Justice. David Pepper (Principal Lecturer in Geography, Oxford Brookes U). London & NY: Routledge, Aug 1993/266p/$49.95;$15.95pb.

"Social justice, or the increasingly global lack of it, is the most pressing of all environmental problems." The UNCED Earth Summit showed clearly that attaining more social justice is prerequisite for combating global environmental problems. Pepper argues that the criticisms of Marxism are partly or wholly wrong. Although it does not constitute a complete eco-socialist theory of itself, the Marxist perspective on the green problematique "can at the very least constantly provided an antidote to the vagueness, incoherence, woolly-mindedness, and occasional vapidity that can invade mainstream and anarchist green discourse." The major principles of eco-socialism, to be recommended to all radical greens, are that it is anthropocentric and humanist, it rejects bioethic and nature mystification (although it does attach importance to human spirituality), and it emphasizes collective control over our relationship with nature. In contrast to anarchism, which is totally opposed to the state and tends to anti-urban decentralist idealism, eco-socialism is ambiguous about the state, confronts capitalism, and emphasizes collective political action.

(eco-socialism)

214

Radical Ecology: The Search for a Livable World. Carolyn Merchant (Prof of Environmental Ethics, UC-Berkeley). NY: Routledge, Aug 1992/276p/$49.95;$15.95pb.

The global ecological crisis results from contradictions between systems of economic production and the environment. The political economy of the First World is legitimated by a mechanistic worldview dominant since the 17th century, and an egocentric ethic assuming that what is best for the individual is best for society. Radical ecology emerges from a sense of crisis in the industrialized world, acting on the perception that domination of nature entails domination of human beings along race, class, and gender lines. It seeks a new ethic of nurturing nature and people, and empowers people to make changes in the world. Chapters discuss the global ecological crisis, environmental ethics, deep ecology, spiritual ecology, social ecology, green politics, ecofeminism, and the sustainability movement. The virtues of radical ecology: 1) it offers a critical standpoint from which to view mainstream society and mainstream environmentalism; 2) it sharpens our understanding of the assumptions underlying Western civilization; 3) visibility of radical movements can make mainstream environmental goals more acceptable; 4) radical actions often raise public consciousness and promote social change. **(eco-radicalism supported)**

215

Ecology and the Politics of Scarcity Revisited: The Unraveling of the American Dream. William Ophuls (ex-US Foreign Service) and A. Stephen Boyan Jr. (Poli. Sci. Dept, U of Maryland-Baltimore County). Foreword by Thomas E. Lovejoy (Asst Secretary, Smithsonian Inst.). NY: W.H. Freeman and Co., May 1992/379p/$14.95pb.

The original edition by Ophuls (1977), which won two awards in political science, has been thoroughly updated by Boyan (with a 7-page Afterword by Ophuls, who despairs that very little has changed in humankind's relationship to its natural milieu since the first edition, despite a vast literature on ecology and environmental problems, several decades of political activism, and considerable government activity). Chapters discuss optimists vs. pessimists (the latter see the situation as urgent, and sci/tech ingenuity unable to keep the ecological wolf from the door indefinitely), the uselessness of reformist politics (a completely new political philosophy and set of institutions is needed), the science of ecology and the limits of ecosystems, the population explosion, limited land and water as the basic agricultural predicament, the limits to resource discovery and substitution, the limits of recycling and conservation in reducing demand, deforestation and the loss of biodiversity, the inevitability of pollution and its costs, the politics of scarcity, market failures in the American political economy, the political functions of economic growth ("growth is the secular religion of American society"), administrative overload and fragmentation as a barrier to effective environmental policy, and ecological scarcity and international politics (ecological degradation may intensify the Hobbesian war of all against all, instead of promoting necessary world government).

Concludes that a new "steady state" world view is needed, that the transition will take several decades, and that the hour is very late, although transition is underway. Characteristics of the steady state include major shifts: 1) from individualism to communalism; 2) from liberty toward authority (although personal and civil rights could be largely retained); 3) from egalitarian democracy to political competence (allowing power to accrue to those who are fittest to rule); 4) from a free market to a planned or designed society; 5) from the values of growth and exploitation to those of frugality and stewardship; 6) from homogeneity and interdependence to cultural diversity and decentralization; 7) from one-dimensional thinking to holism and the systems paradigm as the dominant intellectual mode; 8) from instrumental ethics to genuine morality. "The very success of the industrial stage of civilizational succession has created conditions under which, to avoid a simple relapse into pre-modern civilization, we must move to a new, higher, more mature stage of post-industrial or post-modern civilization that shares many features of earlier civilizations." Political and spiritual wisdom alike urge the adoption of "the minimal, frugal steady state" as the form of a post-industrial society. [**NOTE**: Sophisticated eco-pessimism at its bleakest and most provoking.]

(necessary transition to frugal steady state)

216

Living Within Limits: Ecology, Economics, and Population Taboos. Garrett Hardin (Prof Emeritus of Human Ecology, U of California-Santa Barbara). NY: Oxford U Press, May 1993/339p/$25.00.

Ruminations "on the laws of nature that must be obeyed" by the provocative author of ***The Tragedy of the Commons*** (1968) and various books on population and environment. Topics include the taboo against openly discussing population, the challenge of limits ("four centuries of sedation by the delusion of limitlessness have left humanity floundering in a wilderness of rhetoric"), optimism and pessimism in thinking about population, cowboy economics vs. spaceship ecology (adding to notions proposed by the late Kenneth Boulding in 1966), real and spurious growth, the Malthusian demostat (the mechanism that controls population), nuclear power as a non-solution, denials of carrying capacity, "global pillage" as a consequence of the unmanaged commons, altruism and natural selection, the ubiquity of the "CC-PP Game" of commonized costs and privatized profits, the distinction between birth control and population control, the "Greshamite law of population" (in which having more children lowers quality of life; "loss of freedom is an inevitable consequence of unlimited population growth in a limited space"), and the necessity of immigration control ("to survive, rich nations must refuse immigration to people who are poor because their governments are unable or unwilling to stop population growth"). Concludes that "When the history of civilization is rewritten in the 21st century, it is likely that it will be summarized as consisting first of a revolution from limits to an apparently unlimited world, followed by a counterrevolution back to limits." [**NOTE**: Curmudgeonly and verbose. In an era of infoglut, can't Harden demonstrate word control?]

(population control issues)

217

States of Grace: The Recovery of Meaning in the Postmodern Age. Charlene Spretnak (Berkeley CA). San Francisco: HarperSanFrancisco, Aug 1991/337p/$21.95.

Co-author of **Green Politics** (Dutton, 1984; **FS Annual 1984 #5623**) states that notions of industrialized societies somehow living on top of nature and of humans having no inherent connectedness with one another "are being rejected as ignorant fabrications, which have led to heinous deeds." The assumptions of the cult of modernity have led to widespread ecocide, nuclear arms, the globalization of unqualified-growth economies, and the plunder of the Third and Fourth World (indigenous) peoples' cultures and homelands. One meaning of the "postmodern age" is the current transition beyond the failed assumptions of modernity. But in contrast to "deconstructive postmodernism," which merely stimulates awareness of processes by which conceptualizations are culturally constructed, Spretnak extols "ecological postmodernism" as a constructive transition. The deconstructive-postmodern orientation is critiqued from four perspectives: ecological/cosmological, spiritual, activist-political, and feminist (it is shaped by the traditional concerns of a patriarchal orientation toward life). Kind words are written about "the ecologizing of consciousness," a new politics responsive to the healing and empowering ecospiritual awakening, embracing the core teachings of the wisdom traditions (and their potential to bring a depth of understanding to a multicultural dialogue), and the state of grace when we experience consciousness of the unity in which we are embedded. Grace is "particularly important for a culture that has validated only perceptions of separateness and fragmentation." **("ecological postmodernism")**

218

The Voice of the Earth. Theodore Roszak (Cal. State U-Hayward). NY: Simon & Schuster, June 1992/367p/$23.00.

Author of **The Making of a Counterculture** (1969), **Where the Wasteland Ends** (1972), and **Person/Planet** (1978) argues that environmental problems have become the psychopathology of our everyday life. The "ecological unconscious" has emerged in our time as a deep imbalance: "the epidemic psychosis of our time is the lie of believing we have no ethical obligation to our planetary home." Urban-industrialism is the willful withdrawal of our species from the natural habitat in which it evolved. Roszak's "essay in ecopsychology" seeks to bridge our culture's long-standing, historical gulf between the psychological and the ecological. We need a psychology that is ecologically grounded so as to make the institutional changes that a life-sustaining biosphere requires. Some principles of ecopsychology: 1) the core of the mind is the ecological unconscious; its repression is the deepest root of collusive madness in industrial society; 2) ecopsychology draws upon the findings of the new cosmology of cosmic evolution, and seeks to awaken the inherent sense of environmental reciprocity that lies within the ecological unconscious; 3) it draws on traditional healing techniques of primary people, nature mysticism, the experience of wilderness, and the insights of Deep Ecology; key therapeutic projects include re-evaluation of certain compulsively "masculine" character traits that permeate our structures of political power and which drive us to dominate nature; 4) ecopsychology encourages small-scale social forms and personal empowerment; it is *post*-industrial, not *anti*-industrial in its social orientation; 5) there is synergic interplay between planetary and personal well-being: the needs and rights of the planet *are* the needs and rights of the person. **(ecopsychology principles)**

219

Monocultures of the Mind: Perspectives on Biodiversity and Biotechnology. Vandana Shiva (Director, Research Foundation for Science, Technology and Natural Resource Policy, Dehra Dun). London & Atlantic Highlands NJ: Zed Books; Penang: Third World Network, July 1993/184p/$49.95;$15.00pb.

Five essays by the author of **Staying Alive: Women, Ecology and Development** (Zed Books, 1989), who also serves as science and technology advisor for the Third World Network. Monocultures first inhabit the mind, and are then transferred to the ground. Monocultures of the mind generate models of production which destroy diversity and legitimize that destruction as progress and growth. From the perspective of diversity, monocultures are based on a decline in yields and productivity. They are impoverished systems, highly unstable and non-sustainable. Treatment of biodiversity as mere "raw material" comes from an anti-nature and racist standpoint that treats nature and the labor of Third World people as valueless. Biodiversity does not get value merely from bioengineering performed by white men in white lab coats; it has intrinsic value and also a high use value for local communities. "Diversity as a way of thought and a way of life is what is needed to go beyond the impoverished monocultures of the mind." A final essay critiques the UNCED Biodiversity Convention, which has been distorted into a convention that exploits not protects, and one that risks favoring the US more than the Third World. **(Western monocultures questioned)**

220

Earth Follies: Coming to Feminist Terms with the Global Environmental Crisis. Joni Seager (Geography and Women's Studies, U of Vermont). NY: Routledge, May 1993/332p/$27.50;$16.95pb.

We have degraded the environment, disrupted our biosphere, and endangered physical carrying capacities. If environmental problems are conceptualized in their physical form that leaves causality out of the picture, there is no role for feminist analysis. But the environmental crisis is not just a crisis of physical ecosystems. "The *real* story of the environmental crisis is a story of power and profit and political wrangling; it is a story of the institutional arrangements and settings, the bureaucratic arrangements and the cultural conventions that *create* conditions of environmental destruction." It is a crisis of the dominant ideology. And this is grist for the feminist mill. Militaries, multinationals, governments, the "expert structure" and the eco-establishment are all institutions of men. The institutional culture that is responsible for most environmental calamities is a masculinist culture.

Chapters describe the way in which gender and gender relationships enter into environmental issues. Topics include the military (privileged environmental vandals whose operations are beyond the reach of civil law and protected from public scrutiny), business (the degradation of nature by industrial enterprises stems from the same culture that condones violence against women), government (the military wields enormous influence in most, and most governments are composed largely of men from the wealthy elite), pious globalism, finger-pointing at the Third World, dumping on the poor, the professionalization and bureaucratization of the environmental movements, sexism in the German Green party, the exclusion of minorities from the environmental movement, population and the environment ("blaming population growth for the world's environmental ills is seductively simple"; the biggest problem is not excessive numbers of poor people but excessive pressures of rich countries on the biosphere), the reification of "Mother Earth" (a deceptive paradigm that is not effective for environmental action; "the earth is *not* our mother"), the eco-fringe (deep ecology and eco-feminism), the green consumer movement, and women as the backbone of virtually every environmental group around the world (they generally constitute about 60% of general environmental groups and 80% of small grassroots and animal-rights groups). The grassroots environmental movement "is the story of 'hysterical housewives' taking on 'men of reason'—in the multitude of guises in which they each appear." In the early 1990s, a small sea-change began, as a number of powerful funding agencies and academics started to pay more attention to the idea that non-scientists could contribute to environmental understanding. "It is possible to pursue an analysis of the cultural basis of our ecological crisis without a feminist sensibility—but it will not be insightful, truthful, useful, or even very interesting." [**NOTE**: Illuminating, well-documented, and (pardon the phrase) non-hysterical. Opens a wide new perspective for "men of reason" blinded by gender. ALSO SEE: **Changing the Boundaries: Women**

Centered Perspectives on Population and the Environment by Janice Jiggins (Washington: Island Press, May 1994/291p/$45.00;$22.00pb), and **Feminist Perspectives on Sustainable Development**, edited by Wendy Harcourt (Zed Books and Society for International Development, April 1994/255p/$55.00;$22.50pb), and **Close to Home: Women Reconnect Ecology, Health and Development Worldwide**, edited by Vandana Shiva (New Society Pubs, May 1994/170p/$39.95;$12.95pb).]
(women and environmental problems)

221
Rationality and Nature: A Sociological Inquiry into a Changing Relationship. Raymond Murphy (Prof of Sociology, U of Ottawa). Boulder CO: Westview Press, Dec 1994/294p/$55.00;$21.95pb.

"Sociology has been constructed as if nature didn't matter. It has failed to take the processes of nature into account, perceiving only the social construction of reality. Environmental problems are beginning to send shock waves through this myopic sociological structure. Sociology fabricated as if nature didn't matter constitutes pre-ecological sociology. It was the product of a way of thinking in pre-ecological society at large and is increasingly being exposed as shortsighted by the looming environmental crisis." Murphy seeks to develop an alternative sociology, placing what is distinctively human within the context of what is shared with other forms of life. Chapters discuss rationalization under the premise of the plasticity of nature, the myth of nature's plasticity, the dialectical embrace of rationality and irrationality, sustainable development as rationalization under the premise of greenness, the limits of the sustainable development approach (the very basis of its rhetorical success is the basis of its intellectual inadequacy), the limits of neo-Malthusian research [e.g., the 1972 views of Lester Brown], Murray Bookchin's anarchist social ecology, eco-feminism and its limits, deep ecology and derationalization, the political economy of waste, and the exploitation and attempted mastery of nature by science and technology. Concludes with an extensive bibliography of about 750 items. [**NOTE**: Green sociology, at last, even if mainly of interest to other sociologists. ALSO SEE: **Social Theory and the Global Environment** edited by Michael Redclift and Ted Benton (Routledge, Sept 1994/271p/$17.95pb), on basic social science assumptions challenged by the environmental crisis.]
(green sociology)

222
Biologic: Designing with Nature to Protect the Environment (Revised Edition). David Wann (EPA). Foreword by Frederic Krupp (Exec. Director, Env. Defense Fund). Boulder CO: Johnson Books, April 1994/284p/$14.95pb.

Our mistakes in the past were based on a narrow and disconnected logic. *Biologic* suggests a new environmental ethic and a new way of thinking. Its principles include understanding basic physical concepts (gravity, nutrient cycles, flows of sun/wind/water), using resources sustainably (taking only the "interest" out of our ecological savings accounts), using the right tool for the right job, carefully monitoring inputs, tracing the origins and future route of each physical interaction, acknowledging the uniqueness of each location, using the simplest process or product to get the job done, using software (information) rather than hardware whenever possible, using design solutions that accomplish several things at once, and accounting for costs over the full product lifetime.

"Our economy and our society are defined by the designs we choose to implement." Nature-compatible design includes such strategies as The Strategy of the Backpacker (bringing only what is needed and no more), The Goldilocks Strategy (finding the right fit for each creation), The Strategy of Compost (progressing to a respectful relationship with microbes), The Strategy of the Cactus (getting more mileage from a limited resource), The Strategy of the Lover (our designs are uninspired because we are uninspired), The Strategy of the Rice Paddy (the systems approach used by rice farmers), and The Strategy of the Athlete (going with the flow of nature, rather than against it). Much of the book is organized around four chapters: ***1) Knowing***: risk assessment, biomonitoring, impact assessments, caring capacity; ***2) Choosing***: clarity of desired ends, openness to innovative means, green labeling programs; ***3) Designing***: inventing institutions that nurture smart designs; various strategies [described above]; ***4) Implementing***: playing hardball with the soft path, coevolution of ecology and political will, waste-free lifestyles. [**NOTE**: A green Alvin Toffler, loaded with ideas and presented in a popular style with cutesy chapter sub-heads.] **(nature-compatible design)**

223
Regenerative Design for Sustainable Development. John Tillman Lyle (Prof of Landscape Architecture, Cal State Polytechnic U-Pomona). NY: John Wiley & Sons, March 1994/338p(8x11")/$49.95.

Environmental design is where the earth and its processes join with human culture and behavior to create form. Design in this sense requires reestablishing some connections that began coming loose in the Renaissance and were entirely severed by industrialization. The first connection to reestablish is that between people and nature; next is that between art and science. Chapters discuss sustainability in the neotechnic era, The Center for Regenerative Studies at Cal Poly-Pomona, evolving concepts of nature, strategies for regenerative design (letting nature do the work, aggregating, seeking optimum levels for multiple functions, matching technology to need, using information to replace power, prioritizing for sustainability, etc.), regenerative energy flow and shelter, regenerative water systems and agriculture, waste as a resource, regenerative systems in the social fabric, garden communities, and the economics of regeneration. [ALSO SEE in the Wiley Series in Sustainable Design **Gray World, Green Heart: Technology, Nature and the Sustainable Landscape** by Robert L. Thayer Jr of UC-Davis (Wiley, Feb 1994/352p/$39.95).]
(designing for sustainability)

224
Living Machines: Theoretical Foundations and Design Precepts, John Todd (President, Ocean Arks International, Falmouth MA) and Beth Josephson, *Annals of Earth*, 12:1, 1994, 16-25 (from OAI, 508/540-6801).

Ocean Arks, formerly the New Alchemy Institute, was incorporated in 1982 to disseminate the ideas and practice of ecological sustainability. This article summarizes the epistemological ground for the restorative biology that Todd has practiced for over 20 years, based on the view that the workings of complex natural systems offer a blueprint for technological design. A living machine is a technology with the capacity to do work or provide support services. Within the framework of this new design, complex ecologies can be engineered to undertake work on behalf of human communities. Principles for ecological design/ecological engineering are discussed, especially as applied to waste treatment systems. Viewed as essential are geological and mineral diversity, nutrient reservoirs (mineral diversity provides the long-term foundation for nutrient diversity), steep gradients (a rapid change in

basic properties of adjacent subsystems), high rates of exchange, periodic and random pulsed exchanges, cellular design as the structural model, a law of the minimum, microbial communities, photosynthetic foundations, phylogenetic diversity, sequenced and repeated seedings, and reflection of the microcosmos (Biosphere II in Arizona represents the most prominent experiment in microcosm level ecological engineering to date). "These emergent principles, applied to waste treatment, food production, fuel generation, integrated systems engineering, architecture, and possibly manufacturing, could permit the human community to reduce its negative impact on the planet by as much as 90%." This conclusion is based on the overall efficiency of living machines, and a major reduction in the need for today's vast infrastructure networks. [**NOTE**: Sophisticated, radical bio-thinking. A second part of this article will be in the summer edition of *Annals*. ALSO SEE: **Bioshelters, Ocean Arks, City Farming: Ecology as the Basis of Design** by Nancy Jack Todd and John Todd (Sierra Club, 1984; **FS Annual 1985** #6752).]
(ecological engineering principles)

225
Ultimate Security: The Environmental Basis of Political Stability. Norman Myers (Oxford UK). NY: W.W. Norton, Sept 1993/308p/$25.00.

"Whereas the past 40 years have been dominated by the Cold War, the next 40 years will surely be dominated by environmental conflicts. They will add up to a variant of World War III, a war we are waging against the Earth... It is a war of us all against us all." Moreover, any New World Order must reflect the new situation where the principal threat to security and peace stems from environmental breakdown, plus the need for access to natural resources that are increasingly scarce as more people make greater demands upon them. National security is no longer about fighting forces and weaponry alone: it relates increasingly to watersheds, croplands, forests, genetic resources, climate, and other factors as crucial to a nation's security as military prowess. Not all environmental problems lead to conflict, and not all conflicts stem from environmental problems. But there is ample evidence that environmental factors contribute to security in its all-round sense, security for all, security forever. "It all adds up to a new fact of global living."

Developed nations should be concerned with environmental security in the developing world because of: 1) major interests in exports and imports; 2) the need to maintain political stability; 3) environmental problems that often spill over borders; 4) environmental degradation that triggers mass migrations of people. Chapters are devoted to case studies (each with a negative and a positive scenario) of water in the Middle East, hunger in Ethiopia, food insecurity in sub-Saharan Africa, deforestation and destitution in the Philippines, carrying capacity in India, population growth and ruined environment in El Salvador and other nations of Central America, and population growth and migration in Mexico. Additional chapters describe global population, global warming and ozone-layer depletion, mass extinction of species, environmental refugees (with an estimate of some 150 million future refugees in a mid-21st century world of 10 billion people), synergistic connections, trade-offs with military security (e.g., $12 billion a year, or four days of military spending, could reverse desertification and bring agricultural benefits of $30 billion a year), and ethical aspects of policy. [ALSO SEE: **Redefining National Security** by Lester R. Brown (Worldwatch Paper 14, Oct 1977).]
(environment as "national security")

226
Defining National Security: The Nonmilitary Aspects. Joseph J. Romm (Rocky Mountain Institute). NY: Council on Foreign Rels Press, May 1993/122p/$10.95pb.

For 45 years, the US pursued a national security strategy focused on the single goal of containing the Soviet Union—a doctrine that has now become a victim of its own success. Military security has not vanished as a key element of national security, but it has declined in relative importance. As early as 1974, Gen. Maxwell Taylor stated that "the most formidable threats to this nation are in the nonmilitary field." And in Worldwatch Paper #14 (October 1977), Lester Brown sought to broaden the definition of national security. The definition of national security used here, supplied by Richard Ullman in 1983, defines a threat as actions or events that degrade the quality of life for inhabitants of a state or narrows the range of policy choices available to the government or to private entities within the state. Chapters discuss drug policy, environmental security, energy security, and economic security. Concludes that the two new and pre-eminent post-Cold War nonmilitary national security threats are America's declining competitiveness and global climate change. Moreover, by almost any definition, environmental security and economic security are not just compatible but inseparable, especially because a key component of international trade in the 21st century will be technologies that improve the environment. Japan uses the concept of "economic security" (or comprehensive national security) more than any other country; this doctrine deserves more attention in the US, in that a major point of redefining security is to redefine budgetary priorities. [**NOTE**: Lucid, sensible, and well documented. A briefer and updated version of Romm's **The Once and Future Superpower: How to Restore America's Economic, Energy, and Environmental Security** (William Morrow and Co, Sept 1992/320p/$23.00).] **(broadening "security" definition)**

227
Environmental Change and Violent Conflict, Thomas F. Homer-Dixon (Peace/Conflict Studies, U of Toronto), Jeffrey H. Boutwell (Intl. Security Studies Program, AAAS) and George W. Rathgens (Prof of Political Science, MIT), *Scientific American*, 268:2, Feb 1993, 38-45. (Brief version in *New York Times* Op-Ed, 31 Jan 1993.)

The co-directors of the Project on Environmental Change and Acute Conflict, sponsored by the University of Toronto and the American Academy of Arts and Sciences, warn that human population is likely to exceed 9 billion in the next 50 years, and global economic output may quintuple. Largely due to these trends, "renewable resource scarcities of the next 50 years will probably occur with a speed, complexity and magnitude unprecedented in history." Entire countries can now be deforested in a few decades, and the total area of highly productive agricultural land will drop. Future generations will experience the ongoing depletion and degradation of aquifers, rivers and other bodies of water, the decline of fisheries, and perhaps significant climate change. As such problems become more severe, they may precipitate civil or international strife. These warnings are familiar, but the debate has been constrained by a lack of carefully compiled evidence. A team of 30 researchers was assembled by the Project, with the gathered evidence showing that scarcities of renewable resources are already contributing to violent conflicts in many parts of the world. "These conflicts may foreshadow a surge of similar violence in coming decades, particularly in poor countries." To prevent such turmoil, nations should put greater emphasis on reducing resource scarcities and promoting sustainable development. **(resource scarcity and conflict)**

3. Economics

228

Ecological Economics: The Science and Management of Sustainability. Edited by Robert Costanza (Prof, Coastal and Environmental Policy Program, U of Maryland). NY: Columbia U Press, Sept 1991/525p/$50.00.

Derived from a workshop following the first biannual conference of the International Society for Ecological Economics (ISEE), held May 1990 in Washington DC. In contrast to conventional economics and conventional ecology, ecological economics (EE) "takes a wider and longer view in terms of space, time, and the parts of the system to be studied." It transcends disciplinary boundaries and views the human economy as part of a larger whole. Its domain is the entire web of interactions between economic and ecological sectors. The macro goal of EE is sustainability of the combined ecological economic system.

The 32 papers include **Kenneth E. Boulding** on environmentalism and the integrative power of truth, **Herman E. Daly** on elements of environmental economics, **Garrett Hardin** on the mythical conventional measures of income and wealth and cultural carrying capacity, **Talbot Page** on sustainability and the problem of valuation, **Juan Martinez-Alier** on distributional conflicts, and **Mary E. Clark** on rethinking ecological and economic education. Some of the other topics: sustainability and discounting the future, sustainable resource management, a new scientific methodology for global environmental issues, time and uncertainty in environmental decision-making, the environment as capital, alternative environmental accounting approaches, economic biases against sustainable development, incentives for sustainability, intergenerational transfers, mitigating the impacts of climate change, ecological engineering, integrated agro-industrial ecosystems, and managing tropical forests.

Some policy recommendations from the workshop: 1) institute a consistent goal of sustainability in all institutions at all levels, from local to global; 2) promote long-term thinking and use of a systems approach in decision-making; 3) use a wide variety of policy instruments to assure sustainability (regulation, property rights, permits, marketable permits, fees, subsidies, and bonds); 4) in decisions of scale, individual freedom of choice must yield to democratic collective decision-making by the relevant community; 5) ease up on income taxes (especially on low incomes) and implement fees on the destructive use of natural capital to promote more efficient use; 6) develop transdisciplinary curricula for our overspecialized education system, as well as an EE core curriculum and degree granting programs; 7) develop extension programs that can effectively transfer information among disciplines and nations. [**NOTE**: Important, authoritative, and accessible. For information on ISEE and its journal, *Ecological Economics* (published bi-monthly by Elsevier since 1990), write to R. Costanza, U of Maryland, Box 1589, Solomons MD 20688. ALSO SEE: **Ecological Economics: Energy, Environment and Society** by Juan Martinez-Alier of the Universitat Autonoma de Barcelona (Basil Blackwell, 1987;1990/287p/$18.95pb. with new introduction), a rather dense treatise giving special emphasis to "agricultural energetics," and **Ecology, Economics, Ethics: The Broken Circle** edited by F. Herbert Bormann and Stephen R. Kellert (Yale U P, Dec 1991/233p/$26.50).] **(ISEE on ecological economics)**

229

Environmental Economics: Policies for Environmental Management and Sustainable Development. Clem Tisdell (Prof of Economics, U of Queensland). New Horizons in Environmental Economics Series. Aldershot UK: Edward Elgar (dist in US by Ashgate, Old Post Rd, Brookfield VT), March 1993/259p/$59.95.

Environmental economics is concerned with economic interrelationships between mankind and the environment, both natural and man-made. The two most important factors to be studied in this area are externalities (spillovers between economic entities) and sustainability. In the past two decades, environmental or green economics has become a major focus in economic studies. Most universities in developed countries now include at least one course in this area. The development of environmental economics has opened up new academic vistas and encouraged interdisciplinary work. Diverse new theories and models have been developed, which is likely to foster further progress in economic thought. Chapters discuss relevant and irrelevant externalities, pollution control policies proposed by economists, the law and risk taking, project evaluation, cost-benefit analysis, sustained returns, sustaining intergenerational economic welfare, economists vs. ecologists on sustainable agricultural systems, and resource conservation in LDCs. [**NOTE**: Written for fellow economists. **Ecological Economics** (above) is aimed at a more general audience.] **(ecological economics)**

230

Environmental Economics: An Elementary Introduction. R. Kerry Turner, David Pearce, and Ian Bateman (all Centre for Social and Economic Research on the Global Environment, U of East Anglia and University College, London). Baltimore: Johns Hopkins U Press, Dec 1993/328p/$48.50;$19.95pb.

A textbook with chapters on sustainable economic development, the causes of environmental degradation (how markets and governments fail), cost-benefit thinking, valuing nature, coping with uncertainty, economic control of the environment (using the market, charging for use of the environment, green taxes, trading environmental permits, setting environmental standards), renewable and non-renewable resources, business and the environment, managing waste, climate change, the ozone layer, biodiversity, and international environmental policy. [ALSO SEE: **An Introduction to Sustainable Development: The Developing World** by Jennifer A. Elliott of Staffordshire U (Routledge, Feb 1994/121p/$12.95pb), with emphasis on sustainable rural and urban livelihoods; **Environment, Growth and Development: The Concepts and Strategies of Sustainability** by Peter Bartelmus of the UN Statistical Division (Routledge, June 1994/163p/$15.95pb), on economic accounting for sustainable development, sustainable growth, and global compacts.] **(sustainability: three textbooks)**

*231

The Gaia Atlas of Green Economics. Paul Ekins (Birkbeck Coll., London), Mayer Hillman (Policy Studies Inst., London), and Robert Hutchison (PSI). Foreword by Robert Heilbroner. NY: Anchor Books/Doubleday, March 1992/191p/$16.00pb. (Published in the UK by Gaia Books as **Wealth Beyond Measure: An Atlas of New Economics**.)

As described by Heilbroner, "Green economics is essentially concerned with the consequences of the drive to amass wealth, which has expanded the capacity for survival while at the same time imperiling its prospects; and widened the possibilities for freedom while subjecting large portions of the world's population, women not least, to a diminution of their potential. Presenting the quest for wealth as a moral and environmental problem is the great contribution of this book."

Sections cover: **1) The Need for Green Economics**: industrial society and the rape of Earth, too many people and too much consumption by some, the exploitation of women (who do most of the world's work for little material return), the destruction of community by industrial projects, the explosion of debt, wants and needs ("green economics is the economics of enough"), the market and the state (the green approach sees them as complementary), balancing values (the economy should be viewed as only one of four essential dimensions of the human condition, the others being society, ecology, and ethics); **2) Wealth and Wellbeing**: wealth as anything that makes us better off, needs and satisfiers, the four-capital model of wealth creation (ecological capital, human capital, manufactured capital, and social and organizational capital), economic accounting for the environment, measuring wealth creation, the Adjusted National Product (ANP) as superior to GNP, the nonmonetary economy, social indicators as a necessary supplement to economic data, the failure of capitalism and communism (the Green Mixed Economy is seen as superior); **3) Policies for Wealth**: creating ecocapital with a strategy for sustainability, energy policy for real wealth creation, the greening of industry, transportation and wealth, sustainable agriculture, spiritual resources, stabilizing world population, health promotion, the knowledge industry, education for capability, a Basic Income as a right of all citizens, families and child care, action for community, broadening of business accountability, tax reform (to be shifted away from labor and toward use of resources and the environment), military conversion to peace, global institutions, alternative development policies, building sustainable cities for people, appropriate technology, greening the infrastructure, and conserver tourism. [**NOTE**: Intelligent, informative, and inspiring, with many leading-edge ideas presented in brief essays accompanied with photos, diagrams, and sidebars. As with other books in the outstanding "Gaia Future Series" edited by Norman Myers this is handsomely produced.]

(wealth: green definition)

232

Real-Life Economics: Understanding Wealth Creation. Edited by Paul Ekins (Dept of Economics, U of London) and Manfred Max-Neef (Director, Development Alternatives Centre, Santiago, Chile). London & NY: Routledge, July 1992/460p/$79.95;$23.00pb.

Economics in its mainstream neoclassical form is failing to provide an intellectually coherent explanation of economic reality, especially concerning environmental degradation, the nature of markets, persistent poverty, and household production. The many constructive, potentially complementary, alternative proposals on these and other issues (some from within the economics discipline; others from outside it) are too often left standing in isolation from each other. This book is the main output to date of the international Living Economy Network (42 Warriner Gardens, London SW11 4DU, UK) of economists and other social and natural scientists who are interested in developing new economic thinking that recognizes all economic activity and behavior as profoundly influenced by social, ethical, and ecological realities. The 37 essays include Amitai Etzioni on the new "I & We paradigm," David Pearce on equity and sustainable development, Mark Lutz on humanistic economics, Jane Wheelock on the household in the total economy, Orio Giarini on the modern economy as a service economy, Paul Ekins on a four-capital model of wealth creation, Manfred Max-Neef on development and human needs, Ian Miles on social indicators for real-life economics, Paul Streeten on the evolution of development thought, Severyn Bruyn on a new direction for community development, Robert Chambers on sustainable livelihoods, and Wolfgang Sachs on the transformation of society into a political economy. [**NOTE**: An outstanding collection that pulls together most major critics of industrial-era economics.]

(criticisms of mainstream economics)

233

Alternative Economic Indicators. Victor Anderson (New Economics Foundation, 88-94 Wentworth St, London E1 7SA). London and NY: Routledge, Oct 1991/106p/$59.95;$16.95pb.

The goal of "maximizing economic growth" has become the central long-term policy of most governments, with enormous influence on their thinking and decision-making. In recent years it has been increasingly criticized, especially by supporters of green politics. But the critics have been weak in suggesting an alternative, which is the starting-point for this book. Chapters discuss: **1) The Growth Debate**: the 1972 *Blueprint for Survival*, *The Limits to Growth* and the pro-growth response; **2) National Income Accounting and its Problems**: failure to account for unpaid domestic labor and non-money transactions outside the household, problems of averaging and comparison, problems of stocks and depreciation, the masking of various inefficiencies; **3) Reforms to National Income Accounting**: deductions for environmental depreciation and damage done to human capital; **4) An Enlarged Economics**: greater attention to human economics, natural economics, and the economy as something for which moral evaluation is relevant; describing economic processes in three ways (as monetary flows, as humans beings organized in particular ways, and as the relationship between humans and the natural world); **5) Criteria for a Good Indicator**: information that is easily and cheaply collected, relatively easy to understand, a measure of something important, internationally comparable; **6) Social Indicators**: education and literacy, work and unemployment, consumption, health, and income and wealth distribution; **7) Five Priority Environmental Indicators**: deforestation, CO_2 emissions, annual rate of increase in population, number of nuclear reactors, energy consumption per million dollars of GDP; **8) Promoting Better Indicators**: more comprehensive data collection, publicity for what the priority indicators show, social and environmental auditing for individual companies, monitoring and critique of policies and statements of the IMF and the World Bank, political pressure for changes in national income accounting. [**NOTE**: A cogent primer for a more honest and sophisticated accounting.]

(better economic indicators)

234

For the Common Good: Redirecting the Economy Toward Community, the Environment, and a Sustainable Future (Second Edition, Updated and Expanded). Herman E. Daly (Senior Research Scholar, U of Maryland) and John B. Cobb (Prof of Theology, Claremont Graduate School). Boston MA: Beacon Press, March 1994/534p/$17.00pb

First published in late 1989 (**FS Annual 1991 #10605**), the basic critique against the growth-oriented industrial economy that has led to environmental disaster remains the same. The four parts of the book discuss the failings of economics as an academic discipline, an alternative real-world approach to the economy, policies to promote community in the US, and steps for getting there (university reform, a new Index of Sustainable Economic Welfare to replace GNP, a biospheric religious vision). The major change in this new edition is a 36-page Afterword on "Money, Debt, and Wealth," proposing reforms to restore honesty and accuracy to the function of money in the economic system. Tables of data are updated through 1990, and a few new references added. [**NOTE**: A heavy but thorough critique of industrial era economics. ALSO SEE: **The End of Economics? Ethics and the Disorder of Progress** by U of Brasilia economist Cristovam Buarque (Zed Books, March 1993/172p/$49.95;$19.95pb), who calls for ethics in economics, valuing nature and culture, rethinking "progress," and an ecological economics that includes all the relations of life as part of its study.] **(economics for sustainability)**

235

The Green National Product: A Proposed Index of Sustainable Economic Welfare. Clifford W. Cobb (Sacramento CA) and John B. Cobb Jr (Claremont CA). Co-pub. with the Human Economy Center. Lanham MD: Univ Press of America, June 1994/343p/$65.00;$26.50pb.

In **Through the Looking Glass,** the Red Queen tells Alice that "it takes all the running you can do to keep in the same place." Modern economies practice "Red Queen Economics" by running faster and faster just to stay in place. GNP measures all economic activity as progress, but much of it makes us worse off. A "green national product" is needed to add up the "goods" and subtract the "bads," giving us a picture of our net condition. This new GNP would differ from the old by addressing the long-term health of the planet, making sustainability a central issue. The Index of Sustainable Economic Welfare described here proposes a green national product. A version of the ISEW appeared in **For the Common Good**.

Much of this volume involves eight invited critiques of ISEW by leading thinkers on the measurement of economic welfare: 1) Robert Eisner, author of **The Total Incomes System of Accounts** (U of Chicago, 1989), points to the exclusion of human capital, inadequate treatment of government capital, and estimates of urbanization costs; 2) Allan H. Young and Carol Carson favor the UN System of National Accounts, where nonmarket values are in satellite accounts; 3) Robert Gottfried raises questions of what should be included and how to measure it; 4) Thomas Michael Power points to the problem of ignoring regional differences in the value of per capita income; 5) E.J. Mishan considers issues of distribution, intergenerational equity, and rising land costs; 6) Jan Tinbergen argues that ISEW should include other elements of well-being; 7) Richard Lamm worries that evidence of upcoming crises will not be taken seriously; 8) Hans Diefenbacher offers an ISEW for Germany with similar results to that applied to the US: both show a leveling off of sustainable welfare after about 1970. The Cobbs respond to their critics, make some changes, and update the ISEW to 1990. [**NOTE**: The Human Economy Center (Mankato State U, Box 14, Mankato MN 56002) will be maintaining and revising the Index.] **(Index of Sustainable Economic Welfare)**

236

Ecological Tax Reform: A Policy Proposal for Sustainable Development. Ernst U. von Weizsäcker (President, Wuppertal Institute for Climate, Energy and the Environment; former Director, Institute for European Environmental Policy) and Jochen Jesinghaus (European Statistical Office, Luxembourg). London and Atlantic Highlands NJ: Zed Books, May 1992/90p/$49.95;$15.00pb.

The gap between development requirements under modest assumptions of economic growth and ecologically sustainable levels of consumption is growing ominously wide. Conventional "end of the pipe" environmental protection measures (standards of environmental quality or of maximum emissions per unit of a polluting source) can do very little to narrow this gap. Indeed, classic means of controlling pollution (catalytic converters, recycling, sewage treatment) actually result in increased energy and materials consumption. A drastic reduction in per capita resource consumption in the North (which consumes about 10 times more natural resources per capita than the South) is ecologically imperative. The North must seek a revolutionary increase in resource productivity. To trigger such a revolution, an ecological tax reform can be tailored for social acceptability and business profitability. Such green taxes should primarily be levied on environmentally important input factors, notably energy. The taxes would involve a steady price increase of some 5% per year over some 30-40 years. Revenue neutrality could be secured by reducing other taxes such as VAT, income taxes, and corporation taxes. Over several decades, it should be possible for such new taxes to attain a level of 5-10% of GNP, with other taxes being correspondingly reduced. (One should not talk about "green taxes," but about "ecological tax reform" so as to gain greater political acceptance.) In addition, ecologically unreasonable state subsidies of electricity, coal, and water should be dismantled. Chapters also discuss increasing CO_2 productivity, the polluter pays principle, price elasticity for fuels, objections to ecological tax reform, and the international dimension. Numerous initiatives are underway in Europe for introducing ecological taxes, but difficulties can be expected from the US, whose cultural makeup is linked to unlimited mobility. The relatively high energy prices in Western Europe and Japan have not, however, brought any competitive disadvantage. **(ecological tax reform)**

237

Green Fees: How a Tax Shift Can Work for the Environment and the Economy. Robert Repetto, Roger C. Dower, Robin Jenkins, and Jacqueline Geoghegan (WRI). Washington: World Resources Institute. Nov 1992/96p/$14.95pb.

At present, our taxes fall mostly on activities that make the economy productive: work, savings, investment, and risk-taking. A better system would place more of the tax burden on activities that make the economy unproductive and that should be discouraged, e.g. resource waste, pollution, and congestion. Taxes on these activities would not distort economic decisions, but correct existing distortions. Chapters provide details on: **1) Congestion Tolls**: such tolls on crowded urban highways could generate $40-100 billion and reduce peak congestion by 11-22%; new electronic technology enables easy toll collection with anonymity; **2) Carbon Taxes**: a properly set pollution tax could generate $30-50 billion and improve national economic performance; also discussed is the impact of energy taxes on income class (less burdensome than commonly thought), states and regions, and industries; **3) Solid Waste Charges**: waste collection charges by volume (pay-by-the-bag) could raise $5-10 billion; **4) Other Environmental Charges**: effluent or emissions charges, charges on damaging activities such as recreational user fees and noise charges on airport landings, product charges such as taxes on agricultural chemicals and virgin materials, deposit-return charges on vehicles and tires, reduction of tax benefits and subsidies for grazing rates, timber sales, and irrigation water (together, such charges could contribute another $40-50 billion in revenues). Shifting as much as 10-15% of the total Federal, state, and local revenue base toward environmental charges would help America's economy and environment. [**NOTE**: Technical discussion of a key idea.] **(green taxes)**

238

Governance by Green Taxes: Making Pollution Prevention Pay. Mikael Skou Andersen (U of Aarhus, Denmark). Manchester UK: U of Manchester Press (dist. by St. Martin's Press), Nov 1994/247p/$69.95;$29.95pb.

In the last few years, pollution taxes have moved from economics textbooks to matters of high-level politics. The public has changed its opinion about pollution taxes significantly, with the rise of neo-liberalism and a general market orientation. Andersen examines the role of economic instruments in the water pollution control policies of four European countries: Denmark, France, Germany, and the Netherlands. The Danish water pollution control policy was both expensive and ineffective; the Dutch policy has been quite effective, and the German and French experiences are less clear-cut. Environmental policies are classified along two dimensions in a 2x2 matrix: whether they are imposed or consensus-seeking, and reactive or anticipatory. One can then differentiate four types of environmental programs: ***1) Command-and-Control***: policies that prescribe technology-based "best-available technology" guidelines, as in Germany and the US (reactive and imposed); ***2) Consensus-and-Delegation***: prescribes no specific standards, but defines a process by which bureaucracy and polluters agree on measures to be taken, as in Denmark, Sweden, and UK (reactive and consensus); ***3) Charging Policy***: establishes a system of incentives to charge polluters for their input of raw materials, emissions, or products, thus promoting development of clean technologies, as in Netherlands, France, and Japan (anticipatory and imposed); ***4) Covenant Policy***: emphasis on reaching a voluntary agreement with polluters on targets for pollution control, as in Netherlands and Denmark (anticipatory and consensus).

(new water pollution control policies in Europe)

239

Population, Technology, and Lifestyle: The Transition to Sustainability. Edited by Robert Goodland, Herman E. Daly, and Salah El Serafy (all World Bank). Washington: Island Press, Oct 1992/154p/$19.00.

Largely a retitled version of **Environmentally Sustainable Economic Development: Building on Brundtland**, published in 1991 by the World Bank and by UNESCO [**FS Annual 1993** #11713]. Chapters include Robert Goodland on the case that the world has reached (or exceeded) important source and sink limits, Herman Daly on the transition from "empty-world economics" to "full-world economics," Jan Tinbergen and Roefie Hueting on GNP and market prices as misleading signals that mask environmental destruction, Salah El Serafy on the questionable Brundtland Commission proposal for maintaining an emphasis on growth, Raymond Mikesell on project evaluation for sustainable development, Bern von Droste and Peter Dogsé on why policymakers tend to underestimate the value of environmental investments, Robert Costanza on ecological economics contrasted to conventional economics and ecology, and Lester R. Brown *et al* on transition to sustainability. A new chapter by Goodland and Daly explains the need for the North to stabilize its rate of resource consumption to free resources for the South (an alternative to the traditional view that the North must grow faster to buy ever more resources from the South). Northern growth is not the answer to Southern poverty because of flaws in the GNP measure, the value of self-reliance in alleviating poverty, dysfunctions of imbalanced trade, the insecurity of inequality, etc.

(economics for sustainable development)

240

How Much Is Enough? The Consumer Society and the Future of the Earth. Alan Thein Durning (Worldwatch Institute). Worldwatch Environmental Alert Series. NY: W.W. Norton & Co, Aug 1992/200p/$19.95;$8.95pb.

Consumption is the neglected god in the trinity of issues the world must address if we are to get on a path of sustainable development (the other two issues being population growth and technological change). The world has three broad ecological classes: **1) The Poor**: 1.1 billion people who earn less than $700 a year per family member (2% of total world income), generally have an inadequate diet, and drink mostly unclean water; **2) The Middle-Income**: 3.3 billion people who earn $700-$7500 per family member (33% of world income), and get plenty of calories and protein from their grain and vegetable diet; **3) The Consumer Class**: 1.1 billion people with over $7500 income per family member (64% of world income or 32 times as much as the poor), dine on meat and processed foods, and drink beverages in disposable containers (feeding the meat-eating class takes nearly 40% of world grain, grown on nearly one-fourth of the world's cropland). Unless the rich one-fifth lowers consumption, they will have no authority to object to others despoiling the earth. The future of life on earth depends on whether the rich fifth, having met its material needs, can turn to nonmaterial sources of fulfillment. We can curtail our use of things that are ecologically destructive, and lowering consumption need not deprive us of goods and services that really matter. Voluntary simplicity may be an unattainable ideal, but the philosophy of sufficiency is deeply rooted in the human past. To rejuvenate this ethic, a critical mass of individuals committed to living by it must emerge. A "culture of permanence" will not come quickly; all we can hope for is slow progress against consumerism, punctuated by rapid advances (i.e., cigarettes and ivory).

(over-consumption questioned)

241

Economic Values and the Natural World. David W. Pearce (Prof of Economics and Director, Centre for Social and Economic Research on the Global Environment, University College London). Cambridge MA: The MIT Press, Sept 1993/129p/$35.00;$15.95pb.

Co-author of **Blueprint for a Green Economy** (London: Earthscan, 1989; 1991; 1993) and of **World Without End** (below) focuses on why economists engage in preference measurement in the environmental context, how they do it, and how the results might be used. Critics have nothing to offer in its place, other than "generally unimaginative procedures such as environmental impact assessment." Nor has the world environment fared very well under environmental policies dominated by non-economic considerations of worth and value. If the Earth's resources were available in infinite quantities, and if they could be deployed at zero cost, there would be no economic problem. Choice becomes necessary when resources are finite—when, in the words of Kenneth Boulding, we move from a "cowboy" to a "spaceship" economy. Economic valuation is useful for project appraisal, setting national priorities, and modifying national accounts (GNP) to reflect depreciation of natural capital and damage to human well-being. The goal of sustainable development cannot be interpreted without some idea of the value of environmental assets and services. The revealed economic values for environmental conservation and improvement projects and policies have frequently been found to be large. "Valuation demonstrates that there is an economic case for protecting the environment, in addition to any ethical case." It offers the potential of more cost-efficient public choices, so that limited public income is spent to best advantage. Measuring preferences, however, says little about the distribution of costs and benefits within a time period or between time periods, and thus discriminates against future generations. This problem of discounting the future can be met by imposing an intergenerational criterion of sustainability or making some adjustment to the conventional discount rate in the range of 2-5%. Gives examples for valuing protected areas, wetlands, noise reduction, endangered species, biodiversity, water supplies, and fuelwood planting. [**NOTE**: A brief and authoritative primer for those who can't or won't count.]

(economic valuation of the environment)

242

World Without End: Economics, Environment, and Sustainable Development. David W. Pearce (University College London) and Jeremy J. Warford (Senior Advisor, Environment Dept, World Bank). Published for The World Bank. NY: Oxford U Press, Jan 1993/440p/$39.95.

If poverty is to be reduced and the standard of living improved, economic growth must remain a legitimate objective of national governments and the world community. Limits can be avoided and the world will not necessarily come to an end—if imaginative policies are devised and implemented. Pursuing economic growth without adequate attention to the environment is not sustainable. The issue is how, not whether, to grow. The causes of much environmental degradation lie in the workings of the economy, and in economic distortions of government policies. The major theme in this book is the need for incentives to conserve resources and to change technology, especially in the poorest countries. Chapters discuss environmental economics, the concept of sustainable development, the choice of discount rate, environmental accounting and indicators, evaluating environmental damage and benefits, population growth and carrying capacity, pricing for cost recovery, market failure, environmental taxes, marketable permits, property rights and resource management (privatization can provide incentives to improve land and resources, but can also lead to degradation and externalities; common property management can work with proper incentives), poverty and the environment, world markets and natural resource degradation, transfrontier environmental issues, and managing global resources. [ALSO SEE: **Economic Policy Towards the Environment**, edited by Dieter Helm of New College, Oxford (Blackwell, Oct 1991/326p/$24.95), with 11 essays, including David Pearce on saving tropical forests.] **(environmental economics)**

243

The Growth Illusion: How Economic Growth Has Enriched the Few, Impoverished the Many, and Endangered the Planet. Richard Douthwaite (Westport, Ireland). Foreword by Edward Goldsmith. Tulsa OK 74120: Council Oak Books (1350 E 15th St), June 1993/367p/$16.95pb. (First published in the UK by Green Books in 1992.)

A journalist and economist explains why economic growth is the cause of environmental problems, and why its continuation cannot be part of the cure. We persist in trying to grow because our measure of progress is so defective that it still gives us favorable signals. Moreover, groups that make the investments that generate growth are getting good returns—at the expense of everyone else. And many of the negative effects of growth have yet to reveal themselves, especially those connected with the environment. Rather than striving for rapid growth and ever-larger bundles of unspecified goods and services, we should set specific targets and measure success in terms of progress towards them. Chapters discuss definitions of the quality of life (GNP is merely a measure of the volume of trading), why capitalism needs growth (which leads to change and more profits), the inverse relationship between economic growth and social welfare, loss of jobs from mechanization and international trade, how growth damages family and community life, the ignored warnings of **The Limits to Growth**, the need to lower fossil fuel use, alternative futures for the Irish economy, the myth of sustainable growth (the only sustainable society is one in which all forces are in equilibrium), criticism of David Pearce's **Blueprint for a Green Economy** (e.g., it does not mention population at all), and the need for democratic governments to guide "the invisible hand." Concludes with three key principles of the no-growth age: 1) giving the future equal weight to the present; 2) valuing the interests of other people equally with our own (as demonstrated by the rapidly-spreading "polluter pays" principle); 3) viewing some things as not for sale. [**NOTE**: Well-informed and clearly written, resurrecting the widespread growth/no-growth debates of the 1970s. Contrasts with the ideas of David Pearce, who insists that growth is necessary and that valuation is desirable (see two above items).]

(economic growth questioned)

244

Sustainable Investment and Resource Use: Equity, Environmental Integrity, and Economic Efficiency. M.D. Young (CSIRO, Canberra, Australia). UNESCO Man and the Biosphere Series, Vol 9. Paris: UNESCO & Park Ridge NJ: Parthenon Publishing Group, 1992/176p/$65.00.

Degradation of environment and resources is a widespread problem in rich and poor economies. If present rates of consumption and pollution continue, and human population doubles, we may be vulnerable to dramatic change. The challenge is to establish a new economic order that includes the environment in economic decision-making. We have to find forms of investment that increase production without taking from the environment and the ecological processes that maintain it. Much more quality investment is necessary. Positive investments complement or increase the stock of existing resources without imposing additional costs on future generations. A sustainable future depends on retaining a diverse range of approaches to and definitions of the most appropriate ways to achieve sustainable development. Chapters discuss intergenerational equity, principles for sustainable resource use, the design of resource rights for sustainable investment, shifting the tax base towards resource rent and resource severance taxes, and putting the environment into the market. **(sustainable investment)**

245

Investing in Natural Capital: The Ecological Economics Approach to Sustainability. Edited by Ann-Mari Jansson (U of Stockholm), Monica Hammer, Carl Folke, and Robert Costanza (President, ISEE). Washington: Island Press, June 1994/504p/$35.00pb.

The product of a 1992 workshop following the second biennial conference of the International Society for Eco-

logical Economics in Stockholm. Topics include natural capital contrasted to human-made capital, renewable vs. non-renewable natural capital, why operationalizing sustainable development requires investment in natural capital (by Herman Daly), challenges to ecological economics (by Paul Ehrlich), ecosystem properties as a basis for sustainability, investing in cultural capital (institutions, environmental ethics, ecological knowledge) for sustainable use of natural capital, the economic value of natural ecosystems, a humane and ethical decision system for investment, strategies for environmentally sound economic development, mitigation strategies for sea-level rise, fisheries, and Robert Costanza on three general principles to achieve sustainability: 1) a natural capital depletion tax, passed on to consumers in the price of products; 2) the precautionary polluter pays principle, applied to potentially damaging products; 3) a system of ecological tariffs, to be reinvested in the global environment).
(natural capital and sustainability)

246
Sustainable Livelihoods: Redefining the Global Social Crisis, David Korten (President, People-Centered Development Forum, NYC), *International Society for Ecological Economics Newsletter*, 5:4, Oct 1994, 1,4-5,8.

Author of **Getting to the 21st Century: Voluntary Action and the Global Agenda** (Kumarian Press, 1990; **FS Annual 1992** #11001) argues that the failures of international development efforts have mired the world in a deepening social crisis every bit as severe as the parallel environmental crisis, whereby a major portion of humanity is being marginalized. Both crises result from misplaced priorities, due to confusing means with ends. Rather than developing healthy, sustainable human societies that provide people with secure and satisfying livelihoods, we have defined our goals in terms of growing economies to provide jobs. Consequently, the economic growth of the past 20 years has primarily benefitted a tiny elite. The global economy consumes resources faster than they can be regenerated and thus destroys its resource base; it pays workers too little to buy the products they produce and thus destroys its own markets; it displaces the functions of households and communities and thus destroys the social fabric. A more holistic approach is needed that views the development of healthy societies as the goal, deals with the market economy as one of several means to realize the goal, and returns economic and politic control to people. A good starting point is to focus not on the need for jobs, but on the need for sustainable livelihoods. Assuring everyone the opportunity for a livelihood by which they may obtain the necessities of life, even while significantly reducing traditional welfare assistance programs, is entirely within our collective means. "In light of current realities, it is wholly unrealistic to expect that any available policies will result in providing adequate and satisfying jobs for everyone in the world who might want or need one." [**NOTE**: Briefly and eloquently frames a fundamental issue. The ISEE (Box 1589, Solomons MD 20688) is a rapidly-growing group, listing 1,700 members in their 1994 Directory!]
(jobs vs. livelihoods)

4. Business and Industry

*247
The Ecology of Commerce: A Declaration of Sustainability. Paul Hawken (Sausalito CA). NY: Harper Business, Nov 1993/250p/$23.00.

If every company on the planet were to adopt the best environmental practices of the leading companies, the world would still be moving toward sure degradation and collapse. Although the efforts made by companies to reduce their negative impact on the environment are to be applauded, the fact is that commerce and sustainability are antithetical by design. Our understanding of business is stuck at a primitive level. The ultimate purpose of business is not—or should not be—simply to make money; it should be to increase the general well-being of humankind. "To create an enduring society, we will need a system of commerce and production where each and every act is inherently sustainable and restorative." The economics of restoration is the opposite of industrialization, which separated production processes from land values, and land from people. Business has three interlinked issues to face: what it takes from the environment and how, what it makes and how, and what it wastes. To realize a restorative economy, a set of objectives is needed: 1) reduce absolute consumption of energy and natural resources in the North by 80% in the next 50 years (we already have the technology to make things last twice as long with about half the resources); 2) provide secure, stable, and meaningful employment for people everywhere; 3) be self-actuating, as opposed to regulated or morally mandated; 4) be more rewarding than our present way of life (we must invite people into a world that delivers the goods, not subtracts them); 5) exceed sustainability by restoring degraded habitats and ecosystems ("sustainability is an insufficient objective"); 6) create better systems of feedback and accountability. If we are to endure as a world culture, or a group of local cultures, we must incorporate ecological thinking into every aspect of our lives. [**NOTE**: Fresh, provocative, and well-written.]
(steps to a restorative economy)

248
Changing Course: A Global Business Perspective on Development and the Environment. Stephan Schmidheiny (Chairman, BCSD; World Trade Center, Geneva; and Chairman, UNOTEC). Cambridge MA: MIT Press, May 1992/374p/$35.00;16.95pb.

Schmidheiny, a Swiss industrialist who heads three major holding companies, was asked by Maurice Strong to serve as principal advisor to UNCED for business and industry. This led to forming the Business Council for Sustainable Development; among the 50 members are the heads of ALCOA, Browning-Ferris, Chevron, ConAgra, Dow Chemical, Du Pont, Mitsubishi, Nippon Steel, Nissan Motor, Northern Telecom, Royal Dutch/Shell, and Volkswagen. The BCSD has already organized some 50 conferences, symposia, and issue workshops in over 20 countries. A 3-page Declaration of the BCSD states that business will play a vital role in the future health of the planet, that economic growth and environmental protection are inextricably linked, that "economic growth in all parts of the world is essential... to sustain growing populations," and that new forms of cooperation between government, business, and society are required. The world is moving toward deregulation and private initiatives; this requires corporations to assume more social, economic, and environmental responsibility. "Progress toward sustainable development makes good business sense because it can create competitive advantages and new opportunities."

Chapters discuss "alarming trends" in patterns of development and in environmental degradation, meeting the challenge of "eco-efficiency" by profound changes in the goals and assumptions that drive corporate activities, redefining the rules of the economic game to move from wasteful consumption and pollution to conservation, "full-cost pricing" that includes environmental costs, altering standard national accounts and GNP calculations, making energy markets work by market signals that move toward sustainability, increased energy efficiency, financing sustainable development in emerging capital markets, freer trade and sustainable development, managing corporate change with the vision of sustainable development, the innovation process in pollution prevention and materials substitution, MNCs as primary agents of technology transfer, sustainable agriculture and forestry, and LDC leadership for sustainability. Half of the book is devoted to 38 case studies on making energy conservation pay, the CEO as Chief Environmental Officer, environmental auditing, long-term partnerships, catalyzing improved supplier performance, cooperating with stakeholders, and managing cleaner production and products. [**NOTE**: Cutting-edge thinking for business.]

(Business Council for Sustainable Development)

249

EarthEnterprise Tool Kit: Markets, Technology, Financing, New Business Practices. International Institute for Sustainable Development. Winnipeg: IISD, Dec 1993/129p(8x11″ spiral).

EarthEnterprise is a new IISD initiative linking the business world with new opportunities arising from environment and development concerns. This Tool Kit provides basic knowledge for smaller businesses entering the environmental marketplace. Sections include: ***1) Green Consumers***: sincere in their intentions, but they tend to overstate their actual green behavior and don't want to make sacrifices; few will consciously pay extra for greener products or go out of their way to buy green products; ***2) Building Credibility***: products and services must be credible to cynical and confused consumers; ***3) Be Ready to Compete***: green retailing is headed for a shake-out, with great impact on small-scale retailers and manufacturers that sell exclusively through them; ***4) Green Purchasing***: target retailers who want to sell green products, understand government programs, green your suppliers, forge new alliances; ***5) Demand Side***: understand environmental laws and regulations as technology drivers, assess international agreements as sustainable technology drivers, watch for technology breakthroughs or leaps; ***6) Supply Side***: find and assess new technologies, understand the four generations of environmental technology (for remediation, abatement, pollution prevention, and sustainability); ***7) Financing***: approach capital markets, business planning, identify potential investors and types of financing; ***8) New Business Practices***: in-house waste reduction and pollution prevention, informing employees, networking, and annual sustainable development reports [**NOTE**: Lots of savvy tips and contacts in Canada and the US. ALSO SEE: IISD report on corporate environmental reporting, #257].

(green business Tool Kit)

250

Business and the Environment: A Resource Guide. Allison A. Pennell, Patricia E. Choi, and Lawrence Molinaro Jr (Management Institute for Environment and Business, Washington). Washington: Island Press, March 1992/364p(8x11″)/$55.00.

The first section of the guide provides brief abstracts of about 900 books, articles, and case studies from the 1981-1991 period on environmental aspects of accounting and finance, business/government/society, marketing, production and operations management, and strategic management. The second section provides information on about 180 educators, largely in the US, who write and teach in this area. Includes subject and author indexes. [**NOTE**: Employs a rather arcane numbering system for items, and does not offer any analysis or highlights.]

(business/environment resource guide)

251

Costing the Earth: The Challenge for Governments, The Opportunities for Business. Frances Cairncross (Environment Editor, *The Economist*). Boston MA: Harvard Business School Press, March 1992/c300p/$24.95.

Public interest in green issues may ebb and flow, but politicians and business managers will find some environmental pressure irresistible and lasting. "Environmental issues will not go away." But bad government policies may make even more of a mess than the unfettered market. A better starting point is to look for ways to improve markets, e.g. clear rights of ownership for natural resources and making sure that consumers and producers pay the true costs of the environmental damage they cause. Setting a rough value may be a better basis for policy than none at all. Economic growth cannot be made environmentally benign, "but greener growth is possible." Chapters explain the environmental limits to sustainable growth, discount rates, measuring costs and benefits, anti-environmental farm subsidies, making polluters pay, conserving energy, international environmental management, the challenge to companies (the demand for cleaner products and processes will change the way they think about innovation), green consumerism (the ultimate oxymoron for the most ardent environmentalists), waste disposal, recycling, and effective corporate environmental policy (a clear statement of principles and objectives, regular environmental audits, harnessing the enthusiasm of employees, building environmental considerations into accounting procedures, creating links with local people). Concludes that "a truly green company is unlikely to be badly managed. Conversely, a well-managed company finds it relatively easy to be green." [ALSO SEE: **Green Ink** by Frances Cairncross (Island Press, July 1995/ $17.95) and **Leading Your Business in the Environmental Age** by Steven J. Bennett and Richard Freierman (Wiley, July 1992/320p/$24.95).]

(the greening of business)

252

Beyond Compliance: A New Industry View of the Environment. Edited by Bruce Smart (Senior Fellow, WRI; former US Undersecretary of Commerce and CEO of Continental Group). Washington: World Resources Institute, April 1992/285p/$19.95pb.

Societies worldwide are pursuing two potentially conflicting goals: a cleaner environment and sustained economic growth. Reconciling these goals will require a sweeping technological transformation—essentially an "eco-industrial revolution." Provides case studies of some two dozen companies that have publically stated a determination to move toward environmental excellence: the Pollution Prevention Pays program at 3M since 1975, the Recycled Paper Coalition involving three large San Francisco Bay Area companies, the comprehensive environmental program of Pacific Gas & Electric Co, the Weyerhaeuser intensive forest management strategy, Dow Chemical's Responsible Care initiative, the Business Charter for Sustainable Development sponsored by the International Chamber of Commerce (signed by over 500 companies from 35 countries as of 12/31/91), the Monsanto Pledge, innovation by Briggs Nursery in raising its ornamental nursery stock (an example of a small company improving the environment), cooperation between Allied Signal Inc and Tufts University in forming The Environmental Literacy Institute at Tufts (for training faculty members and integrating environmental issues into all undergraduate, graduate, and professional school curricula), Du Pont's Environmental Leadership Council, and the Royal Dutch/Shell Group's use of scenarios in planning.

Some concluding observations: 1) most major changes are very recent, many responding to events of the 1980s such as Bhopal and Exxon Valdez; 2) perceptions of corporate environmental progress vary ("distinguishing between hypocrisy and the beginnings of real change requires great wisdom"); 3) credible communication is difficult: many companies are frustrated that efforts to communicate their new environmental policies are not very convincing to the public; 4) economic benefits of a forward-looking environmental policy are becoming more widely recognized; 5) educational initiatives and leadership of committed CEOs will continue to be critical.
(corporations and eco-industrial revolution)

253

Green at Work: Finding a Business Career That Works for the Environment. Susan Cohn (Environmental Programs Coordinator, Office of Career Development, NYU School of Business). Washington: Island Press, Nov 1992/222p/$30.00;$16.00pb.

The rise in corporate environmentalism reflects social pressures to hold corporations more strictly accountable for their environmental impacts, as well as opportunities in growing markets for environmentally conscious products and increased employment in environmental industries and functions. Environmentalism has forced companies to rethink all of their decisions. The resulting environmental lens of "life cycle assessment" provides a new model for evaluating products, financial decisions, and innovations. Environmental jobs are available in rethinking consumer products, consulting, finance, nonprofit organizations, environmental services, education, petroleum, chemicals, agriculture and food processing, law, insurance, and government. "Environmental literacy will be to the 1990s what computer literacy was to the 1970s and 1980s... people who are environmentally literate will have a competitive advantage." Chapters on searching for green employment, career profiles (of 12 environmental managers, 5 consultants, 5 investment analysts, and 7 non-profit positions), a 123-page directory of 230 company environmental programs and projects, and a brief guide to books, journals, and organizations. [**NOTE**: Not only for job-seekers, but a useful guide to many corporate programs.] **(environmental programs and jobs)**

254

Management for a Small Planet: Strategic Decision Making and the Environment. W. Edward Stead and Jean Garner Stead (both Profs of Management, East Tennessee State U). Newbury Park CA: Sage Publications, May 1992/211p/$36.00;$17.95pb.

A guide "to prepare managers to make strategic decisions that are both economically successful and sensitive to the Earth's natural environment," seeking to describe national and global environmental issues as they relate to economic activity, explore related social/scientific/psychological/economic concepts, and provide a new decision-making framework to attain long-term economic success. Chapters discuss unprecedented population growth, economic activity as a prime source of environmental ills, the need for a new economic paradigm, the issue wheel (a schematic of issues resulting from population and economic growth), Gaia as a metaphor for future survival, perceptions and values, economics as if the planet mattered, the emerging management paradigm (as suggested by Tom Peters, Robert Waterman, William Halal, Peter Drucker, and Gareth Morgan), attention to stakeholders (with Earth as the ultimate stakeholder), "sustainability" as the transcendent core value for a small planet, instrumental values for achieving sustainability (wholeness, valuing posterity, smallness, quality, community), green consumerism, ethical investors, and sustainability strategies. **(management for sustainability)**

255

Environmental Strategies Handbook. Edited by Rao V. Kolluru (Director of Environmental Health Services, CH2M HILL). NY: McGraw-Hill, Mar 1994/1,030p/$79.50.

Insights and case studies on how industries can develop optimal environmental strategies and integrate them into long-term strategic planning to insure growth and competitive advantage. The 22 chapters cover such topics as corporate environmental excellence and stewardship, the regulatory framework (in the US, Canada, and Mexico), environmental technology and engineering, pollution prevention case studies, "Total Quality Environmental Management," environmental health and safety challenges, developing an environmental action plan, financial issues in environmental liabilities, risk management, communicating risk information, public health and private initiatives, life cycle analysis, environmental marketing, environmental business strategies, environmental management of multinational corporations, trade and the environment, environmental strategies for the developing world, and population and health perspectives. A final chapter by **Walter H. Corson** (Global Tomorrow Coalition) summarizes global environmental issues and strategies, and sketches 30 case studies of corporate progress toward sustainable resource management in such areas as improving energy efficiency, controlling hazardous substances, and managing solid waste. The Handbook also lists regulatory agencies, industry associations, support groups, and consulting firms "that can assist executives in profitably linking corporate ambitions with environmental stewardship and sustainable development."
(environmental strategies for business)

256

Corporate Environmentalism (Focus Issue). *Columbia Journal of World Business* (310 Uris Hall, Columbia U), 27:3-4, Fall-Winter 1992, 297p.

The 30 essays are under six headings: **1) Sustainable Development**: Frances Cairncross on the growth of mass environmental consciousness, Stephan Schmidheiny on the Business Council for Sustainable Development, a politics of hope (connecting a clean and safe environment with a healthy economy); **2) Markets**: market innovation as a powerful tool of environmental policy (tradeable permits, debt-for-nature swaps, the proper allocation of property rights), the modification of capital markets to incorporate environmental costs and benefits and promote long-term investment; **3) Treaty-Making**: the new global negotiations at UNCED, new corporate roles in environmental treaty-making, trade and environmental protection, the corporation as an NGO in international policy-making; **4) Regional Policies**: Japan's "New Earth 21" policy for sustainable development, environmental policy in the EC, the European Environmental Agency and its "Toward Sustainability" program of year 2000 objectives, the 1991 German Packaging Order to reduce and recycle packaging waste, shifting the US tax burden to unproductive activities of resource waste and pollution (by Robert Repetto of WRI), implementing the Montreal Protocol to restore the ozone layer, environmental technology co-operation between transnational corporations and LDCs; **5) Energy and Technology**: S. Fred Singer on the detrimental effects of a carbon tax, the inevitability of increasing carbon emissions, energy management in Eastern Europe and the former USSR, the need for revolutionary restructuring of philosophic and technologic underpinnings; **6) Corporations**: the role of the private sector in environmentally friendly development, getting a strategic edge on environmental innovation, lessons from companies recognized as environmental leaders (the environmental vision must be tied to specific goals carefully tailored to the organization), the increasing accountability of companies to the environmental expectations of stakeholders, corporate environmental communications, steps toward "sustainable communication," environmental option assessment (a new tool to support decision-making on issues of substance life cycle management). [**NOTE**: This is leading-edge thinking, handsomely produced.]

(business and environment)

257

Coming Clean: Corporate Environmental Reporting. Opening Up for Sustainable Development. Ken Beecham (Deloitte Touche Tohmatsu International, London), John Elkington (SustainAbility Ltd, London), and Stephan Barg (IISD). London: DTTI (dist by International Institute for Sustainable Development, Winnipeg), May 1993/64p(8x11")/C$25.00 (204/958-7700).

The first international survey of corporate environmental reporting, analyzing over 70 companies in Europe, North America, and Japan that have produced free-standing environmental reports. A growing number of companies are producing these reports, following such bellwethers as Monsanto and Norsk Hydro. They are doing so because of growing pressure to report on the sustainability of their products and processes, their success in taking advantage of opportunities related to sustainable development, the growing recognition that business must move beyond mere compliance, and support of "corporate citizenship" codes of conduct that promote best environmental practice. Moreover, companies will increasingly be asked to make good on past environmental damage, including adequate accounting provisions for resulting liabilities. These costs for dealing with environmental damage will increase at a rate that will drive them far higher up the agenda of management concerns. Businesses cannot hide the damage they are causing: "sooner or later all will be revealed." Ensuring that environmental liabilities are fully disclosed is a major challenge for accountants. Very few companies give precise targets for improvement in their environmental reports, but most observers see this as the logical next step. The "honeymoon period" in which any report was welcome is almost over; future reports will undergo increasingly rigorous analysis.

(environmental reports by business)

258

EcoManagement: The Elmwood Guide to Ecological Auditing and Sustainable Business. Ernest Callenbach, Fritjof Capra, Lenore Goldman, Rüdiger Lutz, and Sandra Marburg. San Francisco CA: Berrett-Koehler Publishers (155 Montgomery St), Sept 1993/188p/$27.95.

The Elmwood Institute (2522 San Pablo Ave, Berkeley CA 94710; 510/845-4595), founded in 1984 by Fritjof Capra to foster ecological literacy and systemic thinking, has initiated an EcoManagement Program to explore the terrain in which ecology and business interface. Topics covered: 1) trends toward ecologically conscious management (eco-friendly products and services, Total Quality Management for the Environment, mapping stakeholders to expand recognition of interconnections); 2) benefits of ecologically conscious management (human survival, market opportunities, reduced liability risk, reduced costs, personal integrity); 3) methods of ecological auditing (environmental indicators, environmental accounting to show true costs, impact assessment, feasibility calculations, technology assessments, material and energy audits, market and survey research, cost-benefit analysis, social balance sheets, checklists, scenarios); 4) innovative ecomanagement practices and programs worldwide (Smith & Hawken's guidelines for direct mail, the 3M Pollution Prevention Pays Program, Union Carbide's new solvent-free paint, using recycled rubber in the soles of Nike shoes); 5) agents of change (the International Network for Environmental Management, the Social Investment Forum, the German Environmental Management Association, the Eco-Life Center in Japan); 6) the paradigm shift from environmentalism to systemic ecomanagement; 7) managing change to an ecologically conscious corporate culture; 8) checklists for ecomanaging (energy and materials inflows, design and processing operations, sales and marketing outflows, emission outflows, support structures, international business relations). [**NOTE**: Many useful lists, including a brief appendix on how to audit resources.]

(ecological auditing)

259

Environmental Strategies for Industry: International Perspectives on Research Needs and Policy Implications. Edited by Kurt Fischer (Tufts U) and Johan Schot (U of Twente). Washington: Island Press, June 1993/389p/$48.00;$24.95pb.

Essays from the first convening of the Greening of Industry Network, a US-European partnership for studying the behavior of industrial firms around environmental issues (the Network is led by the Center for Environmental Management at Tufts U and the Centre for Studies of Science, Technology, and Society at the U of Twente in The Netherlands). Topics focus on trends in the greening of business, economic analysis of cleaner technology, corporate sustainability strategies, how German companies are dealing with environmental issues, an analysis of two high-performing waste-reduction programs, trust and credibility, managing chemical accident risks, technological responses of industrial firms to environmental problems, information exchange among firms (environmental co-makership), setting requirements for suppliers, and product stewardship (systematic company efforts to reduce product risks or significant segments of a product life cycle). The editors conclude that environmental concerns are becoming progressively more important in the market as regulations are tightened, and that there is growing pressure on companies for environmental credibility; "all these trends appear to be irreversible." Many firms currently follow a defensive environmental strategy to comply with regulations and public pressures, but do not try to identify possibilities to excel in environmental protection to gain a competitive position. Some have moved to a more innovative strategy, but it is hard and challenging work.

The central task of government action should be the deepening and broadening of pressures.
(industry and environmental strategy)

260
Lean and Clean Management: How to Boost Profits and Productivity by Reducing Pollution. Joseph J. Romm (Special Asst for Policy/Planning, US Dept of Energy). NY: Kodansha International, Oct 1994/220p/$23.00.

Former research scholar at the Rocky Mountain Institute asserts that "The trend toward pollution prevention is one of the fundamental shifts taking place in businesses throughout the world... Every company can increase its profits and productivity dramatically by reducing pollution." Institutional problems—not technological problems—are the biggest barriers, but all barriers can be overcome by "lean and clean management." The most compelling reason to embrace pollution prevention is that it pays. A 1992 cross-industry survey of 75 case studies found an average payback for industrial investments in waste reduction of only 1.58 years—an annual return on investment of 63%. The Louisiana Division of Dow Chemical audited 575 energy- and waste-reducing projects since 1982, finding an average annual return on investment of 204%. Energy efficiency and pollution prevention generate jobs for several reasons: 1) money that had been devoted to consuming energy and resources gets shifted to investing in technology and capital equipment; 2) when a company cost-effectively reduces costs for natural resources, waste disposal, and cleanup, it can lower the price of its products, which increases competitiveness; 3) a rich export market awaits the nation that first develops clean technologies [see **Green Gold**, #301]. Chapters discuss the American origins of lean and clean in Henry Ford's thinking, life cycle thinking for managers and workers, setting up a suggestion system (Japanese companies are strong in this area), a case study of Compaq Computer, improved lighting and office equipment, and design for office and factory productivity. **(how pollution prevention pays)**

261
The Next Efficiency Revolution: Creating a Sustainable Materials Economy. John E. Young and Aaron Sachs. Worldwatch Paper 121. Washington: Worldwatch Institute, Sept 1994/58p/$5.00.

The rise of the consumer society, in tandem with rapid industrialization, is at the root of the world's environmental crisis. But the consumer society can no longer avoid confronting the inefficiency of the materials economy. Poor product design, sloppy extractive and processing techniques, and an ethos of casual consumption lead to far more use of materials than necessary. Transition to a sustainable materials economy would be as profound as the movement from the Stone Age to the Bronze Age, or from Bronze to Iron. The next materials revolution will hinge not on a new material, but on far more efficient ways of using concrete, timber, metals, and plastics. It will require not only improvements in technological efficiency, but a shift in our culture and ethics. A necessary first step will be to develop secondary industries in the global economy that facilitate recycling of waste materials and reduce usage of primary materials. Raw material production should be taxed rather than subsidized, with subsidies directed instead toward expansion of recycling, reuse, and remanufacturing. Also needed are efficient design so as to promote sharing and easy repair, new markets for recycled materials, and a new political coalition built around an efficient materials policy. [ALSO SEE: **Rethinking the Materials We Use: A New Focus for Pollution Policy**, edited by Ken Geiser and Frances H. Irwin (Washington: World Wildlife Fund, 1993/137p/$12.50pb), on developing materials, life-cycle assessments, and "sunset" policies to phase out chemicals.] **(the next materials revolution)**

262
Recycling for Profit: The New Green Business Frontier, David Biddle (Executive Director, Public Recycling Officials of Pennsylvania), *Harvard Business Review*, 71:6, Nov-Dec 1993, 145-156.

Despite the proliferation of curbside collection bins and public awareness campaigns, recycling programs around the US aren't working. Modern urban recycling has created a tremendous supply of recycled newspapers, glass bottles, office paper, and other materials. But consumer and business demand for products made from these materials is still outdistanced by supply, and the cost of collecting and processing recyclables far outweighs their value. Myths about recycled products still linger. ***Myth 1: Recycled Products Cost Too Much***: most companies committed to principles of recycling and waste reduction have not paid higher prices; e.g., the computer division of American Airlines saved over $100,000 by converting to 100% recycled paper, and the recycled paper bought by Moore Business Forms is no different in cost than virgin paper; ***Myth 2: The Quality of Recycled Products is Bad***: once a serious concern, this is no longer an issue; "office machinery experts now say that recycled-content paper performs better in modern copiers and laser printers"; plastic lumber made from recycled plastics can cost up to four times as much as its wood counterpart but it shows immense maintenance cost savings over time because it doesn't rot, splinter, or break; ***Myth 3: Recycled Products Aren't Available When Needed***: this was a real problem a few years ago, but most standard business products are now readily available. In sum, stimulating markets for recycled products will make the most difference. In the past two years, many national and local organizations (notably, the Buy Recycled Business Alliance) have initiated "Buy Recycled" campaigns. The 33 steering committee members of the Alliance (including Bell Atlantic, Coca-Cola, and Anheuser-Busch) have alone accounted for $3 billion in purchases of recycled-content products and material. **(overcoming recycling myths)**

263
Industrial Metabolism: Restructuring for Sustainable Development. Edited by Robert U. Ayres (Prof of Environmental Economics, European Institute of Business Administration, Fontainebleau, France) and Udo E. Simonis (Wissenschaftszentrum Berlin für Sozialforschung). Tokyo: United Nations U Press, 1994/376p/$35.00 (UN Sales No.E.93.III.A.3).

The end result of a process that began with a 1988 conference in Tokyo sponsored by UNU, UNESCO, and IFIAS, on the human dimensions of global change. The term "industrial metabolism" was coined in the preparatory stages of the conference, referring to "all the materials/energy transformations that enable the economic system to function." Policy implications of the holistic industrial metabolism perspective are that it is much easier to see how narrowly conceived or short-run (myopic) quick-fix policies are very far from the global optimum; indeed, many such policies can be harmful. Narrowly conceived environmental policies over the past 20 years or so have largely shifted waste emissions from one form and medium to another, without significantly reducing the totals. Essays discuss ecosystem and biosphere as metaphors for human-induced material flows, industrial re-

structuring in rich and poor countries, evolution and sustainability, case studies of industrial metabolism (Sweden, Rhine Basin, Switzerland, US), the precaution principle in environmental management, and transfer of cleaner technologies to developing countries. Concludes with a bibliography of about 300 items.
(industrial metabolism perspective)

264
The Greening of Industrial Ecosystems. Edited by Braden R. Allenby and Deanna J. Richards (National Academy of Engineering). Washington: National Academy Press, Feb 1994/259p/$34.95.

Industrial ecology recognizes humans' unique role in creating complex artifacts and institutions that force changes in materials and energy flows in both natural and industrial systems. Natural ecosystems provide raw materials for economic activity, but also serve as sinks for wastes from producers and consumers. Thinking about industrial ecosystems as a product, an urban area, an industry (such as agriculture), or a material (such as lead) can help to shape practices that are compatible with the environment. The 18 papers herein, from a July 1992 NAE workshop, discuss industrial metabolism (by Robert U. Ayres), energy, input-output analysis (by Faye Duchin), wastes as raw materials, economics and sustainable development, pollution prevention, international environmental law, the role of government in encouraging change, design for the environment, designing the automobile for recycling, the Green Product Realization project at AT&T, resource efficiency and product-life extension, zero-loss environmental accounting systems, implications of industrial ecology for firms, and the role of universities in design for environment. **(industrial ecology)**

265
Industrial Ecology and Global Change. Edited by Robert Socolow (Director, Center for Energy and Environmental Studies, Princeton U), *et al.* NY: Cambridge U Press, Sept 1994/500p.

Papers from a July 1992 Global Change Institute workshop in Snowmass CO, sponsored by the Office for Interdisciplinary Earth Studies at the University Corporation for Atmospheric Research. Industrial ecology seeks to evolve the world's industrial activity into a sustainable and environmentally benign system. The metaphor offers a fresh view of environmental management, and leads to six perspectives: long-term habitability, global scope, understanding the overwhelming of natural systems, recognizing our vulnerability, mass-flow analysis (a unifying analysis based on total flows of materials), and centrality of the industrial firm and the industrial farm as policy-maker. Essays are in five parts: ***1) Vulnerability and Adaptation***: implementing industrial ecology, changing perceptions of vulnerability, the human dimension of vulnerability, a developing country perspective on global industrialization; ***2) Grand Cycles of Disruption and Repair***: human impacts on the carbon and nitrogen cycles, CO_2 emissions, reducing urban sources of methane, energy efficiency in China, biomass energy in sustainable development; ***3) Toxics and the Environment***: vulnerability of soil and biotic diversity, global ecotoxicology, metals emissions, exposure to cadmium and lead, the nuclear industry; ***4) Industrial Ecology in Firms***: product lifecycle management, manufacturing consumer products, design for environment, prioritizing impacts, finding and implementing projects that reduce waste, taking advantage of win-win situations ("free lunch economics" that improves economic growth and the environment); ***5) Industrial Ecology in Policy-Making***: policies to encourage clean technology, linking ecology to economy, military-to-civilian conversion in Russia, raw materials extraction and trade, development and energy efficiency.
(industrial ecology essays)

266
The Handbook of Environmentally Conscious Manufacturing: From Design & Production to Labeling & Recycling. Robert E. Cattanach (William Mitchell College of Law) *et al.* Burr Ridge IL: Irwin Professional Publishing, Jan 1995/300p(8x11")/$75.00.

Manufacturers need not harm the environment to succeed in business. Indeed, market incentives and legal controls increasingly offer rewards for those able to develop cleaner processes and products, and penalties for those unable to eliminate waste. Growing internationalism has also added incentives with emerging standards such as ISO 14000, the environmental counterpart to ISO 9000 standards. Topics include product life cycle, marketing the environmentally conscious product, regulatory evolution, compliance assessments, establishing a proactive corporate management structure, developing an environmental business plan, evaluating existing products and processes, developing ECM process options, new product/process design, the greening of the market, challenges facing green marketers, US and international regulations and guidelines for environmental marketing, and implementing environmentally conscious marketing. Appendices offer a sample pollution prevention plan, a sample environmental management program survey, a hazardous waste/materials plan, guides for environmental marketing claims, responsible environmental advertising, labeling guidelines, and resources and contacts.
(environmentally conscious manufacturing)

267
Greener Marketing and Advertising: Charting a Responsible Course. Edited by Robert Rehak (Sr. VP/Creative Dir., Ogilvy & Mather, Houston). Emmaus PA: Rodale Press, Sept 1993/242p(8x11")/$24.95pb.

Advertising is no better or worse than any other tool; its effect depends on the people who wield it. Advertising can be a powerful way to address environmental problems by changing mass consumer habits, raising awareness of environmental issues, and helping to redefine acceptable standards of environmental behavior. This book begins with six principles issued by the Environmental Marketing and Advertising Council: 1) integrate environmental considerations as part of the advertising message; 2) make sure environmental claims are completely honest; 3) don't overpromise; 4) offer customers hope for solutions and invite their participation; 5) aim for creative leadership by moving people to see products and services in a fresh perspective; 6) produce advertising in an environmentally responsible way. These principles are illustrated with hundreds of examples from the Council's 1993 Environmental Communication Awards. Examples tell customers about a company's environmental record, emphasize quality and durability, promote environmentally sound practices, encourage conservation and recycling, show people finding happiness in simplicity, encourage alternative transportation and responsible consumption, show the relationship between population growth and environmental quality, stress minimal packaging, use recycled materials, employ global thinking, etc. Also includes environmental advertising guidelines and regulations adopted by the US Federal Trade Commission, Canada, 11 states, and the International Chamber of

Commerce. Concludes with statements by leaders of various environmental organizations on how to make both marketing and advertising activities more environmentally responsible. **(greener advertising)**

268

Trade and Sustainable Development: A Survey of the Issues and a New Research Agenda. David Runnalls (Environment and Sustainable Development Program, Institute for Research on Public Policy, Montreal) and Aaron Cosbey (IISD). Winnipeg, Manitoba: International Institute for Sustainable Development, Nov 1992/130p(8x11")/Can$25.00.

Trade is a powerful force, with the potential to frustrate or foster sustainable development on a global scale. In 1990, international trade flows amounted to US$4.3 trillion. Negotiations are underway that will have substantial impact on these flows. But the debates and discussions are fragmented, and the various players often do not address the same issues. The concept of sustainable development is proposed as a framework within which the various positions can be understood in common terms of economy, environment, and human well-being. Chapters discuss the history of the trade-environment issue, effects of trade on the environment, effects of environmental policies on trade (due to uneven international standards for products and production), and obstacles to sustainable development. Concludes with a proposed new research agenda on issues such as harmonizing standards, pollution havens, tough environmental regulation and competitiveness, measuring the ecological content of trade, required institutional change, and the potential of green consumerism. Bibliography contains some 700 items, in six sections: general issues, international organizations, environmental policies, trade regimes and sustainable development, sectoral issues, and theory and methodology. [ALSO SEE: **Trade, Environment & Competitiveness: Sustaining Canada's Prosperity**. Edited by John Kirton and Sarah Richardson (Ottawa: National Round Table on the Environment and the Economy, Dec 1992/391p/free), the proceedings of a November 1991 conference in Toronto. **(sustainable development and trade)**

269

Trade and the Environment: Law, Economics, and Policy. Edited by Durwood Zaelke, Paul Orbuch, and Robert F. Housman (all Center for International Environmental Law, Washington). Washington: Island Press, Dec 1993/318p/$49.95;$24.95pb.

According to Tom Wathen (The Pew Charitable Trusts), "trade policy is becoming the forum for deciding whether the global economy will be based on the ethics of sustainable development or on laissez-faire market principles." Trade and environment issues sprung with little warning to the forefront of public policy debate around 1990. The NAFTA package moves several steps forward in the trade/environment area, but it does not reconcile all issues. The editors assert that "The list of trade challenges to environmental protections continues to grow, as does the list of potential environmental trade barriers." Essays discuss trends since 1972 in the trade/environment nexus, managing through prices and despite prices, harmonizing international and national environmental regulations, the failure of **Agenda 21** to consider trade issues (along with a proposed action agenda for trade policy reform), a business view of the Polluter Pays Principle (it should be applied so as to avoid distortions in trade and investment), foreign investment and environmental protection in the Third World, a steady-state critique of free trade (by Herman E. Daly), trade and the environment as a false conflict (by Jagdish Bhagwati), complementarities between trade and environmental policies (by Robert Repetto), environmental strategies for agricultural trade, and rules for harmonization (inevitable in an increasingly interdependent world). **(environment and trade issues)**

270

Freer Trade, Protected Environment: Balancing Trade Liberalization and Environmental Interests. C. Ford Runge (Prof of Agricultural and Applied Economics, U of Minnesota). NY: Council on Foreign Relations Press, April 1994/146p/$17.95pb.

A report derived from the CFR-sponsored Study Group on Trade and the Environment, which held eight meetings in the 1992-1993 period. The underlying premise is that environmental and trade policies are equally important, and deserve to be considered on equal terms. The two interests can often be pursued jointly, in a manner that protects the benefits that expanded trade can bring, while ensuring that these benefits do not come at the expense of the environment. A new World Trade Organization has resulted from the successful Uruguay Round of trade negotiations. A parallel World Environmental Organization (WEO) is proposed, to advance environmental improvements that are coordinated with increases in market access and expanded trading opportunities. The trilateral Commission on Environmental Cooperation, emerging as a side agreement to NAFTA, provides an "enormously important" prototype for WEO. The WEO must have the capacity to induce nations to upgrade enforcement of environmental laws, provide opportunities for citizen and expert involvement (through a Multilateral Commission on Environment as part of WEO), work with the World Bank and other lending agencies to fund environmental projects, identify trade measures that threaten the environment, and help develop environmental policies that are least burdensome to trade expansion.

(World Environmental Organization proposed)

271

Costly Tradeoffs: Reconciling Trade and the Environment. Hilary F. French (Senior Researcher, Worldwatch Institute). Worldwatch Paper 113. Washington: Worldwatch Institute, March 1993/74p/$5.00.

Trade is neither inherently good or bad. It can bring greater prosperity and improved quality of life, if properly managed, or it can be an engine of enormous destruction. "As the 21st century approaches, it is now clear that the greater economic prosperity to be gained from unrestrained free trade is no longer justifiable—or even possible—as an end in itself." Reconciling the demands of global commerce with the equally compelling needs of a troubled biosphere will not be easy. Global trade issues provide vexing contradictions and tradeoffs at every turn. For example, the free trade zone on the US-Mexican border allows polluting industries to escape enforcement, yet this mobility of investment may give poor countries more revenue to invest in cleaner industry. Unfortunately, in the absence of any common regulation, restrictions imposed in one country can lead to massive degradation elsewhere. "Perhaps the single most important reform needed to place world trade on a more sustainable footing is widespread internalization of environmental costs through tax policy and other measures." For example, if wood from primary forests was priced to reflect the loss of biodiversity suffered when it is harvested, timber from

sustainably managed stands would be more competitive in national and global markets. But this poses a mammoth political challenge, even at the national level. Costs must be internalized not only within nations but between them. "The demands of world trade are thus likely to spur movement toward internationally-agreed measures to internalize costs—be they taxes, regulations, or labeling programs." This means political integration must follow economic union to level the playing field for trade.
(world trade and uniformly internalized costs)

272
Tourism and the Environment: A Sustainable Relationship? Colin Hunter (U of Huddersfield) and Howard Green (Leeds Metro U). London & NY: Routledge, May 1995/212p/$75.00.

Examines the impacts of tourists and the tourism industry on environmental resources, and explains key concepts and techniques for better environmental management of tourism. Chapters discuss the nature of tourism and its huge growth, the environmental impacts of tourism, the emerging view that tourism must become more environmentally conscious, tourism carrying capacity, sustainable tourism development, land use planning for tourism, and environmental impact assessment. A paradox is at the heart of all major tourism developments: that tourism will destroy or degrade the environmental resources that attract tourists in the first place; the tourism industry thus should try to maintain environmental quality. The concept of "sustainable tourism development" not only enables integration of economic/ social/ecological concerns, but encourages a more holistic and multi-disciplinary approach to tourism research.
(sustainable tourism development)

273
Nature Tourism: Managing for the Environment. Edited by Tensie Whelan (VP for Conservation Information, National Audubon Society). Washington: Island Press, Nov 1991/223p/$34.95;$19.95pb.

There are intense economic pressures on people in poor countries to overexploit their natural resources. If the environment is to be saved, people must have alternatives to destruction. If done well, ecotourism can be a sustainable and relatively simply alternative. Tourism is already the second largest industry in the world (comprising 7% of world trade in goods and services), and the World Tourism Organization projects that it will be the world's largest industry by 2000. Ecotourism is a rapidly developing sector of this industry. Most ecotourists are from Europe, North America, and Japan; favored activities are hiking, bird watching, nature photography, wildlife safaris, camping, mountain climbing, fishing, river rafting/canoeing, and botanical study. The most popular destinations are Nepal, Kenya, Tanzania, China, Mexico, Costa Rica, and Puerto Rico. But the ecotourism industry is still in its infancy. The 7000 "protected areas" around the world are often not protected in fact because of lack of funds and local support. Most parks lack entrance fee systems and basic infrastructure. Chapters discuss ecotourism in Kenya and Costa Rica, tourism in the Greater Yellowstone Ecosystem, ecotourism on family farms and ranches in the American West, costs and benefits of nature tourism, local participation in ecotourism projects, recommendations for planning and management (for various ministries, park managers, and tour operators), and a proposed Ecotourism Society to promote "clean" ecotourism and appropriate ethics. **(promoting ecotourism)**

5. Agriculture

*274
Alternative Agriculture. National Research Council, Board on Agriculture. Washington: National Academy Press, 1989/448p/$29.95;$19.95pb.

Report of the Committee on the Role of Alternative Farming Methods in Modern Production Agriculture, stimulated by environmental problems caused by high-tech agriculture: water pollution, soil erosion, pest resistance topesticides, pesticide residues in food, and purchased inputs as a significant part of total operating costs. Alternative agriculture utilizes natural processes, reduces the use of harmful off-farm inputs, and seeks long-term sustainability. It is not a single system of farming practices, but a spectrum of systems known as biological, low-input, organic, regenerative, or sustainable. It utilizes practices such as integrated pest management, crop rotations, management systems to control weeds, and conservation tillage. The major findings of the Committee: 1) farmers successfully adopting these systems generally derive significant sustained economic and environmental benefits; 2) a wide range of Federal policies have served to work against environmentally benign practices and adoption of alternative systems; "government policies that discourage the adoption of alternative practices must be reformed;" 3) a systems approach to research is essential to the progress of alternative agriculture; 4) to make wider adoption possible, farmers must receive information and technical assistance in developing new management skills. [**NOTE**: The authoritative report to promote what is likely to become known of as "sensible" or "post-modern" agriculture.] **(policies for sustainable agriculture)**

275
Sustainable Agriculture, John P. Reganold (Washington State U), Robert I. Papendick (USDA, Pullman WA), and James F. Parr (USDA, Beltsville MD), *Scientific American*, 262:6, June 1990, 112-120.

Since WWII, US farms become highly mechanized and specialized, and heavily dependent on fossil fuels, borrowed capital, chemical fertilizers, and pesticides. "Today, the same farms are associated with declining soil productivity, deteriorating environmental quality, reduced profitability, and threats to human and animal health." A growing cross-section of America is questioning conventional agriculture, and many are seeking alternative practices. Sustainable agriculture embraces several variants of nonconventional agriculture that are often called organic, alternative, regenerative, ecological, or low-input. But organic is not necessarily sustainable. "For a farm to be sustainable, it must produce adequate amounts of high-quality food, protect its resources, and be both environmentally safe and profitable." Sustainable agriculture addresses many serious problems afflicting US and world food production: high energy costs, groundwater contamination, soil erosion, loss of productivity, low farm incomes, and risks to human health and wildlife habitats. It is not so much a specific strategy as a system-level approach to understanding the complex interactions within agricultural ecologies. In 1980, the USDA estimated about 1% of America's farmers were practicing nonconventional agriculture, most of which could now be termed sustainable. Today, some experts estimate that the figure may have doubled or tripled. [**NOTE**: A brief and authoritative overview of what appears to be an inexorable and very positive trend.] **(sustainable agriculture spreading)**

276
Farming in Nature's Image: An Ecological Approach to Agriculture. Judith D. Soule and Jon K. Piper (The Land Inst., Salina KS). Foreword by Wes Jackson. Washington: Island Press, Jan 1992/286p/$34.95;$19.95pb.

Builds on the research and thinking of Wes Jackson's **New Roots for Agriculture** (Friends of the Earth, 1980; **FS Annual 1980-81** #2206), one of the first calls for a sustainable agriculture. The ecological crises of modern agriculture include increasing soil erosion (and a focus on productivity that disguises this fact), resulting pollution of water supplies and the air, loss of genetic diversity, energy dependency, depletion of water resources, and chemical contamination of food and water supplies. A case is made for an ecological and sustainable agriculture that uses natural systems as models. It would incorporate integrated pest management, intercropping, and conservation tillage. Barriers to culturing sustainability include the necessarily long-term and interdisciplinary nature of needed research, competition with biotechnology for funding, and corruption of the term *sustainable development* as it is embraced by the political mainstream (it is coming to mean development that means sustainable material growth). Humility is an essential ingredient in developing sustainable agriculture: a willingness of scientists and farmers to listen to each other, to listen across disciplinary bounds, and to listen to nature. **(sustainable agriculture)**

277
Drought Follows the Plow: Cultivating Marginal Areas. Edited by Michael H. Glantz (National Center for Atmospheric Research, Boulder). NY: Cambridge U Press, Sept 1994/197p/$39.95;$19.95pb.

The 1980s marked the beginning of the "Age of Environmental Enlightenment," as worldwide interest in the environment greatly increased, based on many environmental changes increasingly perceived by the public as global threats. The 1980s also witnessed a sharp increase in concern about global warming. Many researchers speculate that global warming will probably increase drought-related crop failures in areas already subject to drought, and that dry areas will get drier and wet areas wetter. Yet farmers and pastoralists are moving into regions that are less fruitful than the lands from which they emigrated, often as an option to migrating from rural to urban areas.

In the 19th century, there was a belief that "rain follows the plow"—that tree planting, land clearing, cultivation, and fires could alter atmospheric processes so as to increase local and regional rainfall. There is now evidence to suggest that the opposite process is occurring around the globe. "Drought follows the plow" draws attention to the land use changes that are taking place—for the worse. Encroachment of agricultural activities into marginal areas is a process that may now be accelerating due to population pressures. As marginal lands in arid and semi-arid areas are increasingly put under the plow, drought will likely increase, even under normal meteorological conditions. We are thus likely to hear more about droughts in the future, as populations increase and social groups become marginalized. "One could [also] effectively argue that the stork will eventually outrun the plow, as originally suggested by Malthus." [**NOTE**: Perhaps limited by the guiding metaphor. What if there is little or no plowing in the future, but minimum tillage instead, along with new irrigation methods and plastic mulch? (See #384.)]
(drought to increase in marginal areas)

278
World Soil Erosion and Conservation. Edited by David Pimentel (Cornell U; Chairman, IUCN Study Group for World Soil Conservation). Foreword by Martin W. Holdgate (IUCN Director General). NY: Cambridge U Press, 1993/349p/$79.95.

More than 97% of the world's food comes from the land, with an ever-increasing portion from steep land in tropical climates. Soil erosion is one of the most serious environmental problems in the world today. Severe erosion occurs in most of the world's agricultural regions, and the problem is growing as more marginal land is brought into production. It is proceeding largely uncontrolled because the amount of soil that erodes with each rain or windstorm is almost imperceptible. Erosion affects crop productivity by reducing the availability of water—the most harmful effect—as well as nutrients and organic matter, and, as the topsoil thins, by restricting rooting depth. In addition to reducing land productivity, sediments from erosion fill rivers, lakes, and reservoirs, significantly reducing their usefulness. Many effective conservation technologies are available for farmers in all nations, but gains may be offset by other factors, e.g.: the US Soil Conservation Service successfully encouraged farmers to adopt contour planting and no-till technologies; at the same time, farmers abandoned crop rotations and removed hedgerows and tree shelter-belts. Thus, "in the US overall the total amount of soil erosion has remained about the same since 1935." Yet, there is a high payoff for prevention: "$1 invested in soil and water conservation on the farm watershed may save $5-$10 in dredging rivers, building levees, and restoring property damaged by floods." Essays describe soil erosion and conservation efforts in West Africa, Ethiopia, China, India, Australia, Argentina, UK, Poland, and the humid tropics. The principal method for controlling soil erosion is to maintain adequate vegetative cover; concluding essay describes using vetiver grass hedges to solve many soil and water problems. **(soil erosion worldwide)**

279
Sustainable Agriculture and the Environment in the Humid Tropics. National Research Council. Washington: National Academy Press, March 1993/702p/$49.95.

A new attitude of stewardship and sustainable management is required if our global resources are to remain productive. Nowhere is this need more urgent than in the world's humid tropics, where many people will continue to rely on the resource base to meet their needs. Agricultural practices in tropical regions are frequently unsustainable because the capacity of the land is rapidly exhausted—a problem exacerbated by the pressure of poverty. Faced with declining yields, farmers in many areas of the humid tropics typically seek new forestlands to clear for crop production, in turn contributing to further environmental degradation." The 15-member NRC Committee on Sustainable Agriculture in the Humid Tropics arrived at three major findings: 1) degraded lands can be found that have the potential to be restored; 2) many successful land use systems involve integrated approaches to farming and forestry; 3) some locales are successfully shifting from economic growth based largely on forest harvest to a more diversified economy; gains from further conversion of forests in the humid tropics are increasingly marginal. Extensive country profiles are provided for Brazil, Indonesia, Ivory Coast, Malaysia, Mexico, The Philippines, and Zaire. **(farming and forestry in the tropics)**

280
Agricultural Policy and Sustainability: Case Studies from India, Chile, the Philippines and the United States. Edited by Paul Faeth (WRI). Washington: World Resources Institute, Sept 1993/113p/$14.95pb.

Agriculture in many areas of the world is unsustainable. Conventional economic indicators used in the agricultural sector, however, do not include measures of environmental damage, and such evidence rarely makes its way into economic decisionmaking. Accounting systems for business and government include a capital consumption allowance, but accounting practices in agriculture do not capture losses. Soil can be eroded, groundwater depleted or contaminated, wildlife poisoned, and reservoirs filled with sediment. But no depreciation allowance is applied, thus misrepresenting a decline in wealth as an increase in income. Six studies are presented here based on natural resource accounting, finding that "resource-conserving production practices can compete economically and financially with conventional practices when a more accurate picture of agricultural productivity is presented." Each case study shows how official policies presently work against sustainability. Policy proposals: 1) eliminate subsidies that encourage natural resource degradation or depletion; 2) make agricultural pollution subject to the "polluter pays" principle; 3) eliminate agricultural support programs that distort economic signals to farmers and thus create commodity surpluses; 4) give much higher priority to research into sustainable farming practices; 5) drastically improve physical monitoring of agriculture's environmental impacts in LDCs; 6) insure that national economic indicators of agriculture reflect depletion and degradation of natural resources.

(agricultural accounting questioned)

281

Growing Green: Enhancing the Economic and Environmental Performance of U.S. Agriculture. Paul Faeth (WRI). Washington: World Resources Institute (800/822-0504), April 1995/81p(8x11")/$16.95 + $3.50 p&h.

In the last 10 years, the US government paid farmers more than $135 billion in direct income supports. This money supported the prices of a few commodities that occupy most of America's cropland. The benefits to producers are clear, but the benefits to taxpayers and consumers are not, for these commodity programs have substantial economic and environmental costs. A framework for considering sustainability is provided, along with analysis of several scenarios, demonstrating that "a major reduction in agriculture's impact on the environment is both possible and economically advantageous." A win-win outcome is possible: 1) uncouple income support from commodity programs by increasing the percentage of commodity-program land on which farmers can plant whatever they choose; as this "flex rate" increases, costs decline (saving taxpayers about $5 billion a year) and environmental performance improves; 2) use commodity program savings to support farm income directly (providing support directly to the people who need it makes much more economic and environmental sense); 3) support environmental objectives through targeted acreage-diversion programs and agricultural conservation programs (program savings can provide support for farmers to make the transition to more sustainable production practices); 4) increase funding for agricultural research, extension, and technical support enabling conservation of natural resources.

(cost-saving sustainable agriculture)

282

Sustainable Agriculture in the American Midwest: Lessons from the Past, Prospects for the Future. Edited by Gregory McIsaac (U of Illinois) and William R. Edwards (Illinois Natural History Survey). Urbana IL: U of Illinois Press, Sept 1994/291p/$32.95.

Essays on the dynamics of past changes in the natural environment and agriculture of the American Midwest, in order to assess and suggest what a more sustainable agriculture could or should be. Topics include sustainable practices of the Amish and Native Americans, cultural beliefs about sustainable agriculture (contrasting the stewardship of German Yeomen vs. Yankee Entrepreneurs in Illinois), sustainable cropping systems, insects and pest management, implications of warmer future climate for Midwest agriculture, and a proposal for large-scale biomass energy farming to cope with America's excess production capacity. In the introduction, McIsaac notes the problem of conflicting definitions of "sustainability," and disagreement about what can and should be sustained, and for how long. [ALSO SEE: **Food for the Future: Conditions and Contradictions of Sustainability**, edited by Patricia Allen of UC-Santa Cruz (NY: Wiley, 1993/328p/$39.95pb), concluding with the problem of forging broad agreement on the meaning of sustainable agriculture.] **(sustainable agriculture essays)**

283

Future Harvest: Pesticide-Free Farming. Jim Bender (Cass County NE). Our Sustainable Future, Vol 5. Lincoln NE: U of Nebraska Press, Feb 1994/159p/$21.00.

Bender returned to Nebraska from graduate studies in the humanities in 1975, to help operate a four-generation family farm of 642 acres. His book offers a case study of transition to stewardship agriculture, arguing that many of the techniques and strategies are applicable well beyond the Midwest. The cropland on Bender's farm has been free of herbicides, insecticides, and fungicides since 1980, and it has received only alternatives to synthetic fertilizers since 1987. A wide array of conservation structures are in place, and conservation practices followed. The key to successful transition is a gradual, rather than immediate and complete, conversion. Unfortunately, trends in contemporary agriculture (separation of livestock from farms, increasing farm size, separation of land ownership from farming, the increase in part-time farming) and myths surround pesticide-free farming (that alternative systems revert to a more primitive agriculture, and that chemicals are more convenient) create a context that inhibits change. Moreover, Federal farm programs penalize crop rotation, agricultural research has largely ignored organic methods of cultivation, and opponents of alternative agriculture make compelling but questionable assertions (e.g., that the key to using pesticides safely is simply following label directions). Compared to conventional practices, organic farming is more conducive to soil conservation, no less convenient than chemical weed control, reduces risk, and need not be less productive. [ALSO SEE: other volumes in the Our Sustainable Future series: 1) **Ogallala: Water for a Dry Land** by John Opie; 2) **Building Soils for Better Crops** by Fred Magdoff; 3) **Agricultural Research Alternatives** by William Lockeretz and Molly Anderson; 4) **Crop Improvement for Sustainable Agriculture** edited by M. Brett Callaway and Charles A. Francis.]

(transition to organic farming: case study)

284

Harvest of Hope: The Potential of Alternative Agriculture to Reduce Pesticide Use. Jennifer Curtis. NY: Natural Resources Defense Council, May 1991/124p (8x11")/$19.95pb.

The widespread use of agricultural chemicals has serious environmental, social, and economic costs. For example, nitrate (one of the most common contaminants of groundwater) is present in 52-57% of community and private wells nationwide. Various state and Federal laws have been enacted to protect water quality, but they have generally failed to address agriculture's role in polluting water supplies. Source reduction, as in other industries, is a practical strategy for preventing the environmental problems associated with agriculture. Alternative (or sustainable) agricultural practices hold the greatest potential for reducing use of pesticides and fertilizers. Barriers to alternative agriculture include underfunding of research, inadequate dissemination of information, subsidies that penalize crop rotation and promote surplus production, unnecessary cosmetic criteria, lack of incentives for water conservation that could facilitate reductions in pesticide use, and pesticide prices that fail to include externalities (the costs imposed on human health and the environment). Policy recommendations are made for agricultural research, Federal farm programs, Federal and state marketing policies, improved development and approval of biological pest control methods, and pricing of water and agricultural chemicals. **(sustainable agriculture)**

285

The Pesticide Question: Environment, Economics, and Ethics. Edited by David Pimentel (Prof of Entomology, Cornell U) and Hugh Lehman (Prof of Philosophy, U of Guelph). NY: Chapman & Hall (div of Routledge), Feb 1993/441p/$45.00.

Worldwide, about 2.5 million tons of pesticides are applied each year, with a purchase price of $20 billion. In the US, about 500,000 tons of 600 different types of pesticides are used annually at a cost of $4.1 billion, saving about $16 billion in US crops (pests still destroy 37% of all potential food and fiber crops). The indirect environmental and social costs of pesticide use, however, are at least $8 billion a year, perhaps much greater. Pesticides have contributed to impressive agricultural productivity, but have also caused serious health and environmental problems. A growing number of scientists and government officials have been viewed as primarily concerned with promoting commercial interests rather than the public welfare. Serious ethical investigations of policies related to pesticide use must consider costs and risks as well as benefits. These papers discuss the impact on the environment, pesticide use and "cosmetic standards" for foods, health effects, social implications of reducing agricultural chemical use in the US, Swedish pesticide policy, government policies encouraging US pesticide use, pesticide usage (stabilized since the early 1980s), the dangers of Alar on apples, growing public concern over pesticides, ethical choices in entomology. and a reprint of an important essay by Pimentel *et al.* **(pesticide costs and benefits)**

286

Environmental and Economic Effects of Reducing Pesticide Use, David Pimental (Prof of Insect Ecology, Cornell U) et al., *BioScience*, 41:6, June 1991, 402-409.

American farmers use an estimated 700 million pounds of pesticides annually (69% herbicides, 19% insecticides, and 12% fungicides), costing about $4.1 billion. Dollar returns for the direct benefits to farmers have been estimated to range from $3 to $5 for every $1 invested in use of pesticides—but these figures do not reflect indirect costs such as human pesticide poisonings (about 20,000 per year), reduction of fish and wildlife populations, livestock losses, destruction of susceptible crops and natural vegetation, honeybee losses, destruction of natural enemies, evolved pesticide resistance, and creation of secondary pest problems. A conservative estimate of these costs is $2.2 billion annually, and the actual cost is probably double this amount. The authors confirm previous reports that it is feasible to reduce pesticide use by 50% with no decline in crop yield. Total price increase in purchased food due to the extra cost of alternative controls is estimated at only 0.6%, and the public is probably willing to pay this amount. The public would also accept some reduction in cosmetic standards if it meant less pesticide contamination of food. Denmark, Sweden, and the Netherlands already have programs to reduce pesticide use by 50% in the next few years. [ALSO SEE: ***Research on Biological Pest Control Moves Ahead*** (*Science*, 12 April 1991, 211-212), reporting that "researchers are confident that biocontrols are at last poised to make significant inroads in the pesticide market."] **(feasibility of lower pesticide use)**

287

The Pesticide Hazard: A Global Health and Environmental Audit. Compiled by Barbara Dinham (The Pesticides Trust & PAN). London and Atlantic Highlands NJ: Zed Books, Feb 1993/228p/$55.00;$19.95pb.

Rachel Carson alerted the world 30 years ago to the environmental hazards of pesticides. Since then, new pesticides have been developed that are less persistent in the environment, but which bring health hazards to pesticide users, particularly in Third World countries. In 1985, the UN's Food and Agriculture Organisation introduced an International Code of Conduct on the Distribution and Use of Pesticides, and in 1989 it incorporated provisions for governments to refuse to consent to hazardous imports. Still, the World Health Organisation estimates that there are a minimum of three million acute severe cases of pesticide poisonings and 20,000 unintentional deaths each year. This excludes cases of chronic pesticide poisoning ("a 1990 study estimated that 25 million Third World agricultural workers are acutely poisoned every year"). This book provides evidence, gathered by NGOs and researchers around the world, of health and environmental problems of pesticide use that continue in spite of the FAO Code. Many of these NGOs have collaborated through the Pesticides Action Network, an international coalition of over 300 groups and individuals who oppose misuse of pesticides and support safe and sustainable pest-control methods. Chapters discuss issues at stake and country reports from Brazil, Costa Rica, Ecuador, Paraguay, Venezuela, Egypt, South Africa, India, and Malaysia. **(pesticides in Third World)**

288

Beyond Beef: The Rise and Fall of the Cattle Culture. Jeremy Rifkin (President, Foundation on Economic Trends, Washington). NY: Dutton (Penguin Books), March 1992/353p/$21.00. [Brief versions in *The New York Times* Op-Ed (23 March 1992, A17) and *The Utne Reader* (Mar-Apr 1992, 96-109).]

There are currently 1.28 billion cattle populating the Earth, taking up nearly 24% of the planet's landmass. Their combined weight exceeds that of the world's human population. Cattle and other livestock consume about one-third of the world's total grain harvest, while as many as a billion people suffer from chronic hunger and malnutrition. Cattle are castrated, drugged, and made docile. They spend long hours at the feed troughs consuming an array of feeds saturated with herbicides. The modern cattle complex is pervasive; its activities have had a major effect on the deterioration of the world's environment. This cattle complex is "a new kind of malevolent force in the world"—*cold evil* inflicted at a distance, concealed by layers of technological and institutional garb, bound by rational organizing principles with only market forces and utilitarian goals to guide decisions. Chapters describe cattle and the making of Western civilization (the eating of fatty beef was a privilege of the rich in England, and soon became an aspiration of the middle and working classes), the conquering of the US West to meet the demand for beef and hides, the industrialization of beef, the sociology of fat, the global environmental crisis (cattle raising destroys rain forests and rangelands, consumes scarce water supplies, emits methane that warms the planet), and the consciousness of beef-eating cultures (the ubiquitous hamburger represents the final deconstruction of modern meat—a testimonial to the utilitarian ethos).

Concludes that "to go beyond beef is to transform our very thinking about appropriate behavior toward nature." The elimination of beef will be accompanied by an ecological renaissance, a grand restoration of nature on every continent. In the rich countries, the artificial protein ladder will begin to collapse, and dramatically reduce the incidence of heart disease, cancer, and diabetes. "Liberating the land to grow grain for human beings will trigger a large-scale human migration out from the crowded urban shanty towns back to the countryside." Moving beyond the beef culture is a sign of our willingness to reconstitute ourselves, to make ourselves whole—essential elements of a new postmodern sensibility of earth-centered awareness. [**NOTE**: Original, wide-ranging, one-sided, and *very* provocative.] **(beef culture as major environmental evil)**

289

Microlivestock: Little-Known Animals with a Promising Economic Future. National Research Council, Board on Sci/Tech for International Development (Noel D. Vietmeyer, Study Director). Washington: National Academy Press, 1991/449p/$29.95pb. (Brief version in *The Futurist*, May-June 1992, 37-40.)

"Microlivestock" includes species that are inherently small (rabbits, poultry) and microbreeds of cattle, sheep, goats, and pigs that are less than half the size of most common breeds. This survey of more than 300 animal scientists in 80 countries reports on miniature animals with a promising future—animals that are seldom considered in the broad picture of livestock development. Lack of space and time precludes discussion of small aquatic species and invertebrates (edible insects, snails, worms, turtles, bats), which could be in future volumes. Chapters discuss microbreeds (which may be the only practical livestock in many settings), poultry (including wild species that might be domesticated), rabbits (which can utilize almost any type of edible vegetation and collectively produce as much meat per unit of forage as large livestock), tasty and clean-living rodents (agouti, capybara, giant rat, grasscutter, guinea pig, paca, etc.), deer and antelope (species that are smaller than most dogs), lizards (iguanas are highly prized in Central America), and honeybees (present almost everywhere). Concludes that such small animals are likely to become increasingly important, as human populations continue to rise and space for growing forage decreases—especially in urban areas, where land is at a premium. Raising personal livestock on weeds and table scraps in cages beside the house or boxes under the bed will, in many instances, get quality protein to the most poverty stricken more effectively than raising large livestock on pastures. [ALSO SEE: earlier articles by Vietmeyer on "livestock for the landless" (*Ceres*, March-April 1984; **FS Annual 1985** #6745) and on "lesser-known plants of potential use in agriculture and forestry" (*Science*, 13 June 1986; **FS Annual 1986** #7621).]

(microlivestock: vast potential)

290

Neem: A Tree For Solving Global Problems. National Research Council, Board on Sci/Tech for International Development (Noel D. Vietmeyer, Study Director). Washington: Natl. Academy Press, 1992/141p/$19.00pb.

Native to India and Burma, neem is a botanical cousin of mahogany. It can grow up to 30m tall, and bears masses of honey-scented white flowers like a locust. Its dense, complex, evergreen foliage resembles that of walnut or ash, and its fruits look much like olives. It grows vigorously in difficult sites, and has been introduced into Saudi Arabia, the Caribbean (it is already a major tree species in deforested Haiti), and many countries of Central and South America. Neem has many benefits: 1) it often grows rapidly, and can be cut for timber after just 5-7 years; 2) weeds seldom affect growth (in fact, the trees themselves may become "weeds"); 3) extracts from its bitter seeds and leaves have remarkable powers for controlling insects that could usher in a new era of safe, natural pesticides (neem materials can affect more than 200 insect species, but appear to have little or no toxicity to warm-blooded animals); 4) some neem products may benefit human health (most people in India have bright and healthy teeth attributed to neem "chewsticks," and neem is now used in popular toothpaste in Germany and India); 5) neem may play a part in controlling population growth, in that materials from the seeds have contraceptive properties; 6) it is promising for planting in rural areas now suffering from desperate fuelwood shortages, useful as a windbreak, and exceptional as a city tree for shade; 7) large-scale neem production promises to help alleviate deforestation, desertification, soil erosion, and perhaps even global warming. Concludes that, despite some unresolved questions, enough is already known that exploiting certain neem uses can begin immediately. An orderly creation of plantations and markets could see neem rise to become one of the most widely grown trees in the world. Centering rural industries in the warmer parts of the world around neem extraction could be a key to sustainable, self-reliant, and decentralized rural development. **(neem trees: vast potential)**

291

The Greening of Agricultural Policy in Industrial Societies: Swedish Reforms in Comparative Perspective. David Vail (Prof of Economic, Bowdoin College), Knut Per Hasund (Swedish U of Ag. Sciences), and Lars Drake (SUAS). Ithaca NY: Cornell U Press, June 1994/300p/$45.00.

As international interest about sustainable agriculture intensified in the 1980s, Sweden, with little fanfare, implemented a broad range of green initiatives. Several were quite innovative: taxes on polluting inputs, financial assistance for converting to organic farming, measures to improve the well being of farm animals (responding to demands of animal rights groups that livestock be reared in "more natural conditions"), prototype development of renewable agricultural bioenergy, and safe recycling of municipal wastes on farmland. In 1990, Sweden quietly reformed its heavily regulated agricultural regime, becoming the first highly protectionist nation to set a firm timetable for deregulation of agricultural markets. The new food policy has goals of "minimizing the negative environmental effects of agriculture," using directly targeted instruments such as reducing nitrogen and biocide use and encouraging economically submarginal land to be abandoned or afforested. (The authors assess the environmental effects as "pale green.") Swedish greening is well in advance of most other nations in several areas, although other countries have influenced Swedish policy. "The industrial nations are still at a relatively early stage in designing instruments to internalize agriculture's positive and negative externalities. None have taken the qualitative leap beyond piecemeal greening, to articulate a far-sighted and holistic vision of sustainable agriculture or to translate such a vision into policy." Even the array of Swedish initiatives was not inspired by a clear, holistic vision of the agricultural future, nor do they sum up a coherent green strategy. But, as knowledge about which green measures are effective spreads internationally, further convergence toward the most effective policy is likely. And "there is certain to be a powerful political pressure for—and heated contention over—international harmonization of agro-environmental standards and policies." Sweden's agriculture is moving toward environmental goals, but the authors worry about the effect of liberal markets and new technology on the social and economic sustainability of farming. "An agriculture that was environmentally sustainable but had few farms and no farm communities would indeed be a pyrrhic victory." [**NOTE**: Thorough and sophisticated analysis of bellwether policies.] **(greening of Swedish agriculture)**

292

Sustainable Agriculture: The Research Challenge. Science Council of Canada. Report #43. Ottawa: Minister of Supply and Services, July 1992/46p(8x11"). Catalog #SS22-1002/43E.

The Canadian agriculture-food industry directly or indirectly employs 14% of Canada's workforce. The industry is being buffeted by shifting world trade patterns, rising costs, and mounting pressure over practices that lead to environmental degradation, resource depletion, and safety concerns. "There is a growing realization that for Canadian agriculture it can no longer be business as usual." Increasingly, agriculture must be integrated with environmental management. New policies and programs must be developed, and academic, government, and industry institutions related to agriculture must be recast. Recommendations include redesigning the research system to make it more responsive to changing priorities, more collaborative research and training (multidisciplinary programs with an emphasis on agricultural systems), improving the knowledge base (developing measures of soil and water quality, helping farmers collect data), developing physical and biological indicators, decoupling subsidies from specific production practices, redirecting subsidies so as to maintain the rural landscape, a massive public education program to alert consumers and farmers to the benefits of sustainable agriculture, school curricula related to modern agriculture and food production, developing codes of agricultural practice (and ways to measure compliance) to ensure that farms are operated in an environmentally sound manner, and mechanisms to help farmers and food processors enter rapidly emerging niche markets. **(sustainable agriculture in Canada)**

293

Sustainable Farming: Possibilities 1990-2020. Ottawa: Science Council of Canada (100 Metcalfe St), Sept 1991/97p/free. (Brief version published in *The Futurist*, Sept-Oct 1992, 33-37.)

The first of several discussion papers to be issued by the Science Council's Committee on Sustainable Agriculture. Scenarios are provided by **Marc Zwelling** (President, Vector Public Education, Toronto; Publisher, *John Kettle's FutureLetter*) on various paths to make sustainable agriculture a reality: **1A) Continuing Trends, Deregulation**: steadily reduced government support, growing farm specialization, more investment in aquaculture, more brand names on generic products, market dictates a slow change away from conventional agriculture; **1B) Continuing Trends, Re-Regulation**: growing consumer loyalty to green products, expanded government support for agriculture, agricultural chemicals phased out in the early 2000s, government dictates a slow change away from conventional agriculture; **2) All-Import Agriculture**: most farm land becomes too valuable to farm, assistance to farmers promotes alternative employment and relocation, new technology enables better food transportation and storage, farming largely disappears by 2020; **3) Desktop Agriculture**: public views food from traditional sources as unsafe, demand for meat reduced to near zero by 2020, aquaculture fails because of great demands on energy sources, imported food seen as unreliable because of threats by terrorists, consumers accept new non-food foods which taste better than the dirty food they replaced, almost everything is now produced in "farmories" (farm-factories) near major population centers, Canada reaches its environmental objectives because agriculture is not sustainable; **4) High-Tech/Export-Based Agriculture**: Canada becomes a world leader because its farmers are the first to learn really high-tech sustainable agriculture, pesticides and growth hormones outlawed, home testing kits allow consumers to evaluate food, massive topsoil reclamation and large-scale water diversion projects assure basic resources, share of national employment in all agribusiness rises from 15% in the 1980s to 20% in 2020; **5) Clean Agriculture**: regulators ban all unnatural inputs to the food chain, bottom-up rebuilding of the food chain begins in 2010, return of the small family farm, flourishing of backyard gardens and rooftop farming in cities, failure of biotech companies because products are deemed dangerous, increased cost of food requiring new government welfare programs. Each of the six scenarios and their probabilities are analyzed in seven essays by various experts. A summary is provided by Zwelling, who concludes that Scenario 1A is seen as most likely. The least that will happen is growing demands for environmentally-friendly farming in the 1990s, organic and other "pure" foods will have 20% of market share by 2010 and 30% by 2020, and hydroponic agriculture and aquaculture will take off in the 2010s. [**NOTE**: Rich and provocative scenarios showing several paths to safe food and sustainable production. ALSO SEE: **It's Everybody's Business: Submissions to the Science Council's Committee on Sustainable Agriculture** (Jan 1992), which shows that "although the principle of sustainable agriculture is broadly accepted as a long-term goal, opinions are divided on what the goal really means and how to achieve it."] **(paths to sustainable agriculture in 2020)**

6. Science and Technology

294

Science and Technology for Eight Billion People: Europe's Responsibility. Edited by Peter H. Mettler (Wiesbaden Polytechnic U). Adamantine Studies on the 21st Century. London: New European Publications & Adamantine Press Ltd, July 1995/333p.

EUROPERSPECTIVE I (Paris, 1987) presented various results of top-level European long-range forecasting studies. EUROPERSPECTIVE II (Belgium, 1991) enabled discussion of the future of European peoples and countries. The EUROPERSPECTIVE III conference, organized by the FAST (Forecasting and Assessment in Science and Technology) Programme of the Commission of the European Union, was held in Wiesbaden in June 1993, bringing the 8 billion people of the planet by 2020 into focus. The general policy recommendations of the conference are that the 8 billion people and a sustainable economy are everyone's responsibility, and that Europe's contribution could be decisive. The conference also proposed that the EU become more active in four S&T priority areas: human settlements and housing, communications, education, and the conversion of today's economy to a non-carbon basis.

A 70-page report, ***Living Together***, presents the results and recommendations from the FAST study **Global Perspective 2010: New Tasks for S&T,** carried out in 1991-1992 by a group of research institutes from various EU countries. It discusses: 1) the present global S&T agenda oriented to the needs of the rich countries; 2) alternative world scenarios for the next 30 years (apartheid world, survival world of extreme fragmentation, pax triadica in a reasonably stable world order, sustainable global integration, a multi-regionalized world, universal GATTism); 3) 47 critical areas for application of S&T (boiled down to 16 crucial interrelated areas); 4) 25 barriers and obstacles to applying S&T to the target of 8 billion people; 5) protocols for action: a World Forum on S&T, the shelter initiative, reforming global financial institutions, etc. Also includes documents from Canada (the International Development Research Centre on science for development), India (on science, population, and development), The Japan Society for Technology (towards building a harmonious global society), the US Carnegie Commission on Science, Technology and Government, and the Third World Academy of Sciences. [**NOTE**: A great book title, with some interesting ideas; regrettably, the book is poorly produced and remarkably confusing in its organization.] **(EUROPERSPECTIVE III conference)**

295

Perspectives on the Role of Science and Technology in Sustainable Development. Office of Technology Assessment, US Congress. Washington: USGPO (202/512-1800), Sept 1994/101p/$7.00. (S/N 052-003-01396-7)

"Sustainable development (SD) has emerged as a new goal for international development," driven by industrial development patterns, resource degradation and depletion, widening economic gaps, and resource conflicts. The role of sci/tech in SD is viewed in a new way as well, with an emphasis on technologies for empowerment and for environmental sustainability, and a focus on whole systems. The report discusses five paradigms of environmental management development, challenges in SD (intergenerational equity, substitution of resources, uncertainty and irreversibility), measuring SD progress, global trends and consumption patterns, developing countries and the world economy, and various areas of needed technology: ***1) Agriculture***: technologies to improve use of local natural resources, improve soil fertility, improve water availability and efficiency of use, reduce food losses, foster genetic improvement in plants and animals appropriate to local conditions; ***2) Energy***: technology that stimulates employment and is small-scale with short lead times, reliable for rural areas, and less damaging to the environment; ***3) Environment***: technologies to help in providing basic water, sewer, and refuse disposal services; ***4) Biomedicine***: technologies to reduce tropical diseases; ***5) Local Development***: resource-efficient residential architecture, solar greenhouses, solid waste resource recovery; ***6) Underpinnings for Sustainable Development***: information sharing, education, infotech. Policy options for Congress include an Interagency Development Coordinating Council, a special Joint Committee on Sustainable Development in Congress, revising GNP reporting, revising US tax laws to encourage industry to internalize full production costs, SD demonstration sites, an Endowment for Sustainable Development, and an Electronic Peace Corps to facilitate information flow. [**NOTE**: In many ways similar to the FAST study (above).]

(sustainable development and sci/tech)

296

Science and Technology Policy in the United States: Trading in the 1950 Model, Michael M. Crow (Prof of Public Policy and Vice Provost, Columbia U), *Science and Public Policy,* 21:4, Aug 1994, 202-212.

Present sci/tech policies and institutional designs of the US government are inadequate for the challenges that lie ahead. Based on classifications proposed by Christopher Freeman (SPRU/U of Sussex), Crow points to three sci/tech policy phases that industrialized nations have gone through since the 1940s: ***Phase I*** is directed to military purposes (globally, the vast majority of scientific effort since the mid-1930s has been directed to this end); ***Phase II*** goes beyond weapons systems to include economic growth, with a national strategy of targeting specific industries (Japan and Germany have had much experience here); ***Phase III*** moves science to an even broader focus to make life better for people and to use sci/tech for sustainable growth and environmental quality. No nation has fully begun the Phase III transition to "Sustainable S&T Policy," although Sweden and Denmark are moving toward it. The US will likely enter Phase II in the next decade.

Some proposals to facilitate this transition and position the US to enter Phase III after 2005: 1) establish a new Office of Science and Technology Planning and Development with coordination authority similar to OMB; 2) convert selected national laboratories into national technology centers; 3) redesign and recharter the National Science Foundation to lay the foundation for a Phase III sci/tech environment; 4) establish national science centers under NSF in areas of future sci/tech activity; 5) establish regional technology development and transfer centers; 6) create a national network to support advanced research, treating sci/tech information as a freely available public asset (the present National Technical Information Service is wholly inadequate to the task); 7) set up the American Technology Corporation to identify high-risk or economically uncertain technologies; 8) establish a National Sustainability Research Center to rethink the basic design of sci/tech development processes with an emphasis on long-term quality-of-life issues.

(new phase of US sci/tech policy needed)

297

Technology for a Sustainable Future: A Framework for Action. Environmental Technology Strategy Staff, OSTP. Washington: Office of Science and Technology Policy (Old Executive Office Bldg; etcomments@nsf.gov), July 1994/155p.

Environmental technologies are the bridge to a sustainable future, offering a win-win opportunity for the US and the world. If well-engineered, they will facilitate attainment of our long-term environmental, energy, and economic goals. "Our vision is of an integrated system of social, economic, and environmental values that reinforce each other." This document seeks to catalyze a dialogue with Congress, state governments, industry, NGOs, academia, and interested citizens, toward advancing the development and diffusion of environmental technologies. Four dimensions are highlighted: ***1) Facilitating Inno-***

vation: fostering R&D to create jobs and protect the environment, improving testing and demonstration; ***2) Encouraging New Approaches***: overcoming unintentional regulatory constraints to innovation and commercialization, market stimulation in such areas as consumer labeling and government procurement, fiscal and tax policies that encourage private R&D investments; ***3) Engaging the World***: overcoming barriers to exporting environmental technologies, promoting environmental stewardship as a central goal of foreign aid; ***4) Learning and Working Together***: strengthening public-private partnerships, cultivating environmental awareness and expertise in the workforce, including sustainable development concerns in school and college curricula, enhancing information flows from government to the private sector. Concludes with favorable comments on the new paradigm of industrial ecology, and the value of life cycle analysis and a long-term perspective.

(environmental technology promotion)

298

Technology For A Sustainable Future: A National Dialogue on Environmentally Advanced Technology. Robert L. Olson (Research Director, IAF). Alexandria VA: Institute for Alternative Futures (108 N Alfred St; 703/684-5880), April 1994/63p(8x11")/$35.00.

Summarizes a dialogue between 40 business leaders, environmentalists, and government officials around three major themes: ***1) Benefits of Creating an Environmentally Advanced Technology***: to reconcile environmental and economic goals, set a global example, increase economic efficiency and competitiveness, improve overall quality of life (promote health, create better jobs, revitalize local communities), encourage a redefinition in the role of government (to become less intrusive and controlling, but more innovative and facilitative); ***2) What An Environmentally Advanced Technological Infrastructure Will Be Like***: it will involve transformation of our entire technological infrastructure, based on key design principles of increasing eco-efficiency and moving toward industrial ecology, with infotech playing a critical role; ***3) Six Major Dimensions of the Challenge***: eco-efficiency and industrial ecology, total quality design, a dematerialized and knowledge-based economy, better-informed consumers (relying on certification for products, ecological indexes, seeking quality), true cost pricing, and integrated policy reform (based on a new national vision and strategy of sustainable development).

The Report also includes five presentations: Darryl Banks (World Resources Institute) on environmentally critical technologies, Michael Nelson (White House Office of Sci/Tech Policy) on creating an environmental technology policy, David Rejeski (EPA) on knowledge networking for pollution avoidance, Dick Holt (Department of Energy) on impact reduction through infotech, and Skip Stiles (House Committee on Sci/Tech) on the challenge of greening technology. [**NOTE**: Upbeat and leading edge, with an emphasis on collaboration. This report is the first product of IAF's new program on *Technology for a Sustainable Future*. Several possible "next steps" are listed: greening of industry projects, mapping opportunities for collaboration, multi-industry conferences to familiarize environmental leaders with new and emerging environmentally advanced technologies, consumer-corporate dialogues on true cost pricing, using the Internet to gather views on technology for local development, and reinventing the EPA by shifting emphasis from regulatory fault-finding to promoting superior technologies.]

(environmentally advanced technology)

299

The Greening of High Tech, Robert L. Olson (Institute for Alternative Futures, Alexandria VA), *The Futurist*, 25:3, May-June 1991, 28-34.

Both environmentalists and techno-optimists are often trapped in misconceptions: that a truly advanced technology exists (rather, we are only at the beginning of many revolutions); that advanced technology is always complex and environmentally destructive (photovoltaic technology is a dramatic example of an advanced technology that is small, simple, safe, and nonpolluting); and that there is only one kind of advanced technology (rather, technology might evolve along sharply different alternative paths). Truly advanced technology can and must be shaped by environmental values. Characteristics of an *environmentally advanced technology* include: 1) sustainability (it must satisfy present needs without jeopardizing prospects of future generations); 2) use of a safe and inexhaustible supply of energy such as fusion; 3) high efficiency in the use of energy and other resources; 4) high efficiency in recycling and use of byproducts; 5) increasing intelligence (our tools are evolving to a high level of information-processing capability); 6) increasingly alive (e.g. "living machines" as proposed by John Todd, or "green technology" as proposed by Freeman Dyson). Transformation of technology will be pushed by accelerating environmental deterioration, and pulled by positive images of possible futures, as well as the profit motive. "Creating an environmentally advanced technology will provide enormous business opportunities in the 1990s and over the next 50 years," e.g. redesigning manufacturing processes, reusing waste materials, renewable energy, sustainable agriculture, etc. **(environmentally advanced technology)**

300

Backs to the Future: U.S. Government Policy Toward Environmentally Critical Technology. George R. Heaton Jr (Worcester Polytechnic Institute), Robert Repetto (VP, WRI), and Rodney Sobin (OTA). Washington: World Resources Institute, June 1992/34p(8x11")/$12.95.

In the Foreword, James Gustave Speth (President, WRI) states that "Without a dramatic technological transformation throughout the economy, economic and population growth will create increasingly severe environmental pressures." New technologies are needed that increase the efficiency with which energy and raw materials are used, and that eliminate or reduce pollution. Many industrial countries are now devoting larger R&D resources to the search for such technologies, not only to solve environmental problems but to strengthen their competitive position. The US, however, is still spending 60% of its Federal R&D dollars on defense-related research. By 1991, the concept of "critical" technologies (those that enable related developments, confer a strategic economic lead, or dominate the technological future) had a central place in US technology policy discussions, and lists of critical technologies were issued by the White House Office of Sci/Tech Policy and the Council on Competitiveness (compared in Table 1, along with lists from the EC and Japan's MITI). Environmental technology was not prominent in these lists, and is narrowly seen as pollution treatment.

This report provides a long list of environmentally critical technologies that meet four criteria: bringing large and cost-effective reductions in environmental risk, embodying significant technical advance, having broad applicability at the precompetitive stage of development, and involving a high ratio of social to private returns. The 12 categories include: 1) energy capture (the "enormous potential" of photovoltaics, geothermal, solar thermal electricity, nuclear fission); 2) energy storage and application (batteries, superconductors, hydrogen storage, heat storage, fuel cells); 3) special energy end-uses (transportation, buildings); 4) agricultural biotechnology; 5) improved agricultural techniques (alternative agriculture, precision agriculture); 6) manufacturing modeling/monitoring/control systems; 7) catalysis in industrial processes and effluent treatment; 8) separation technologies (for distillation, drying, cleaning, degreasing, evaporation); 9) precision fabrication (greatly reducing the use of natural resources and industrial emissions); 10) materials design and processing; 11) information, communications, and computing (aiding the process of dematerialization by substituting information for natural resources; also computerized effluent monitoring and process control); 12) contraception (the scope of potential developments is very large).

Concludes that new technology is central to creation of an environmentally sustainable economy, and "the only viable long-term basis for continued economic growth and competitiveness." The breadth of items on the above list suggests that an environmental technology policy must employ a portfolio approach that promotes a wide range of new developments. Any or all of these proposals would move environmental technology forward: a new Federal Institute for Environmental Technology, environmental funding criteria for R&D, new missions for the national laboratories, inter-firm cooperation in R&D, international collaboration, reorienting regulatory policy to support technological innovation, and reorienting ongoing Federal programs that support industrial technology. [**NOTE**: A very important report, pointing to a critical realm in which the US is losing competitiveness.]

(environmentally critical technologies: US neglect)

*301

Green Gold: Japan, Germany, the United States, and the Race for Environmental Technology. Curtis Moore (Washington; former Counsel, Senate Committee on Environment) and Alan Miller (Director, Center for Global Change, U of Maryland). Boston: Beacon Press, Aug 1994/ 297p/$25.00.

"The US is confronted with a threat to its viability as a major industrial power in the coming century because it is failing to adapt to new environmental and technological imperatives." Propelled by these imperatives, the world is rapidly moving into a new technological era, a second Industrial Revolution as important as that of two centuries ago. The US was the first to develop many of these technologies in its massive defense and space programs, but lax environmental and industrial policies allowed them to fall into the hands of Japan, Germany, and other competitors. These nations view environmental protection as a strategy for enhancing economic competitiveness, and seek a major share of the global environmental market, estimated to grow from today's $200 billion/year to as high as $600 billion/year by 2000. Through Germany's policy of *Dynamisierungsklausel* (dynamic, state-of-the-art regulation), constant review assures the newest and best pollution control technologies and maximum possible emissions reduction. Japan has established the New Energy and Industrial Technology Development Organization in 1980 and the Research Agency of Innovative Technology for the Earth in 1990. Industries in Germany and Japan are attuned to opportunities to lesson expenses, in part because they lack a large domestic energy industry (in the US, the energy industries are a major lobby against change). Necessity and foresight also explain why energy and environment policies in California are different from those set in Washington and the other 49 states. California has nurtured alternative energy sources, and set the pace on emissions standards.

Chapters describe policies in Germany, Japan, and California, as well as zero-emission vehicles, clean power technologies, cleaner fuels, and zero-polluting energy sources. Concluding policy proposals: 1) announce commercialization of important environmental technologies as a clear US national goal by a specific time (say, 2000), comparable to the vow to put a man on the moon (a list of 24 such technologies is provided, including cogeneration, advanced scrubbers, fuel cells, and hydrogen use); 2) set specific targets and develop a phased program for reducing CO_2 emissions; 3) tax coal and oil, and favor clean forms of energy with rebates; 4) systemically replace lights and other high energy-consumering devices; 5) convert the entire utility system to zero-polluting sources of electricity. There are many options, but what's important is to agree upon goals, and do so quickly.

(US lag in new technology era)

302

Japan Bids for Global Leadership in Clean Industry, Frederick S. Myers (Tokyo), *Science*, Vol 256, 22 May 1992, 1144-1155.

In 1990, Japan's Ministry of International Trade and Industry (MITI) announced "The New Earth 21—Action Program for the Twenty-First Century," a long-term strategic plan that seeks "to undo the damage done to the earth over the past two centuries, since the Industrial Revolution." Three related developments: **1) RITE**: MITI has created the Research Institute of Innovative Technology for the Earth to develop technologies needed in the next century (rather than searching for ways to improve energy efficiency, RITE is aiming toward schemes to strip CO_2 out of industrial emissions and to recycle the carbon); **2) ICETT**: MITI is also backing the International Center for Environmental Technology Transfer, which seeks to train 10,000 people over the next ten years (mostly from LDCs) in energy conservation, pollution control technology, and environmental protection regulations; **3) UNEP/ ITEC**: a new UN Environmental Program International Environmental Technology Center will be built in Kansai Science City and should be operating by late 1992, providing an added boost to Japan's technology-transfer activities. [**NOTE**: An important development to watch.]

(Japan's "New Earth 21" program)

303

Regreening the Earth: Japan's 100-Year Plan, Yasuhiko Ishida (Global Industrial and Social Progress Research Institute, Tokyo), *The Futurist*, 27:4, July-Aug 1993, 20-24.

Japan's MITI proposed "New Earth 21" in 1990, a 100-year plan for improving the environment, principally by reducing CO_2 emissions. Ishida summarizes the MITI proposal and the assessment of GISPRI. The first 50 years will be devoted to five decade-long tasks: **1) First Decade**: improving and implementing energy efficiency through increased use of existing technology and intensified R&D in new, efficient technologies; enhancing CO_2 absorption sinks by increased reforestation and improved forestry management; **2) Second Decade**: promotion of clean energy such as renewable energy sources and inherently safe nuclear power (nuclear operating capacity in 2050 will be about six times that of today); **3) Third Decade**: spread of non-greenhouse-gas substitutes for CFCs, low-energy production processes, and CO_2 reutilization technologies; **4) Fourth Decade**: afforestation of deserts using biotechnology; new technology to enhance oceanic sinks; **5)**

Fifth Decade: the dawn of a new generation of technologies such as fusion, orbiting solar-power plants, magma electricity generation, and superconductors. All of these programs will be implemented in a global partnership with all countries. In the second 50 years, the results of the first five decades will be integrated to complete the work of regreening the planet. Concludes with suggestions for restructuring social systems to conserve energy by preventing overpopulation in urban areas, changing business evaluation (from the standard of sales and profit to standards that indicate a beneficial role in society), and changing national goals from economic to cultural prosperity. **(Japan's "New Earth 21" plan)**

304

The Role of Engineering in Sustainable Development: Selected Readings and References for the Profession. Edited by Monica D. Ellis (WEPSD). Washington: AAES (1111 19th St NW; 202/296-2237), March 1994/106p.

This collection is sponsored by the American Association of Engineering Societies and the World Engineering Partnership for Sustainable Development (7010 Little River Turnpike, Annandale VA 22003; 703/750-6401). WEPSD is a global initiative formed in 1992, "dedicated to unifying the world engineering community to facilitate sustainable development" by creating a global information-sharing network. It views sustainable development as the dominant economic, environmental, and social issue of the 21st century, and places engineers at the nexus of this debate. Documents include WEPSD goals and priority projects, the World Federation of Engineering Organizations "Code of Environmental Ethics for Engineers," the International Federation of Consulting Engineers "Environmental Policy Statement," and "Environmental Principles for Engineers" formulated by the Australian Institution of Engineers. Essay topics include the significance of the UNCED Earth Summit (by Thomas F. Malone of North Carolina State U), the challenge of sustainable development (by Henry J. Hatch, former Commander of the US Army Corps of Engineers), a global strategic plan for engineering (by David Thom of New Zealand), and the need to pattern the human ecosystem after natural processes (by Don V. Roberts, President of WEPSD). Concludes with 36 annotated references, most of them adapted with permission from *Future Survey*. [**NOTE**: An important boost for sustainability. Between 1989 and 1992, enrollment in the 100 environmental engineering programs in US higher education jumped more than 25% (***The Greening of the Engineer***, *U.S. News & World Report*, 21 March 1994, p90).] **(engineering and sustainable development)**

7. Energy

305

Energy and the Environment in the 21st Century. Edited by Jefferson W. Testor (Director, MIT Energy Lab), David O. Wood (Director, MIT Center for Energy Policy Research), and Nancy A. Ferrari (Energy Lab). Cambridge MA: MIT Press, May 1991/1,006p(8x11")/$50.00.

Proceedings of a March 1990 MIT conference, with over 100 contributions. Plenary presentations include **Sen. Albert Gore Jr** on his Strategic Environment Initiative (whereby the US takes the lead in accelerating and disseminating new technologies), **Stephen H. Schneider** on predicting future climate change, **John H. Gibbons** on the interface of technology and environmental science, **William C. Clark** on strategic perspectives for policy design, **José Goldemberg** on energy and environmental policies in developed and developing countries, **Nazli Choucri** on population and the global environment, **William D. Nordhaus** on economic policy in the face of global warming, and a panel on policy strategies for managing the global environment. The editors conclude that the world needs another technological revolution if it is to insure the efficient and sustainable use of energy and other resources for social well-being. A new perspective is needed on our energy/environmental future, viewing it not as a series of isolated problems, but as a set of interconnected trade-offs that must be made between sufficient energy and bearing the environmental and other costs.

Specialized sessions focus on six topics. ***1) Transportation Systems***: role of the automobile in urban air pollution, improving vehicle fuel economy, alternative transportation fuels (by **Daniel Sperling**), the slowing growth rate in US vehicle use, meeting transportation aspirations of LDCs, etc.; concludes that "unsurprisingly, there is no panacea", and that we still lack the understanding needed for good economic evaluation of transport options as if the environment mattered. ***2) Industrial Processes***: modeling energy intensity and CO_2 emissions in US manufacturing, energy conservation in the industrial sector (by **Robert U. Ayres**), cogeneration applications, and trends in various industries (cement, glass, chemicals, pulp and paper). ***3) Building Systems***: energy-efficient buildings, indoor air quality, the slow change in the fragmented building industry (major company alliances might provide critical mass for implementing new technology), buildings in the 21st century (radically improved components are expected, enabling minimal energy use), progress in building materials research, appliance efficiency, barriers to building energy conservation, encouraging new solutions. ***4) Electric Power Systems***: power for developing nations, clean coal technology, pollution control, natural gas to 2020, nuclear power development, large-scale solar/wind systems, future photovoltaics, geothermal opportunities, networks of the 21st century, energy conservation. ***5) Economics and Policy***: economic analysis of greenhouse effects, least cost analysis of CO_2 reduction options, reducing the risks of rapid climate change (by **Irving M. Mintzer**). ***6) Advanced Energy Supply Technologies***: the prospect for fusion energy, the role of renewable energy (expected to increase from 8% of today's US energy use to 15-35% by 2030), the large worldwide potential of hot dry rock geothermal energy that utilizes natural heat in the earth's crust, magnetohydrodynamic power generation from fossil or nuclear fuels (coal-burning central station MHD plants promise to generate power at up to 50% greater efficiency, with far less environmental intrusion), and fuel cells. [**NOTE**: An authoritative near-encyclopedia.] **(energy/environment mega-anthology)**

306

Cool Energy: Renewable Solutions to Environmental Problems (Revised Edition). Michael Brower (Research Director, Union of Concerned Scientists, Boston). Cambridge MA: MIT Press, Sept 1992/219p/$12.95pb.

The US is rapidly approaching a new energy crisis. Consumption of energy is growing, while Americans are increasingly demanding that energy come from clean and safe sources. Renewable energy sources can provide a way out of this emerging crisis. It is not only desirable but practical to make the transition from fossil fuels to "cool" renewable energy. By doing so, we will help preserve the

environment and sustain our economy. Conventional wisdom holds that renewable energy sources can make only a limited contribution because of high cost, dispersed resources, and the variability of wind and sunlight. Typical of this view is the Bush Administration's National Energy Strategy, released in February 1991, which envisions renewables as supplying only 18% of America's 129 exajoules total consumption in 2030. In contrast, a 1991 study by the UCS and three other groups (**America's Energy Choices: Investing in a Strong Economy and a Clean Environment**) concludes that US energy requirements in 2030 could be cut nearly in half from the Business-as-Usual case, with renewable energy sources providing 53% of the supply, and consumers receiving a net monetary savings of $2.3 trillion over the next 40 years. Chapters describe the potentials of solar energy, wind energy (generating cost is now a fourth of what it was a decade ago), biomass, rivers and oceans, geothermal, and storage alternatives for thermal and electricity (including the vision of producing hydrogen from solar- or wind-generated electricity or from biomass). A pollution-free vehicle powered by hydrogen fuel cells was demonstrated in 1991, and others are likely to follow. Concludes with a chapter on policies to promote renewable energy sources: increasing taxes on fossil fuels, financial incentives to lower initial costs, inexpensive changes in regulations (e.g., requiring the principles of "least-cost planning" to be applied to all utility investments), public education and technical assistance programs, and revised priorities in Federal R&D funding. [**NOTE**: A good introductory overview.]
(renewable energy potential)

307

Powering the Future: Blueprint for a Sustainable Electricity Industry. Christopher Flavin (VP, Worldwatch Institute) and Nicholas Lenssen (Senior Researcher). Worldwatch Paper 119. Washington: Worldwatch Institute, June 1994/74p/$5.00pb.

Electricity provides 13% of the world's end-use energy and 17% in industrial countries. Generating and distributing electricity is one of the world's largest businesses, with estimated annual revenues of more than $800 billion—about twice the size of the world auto industry. The power industry also generates some of the world's most serious environmental problems. It is the leading consumer of fossil fuels, particularly coal, and thus accounts for nearly one-third of global CO_2 emissions. Governments are deeply involved in the power systems of all countries, either owning or heavily regulating most of the companies involved. To build an environmentally sustainable power system, the structure of the industry must be revised—a process that has already begun in several nations. Although no one model will work for all nations, there are key elements of a system that can meet both economic and environmental needs: 1) a competitive market for wholesale electricity generation that encourages private generators to enter the arena (new technologies such as fuel cells and solar generators will enable this decentralization); 2) an open-access transmission system; 3) incentives for relying on diverse power sources that take into account environmental differences among them; 4) development of a service-oriented local distribution system committed to demand-side management and integrated resource planning (the most comprehensive model of the future distribution utility is in Sacramento, California). "Like its cousin the telecommunications business, electric power will create large, competitive markets for new, decentralized technologies, and open up a host of business opportunities." The structure and regulation of the new industry will take at least a decade to sort out, with the end-result likely to be far more attractive than the messy transition process suggests.
(electricity industry restructuring underway)

308

Power Surge: Guide to the Coming Energy Revolution. Christopher Flavin and Nicholas Lenssen. The Worldwatch Environmental Alert Series. NY: W.W. Norton, Oct 1994/382p/$23.00;$10.95pb.

The world energy economy is poised for a sweeping shift away from imported oil and environmentally damaging coal during the next few decades. Pushed by the need to stabilize the earth's climate, and pulled by investment opportunities, global energy markets are beginning a rapid move to more efficient, decentralized, and cleaner systems, echoing the shift from mainframe to personal computers in the 1980s. Among the expected changes: a new generation of lightweight and super-efficient electric cars, rapid conversion of coal and nuclear plants to efficient gas turbines, a new generation of mass-produced wind and solar generators that are cost-competitive with the most advanced fossil plants, tiny fuel cells and rooftop solar panels allowing people to generate their own electricity, and the gradual appearance of non-polluting hydrogen as the world's main energy carrier, supplanting oil and natural gas. Many giant oil, auto, and utility companies may find themselves in the current position of IBM—overtaken by competitors better able to anticipate the coming revolution. However, "energy price reform is a prerequisite to the development of a sustainable energy system." Governments have begun to remove some heavy subsidies to traditional energy sources, occasionally with spectacular results. R&D programs also need to be fairly balanced. **(energy revolution ahead)**

309

Energy for Tomorrow's World: The Realities, the Real Options and the Agenda for Achievement. World Energy Council, WEC Commission. London: Kogan Page & NY: St. Martin's Press, Dec 1993/320p/$45.00.

The WEC, founded in 1924, has Member Committees in some 100 countries, representing most major energy interests. The WEC Commission was created in 1989 to look at future energy developments and "practical strategic options" on a global basis and for nine regions. Four energy Cases (scenarios) were developed to illustrate future possibilities in a world of 8.1 billion people by 2020, with global demand ranging from 17.2 gigatons oil equivalent (Gtoe) in the "high growth" scenario to 11.3 Gtoe in the "ecologically driven" scenario (contrasted to 8.8 Gtoe in 1990 and 3.3 Gtoe of energy in 1960). Major concerns are reflected in the proposed Agenda for Action: 1) reducing "energy poverty" by extending access, improving efficiency, and reducing environmental impacts; 2) expanding supply availability of acceptable traditional forms of energy, and increasing availability of non-fossil fuels; 3) reducing demand by a package of measures to promote technological innovation and its diffusion; 4) curbing harmful emissions from energy provision and use by a variety of measures; 5) removing institutional rigidities that act as barriers to change; 6) ensuring effective functioning of energy markets to meet demand in the most cost-effective way; 7) developing more energy-efficient vehicles and modes of transport; 8) recognizing the wide variation in regional concerns while finding a path to sustainable energy development. Beyond 2020, the magnitude of supply problems could expand, especially if higher global energy demand occurs and too little is done to develop alternatives. "Action needs to start now, on a rapidly widening multi-national basis." [**NOTE**: Good global overview, but too generalized.] **(energy scenarios for 2020)**

310

Fueling a Competitive Economy, Joseph J. Romm and Amory B. Lovins (both Rocky Mountain Institute), *Foreign Affairs*, 71:5, Winter 1992-93, 46-62.

America's energy and economic policies remain tied to Cold War concepts of national security. A coherent approach is now needed to national and global security—one that is profitable and market-oriented. An ultimate goal for the US should be an "industrial ecosystem" that minimizes both inputs of energy and materials and outputs of waste products and pollutants. "With its emphasis on resource efficiency the industrial ecosystem should become the dominant manufacturing approach of the 21st century." Such an approach frees up capital by reducing energy waste that bleeds the economy, e.g.: oil imports alone have accounted for nearly three-fourths of the US trade deficit since 1970, or $1 trillion transferred to OPEC nations. Rather than pursuing energy supplies, which has led the nation astray, we should seek efficiency. Germany and Japan use half as much energy per dollar of GNP as does America, mostly because of smarter technologies. "Becoming equally efficient could cut almost $200 billion a year from America's energy bills." Some examples: 1) technological improvements already in widespread commercial use could make light vehicles in the US 50% more efficient by 2005; 2) remarkable recent advances (in materials, electronics, motors, and energy storage) "could bring a 120 to 150 mile-per-gallon, safe, comfortable, and affordable station wagon to market early in the next decade"; 3) technical improvements could save as much as four-fifths of all electricity now used for lighting; 4) "increasing R&D by a mere $160 million a year for 20 years could by 2030 enable renewable energy sources to provide cost-effectively about half the total energy and all the electricity used in the US in 1989"; 5) increased investment in efficiency could net 1.1 million new jobs by 2010. [**NOTE**: Not to be dismissed: Lovins has been the widely-respected leader in shifting energy thinking from supply to efficiency.]

("industrial ecosystem" for US competitiveness)

311

Energy Efficiency Policies. Victor Anderson (New Economics Foundation, London). London & NY: Routledge, May 1993/91p/$69.95;$18.95pb.

Conventional economics ignores the fact that the atmosphere is a basic resource, essential to life and livelihoods. New Economics challenges this outlook. This book applies the New Economics approach to the issue of government policies to promote energy efficiency. Chapters discuss: **1) The Rise of the Carbon Crisis**: the fundamental problem underlying the crisis is that the natural world has been regarded as "free"; the two ways out are switching to a nuclear economy or a renewable economy; **2) Reasons for Energy Efficiency**: saving money, reducing energy dependence, avoiding a worsening of the greenhouse effect, aiding the shift to sustainability, promoting intergenerational and international equity; **3) Reasons for Energy Inefficiency**: poorly informed consumers, invisibility of some costs, emphasis solely on energy supply, poor organization of the energy efficiency industry, lack of financial resources for some consumers and small businesses, short payback periods, discriminatory taxes; **4) Policies for Energy Efficiency**: publicity campaigns, regulations and standards, R&D, government as example-setter, taxation, tax incentives and loans; **5) UK Energy Policy**: promotion of energy efficiency has not played a central role because the UK has large energy reserves (coal, oil, gas), powerful vested interests favor expansion of supply, "free market" ideas discourage government intervention, and green politics is relatively weak; **6) Least Cost Planning in the US**: this approach to reducing demand has generated many useful policy ideas to overcome the "payback problem"; **7) Policy Recommendations**: a radioactivity tax along with a carbon tax (so that the latter does not produce a switch to nuclear power), providing fuller information to energy users, minimum efficiency standards in energy use, improved public sector energy efficiency, help for poorer consumers, a Progressive Carbon Tax (similar to the progressive income tax), implementing a Carbon Tax on a large scale (perhaps by agreement between the EC and US), charging a Carbon Tax on imports from countries without such a tax. [**NOTE**: Anderson is also author of **Alternative Economic Indicators** (#233). Both books are notably concise, yet packed with intelligent ideas.]

(energy efficiency: what, why, how)

312

Energy Management & Conservation. Frank Kreith and George Burmeister (both NCSL). Denver CO: National Conference of State Legislatures (1560 Broadway), Jan 1993/363p/$55.00pb.

During the past 10 years, primary responsibility for energy regulation and planning has shifted from the Federal government to individual states. There is also a growing realization at the state level that energy must be used efficiently, and that state laws and state agency programs can be used toward this end. "As state budget constraints increase, energy efficiency and conservation are expected to increase in importance." Chapters discuss the US National Energy Strategy proposed in 1991, projection of US energy consumption to 2000, the 1992 NCSL state survey of energy issues, utility regulation and incentives, integrated resource planning (used in 1990 by 23 states, with another 19 developing or considering IRP frameworks that examine all energy saving and seek options to minimize total costs), externality studies of energy's social costs, integrating social costs in IRP, criteria for conservation measures (fiscal fairness, environmental impact, ease of implementation, probable success), and opportunities for state action (in transportation, lighting, building standards). About half of the book is devoted to six case studies of energy efficiency measures in Florida, Iowa, Nevada, New York, Washington, and Wisconsin. Concludes that many state programs to improve energy efficiency and reduce consumption have been introduced in the past few years, and "the pace of legislation is expected to increase." Unfortunately, many programs have been carried out without evaluation. Although reliable data are meager, the results of those programs that were analyzed confirm forecasts that "the cost of conserved energy is below that of energy-supply options, particularly electric power." In the future, it will be important to record the total effect of conservation programs and their implementation costs, share the experience gained from various experiments and demonstration projects, and provide more funding for state energy offices.

(state energy efficiency programs)

313

Industrial Energy Efficiency. U.S. Congress, Office of Technology Assessment. Washington: USGPO, April 1993/152p/$9.50. (S/N 052-003-01336-3)

Using energy more efficiently is one of the most economical ways to reduce the great financial, environmental, and security costs that energy use exacts on society. Industry (comprising manufacturers, mines, farms, and construction) accounts for about 30% of US energy consumption, and thus has a major role in making the US more energy efficient. Despite efficiency gains in recent years, many additional technologies and operating practices could further improve efficiency. A few technologies, such as high-efficiency motors and computerized process controls, can be applied throughout industry. But most technologies have narrow, industry-specific uses and are often highly site specific. Another problem is that saving energy is a low priority in most businesses, in that energy accounts for 5% or less of production costs for 88% of industrial output. Increases in efficiency may come more from general modernization, equipment turnover, and maintenance than from direct efforts to reduce consumption. "This makes economic growth and general capital investment two of the most important routes to improved energy efficiency." The Federal government can assist by providing technical assistance, supporting education and advertising programs, offering financial incentives to alter investment patterns, and boosting energy efficiency R&D. **(energy efficiency in US industry)**

314

Building Energy Efficiency. U.S. Congress, Office of Technology Assessment. Washington: USGPO, May 1992/ $8.50 (S/N 052-003-01280-4).

About $170 billion/year is spent on energy used in US residential and commercial buildings. Energy-efficient technologies such as improved furnaces, air conditioners, and lights are available, and resulting energy savings typically repay the extra capital cost in only a few years. Implementing all cost-effective energy-efficient technologies could hold building energy roughly stable through 2015, saving about one-third of the "business-as-usual" projection. But these technologies are underused because equipment purchasers are often different from those who pay to operate the equipment (thus undermining efficiency incentives), energy costs are often low relative to other costs such as personnel (and are thus ignored), and information is often incomplete or unavailable. Strategies to encourage new technologies: enhance financial incentives, Federal leadership through procurement and demonstration, changes in R&D, mandate efficiency, and improve awareness. **(promoting energy-efficient technology)**

315

A New Energy Path for the Third World, Nicholas Lensson (Senior Researcher, Worldwatch Institute), *Technology Review*, 96:7, Oct 1993, 42-51.

An adaptation of **Empowering Development: The New Energy Equation** (Worldwatch Paper 111, Nov 1992/57p/$5.00), noting that developing countries have more than quadrupled their energy use since 1960, and doubled their per capita use. Yet they are staggering from oil price shocks, struggling with foreign debt, and suffering from serious environmental and health problems—while still facing severe energy shortages. Moreover, conventional energy development has also increased the threat of global warming. Although industrial countries are responsible for 79% of fossil fuel-derived CO_2 emitted since 1950, and still accounted for 69% of the total in 1990, future growth in fossil-fuel use will be greater in the South than in the North. China alone is expected to emit more CO_2 by 2025 than the current combined total of the US, Canada, and Japan.

"An energy strategy based on clean and efficient sources is a cornerstone of sustainable development, and is essential if the countries of the South are to improve their living standards." Over the next 35 years, $350 billion invested in efficiency improvements could eliminate the need for $1.75 trillion worth of power plants, oil refineries, and other energy infrastructure by reducing growth in demand. This would free up money for vastly larger investments in food production, health, education, and other neglected needs. The Third World must also develop alternatives to costly oil and polluting coal. Many developing countries have extensive untapped reserves of natural gas, and all have enormous potential to rely on solar, wind, biomass, or geothermal energy. By combining efficiency and alternative energy sources, developing countries can "leapfrog" to the advanced technologies being commercialized in industrialized countries today, avoiding billions of dollars of misdirected investments. Integrated Resource Planning, originally pioneered in the US, is now spreading to Canada, Japan, and Western Europe. It was recently adopted in Thailand, where utilities expect as a result to save at least $180 million by 1997. To encourage such policies, lending agencies must shift their priorities. A step in the right direction is the Global Environmental Facility, which makes grants to slow global warming. But the GEF budget is not large enough. A new UN agency could take a lead role in promoting efficient energy systems. [ALSO SEE: **Fueling Development: Energy Technologies for Developing Countries** (US Congress, Office of Technology Assessment, April 1992/$15.00; summary free from OTA).] **(energy efficiency for Third World)**

316

A New Power Base: Renewable Energy Policies for the Nineties and Beyond. Keith Lee Kozloff and Roger C. Dower. Washington: World Resources Institute, Dec 1993/196p/$24.95.

Encouraging development and use of renewable energy technologies has been part of US national energy policy for the last 20 years. Yet renewables currently meet only 8% of national energy needs, and, under current policy and market trends, this share will not rise beyond 9% by 2010. The gross disparity between current and potential contributions of most renewable energy sources suggests foregone social benefits: environmental protection, economic sustainability, and energy security. These benefits are a major rationale for policies that go beyond current interventions. Four primary barriers divert investment away from renewables: market energy prices that do not fully capture environmental and other social costs of various energy sources, utility regulation that biases energy choices toward fossil fuel technologies, high costs for consumer information acquisition, and high perceived investment risks. A National Renewable Energy Strategy is proposed to coordinate the selection and implementation of policies to overcome these barriers. Three shifts in policy are needed: 1) reforming energy prices and tax policies that distort energy consumption or investment decisions (e.g., through pollution taxes); 2) reducing nonprice incentives that cause utilities and energy users to favor conventional sources; 3) government initiatives to promote private investment in commercializing renewable technologies. **(renewable energy strategy proposed)**

317

The Great Energy Harvest, Helena Li Chum, Ralph Overend, and Julie A. Phillips (all National Renewable Energy Laboratory, Golden CO), *The Futurist*, 27:3, May-June 1993, 34-40.

After nearly 20 years of R&D, biomass technologies for producing power, fuels, and chemicals are just now entering the commercial arena or are ready for large-scale experiments, both in the US and Europe. During the 1990s, the US Dept. of Energy would like to participate in joint ventures linking farmers, foresters, and energy industries. These projects will test energy crops as diverse as elephant grass, eucalyptus, bagasse, black locust, poplars, switchgrass, sorghum, sugar cane, and sycamores. Such crops retard soil erosion and build productive soil, while removing as much CO_2 from the atmosphere when they are grown as the fuels emit when they are burned. Many other projects are already under way: 1) biomass (wood and other plant-derived materials) produced as much electricity in the US as six nuclear power plants, and new technologies are on the horizon (e.g., a Minneapolis firm is developing a new process for burning whole trees to make electricity); 2) a number of companies plan to develop gasifier technologies for producing methanol; 3) corn crops in the US produced about a billion gallons of ethanol in 1992; Brazil currently produces more than 4 billion gallons of ethanol a year from sugar cane; 4) NREL has developed a promising new technique to convert energy crops and cellulose-rich wastes to biocrude oil, the raw material to produce a range of high-value chemicals. With a major commitment to producing biofuels, the $50 billion a year now spent on oil imports could be redirected into US farms, tree plantations, regional energy firms, and rural infrastructure. "Within about 40 years, the world's farmers could grow all the food needed, with enough capacity left over to easily satisfy up to 30% of global energy requirements." **(energy crops potential expanding)**

318

Tilting Toward Windmills, Jon G. McGowan (Prof of Mechanical Engineering & Co-Director, Renewable Energy Research Lab, U of Mass-Amherst), *Technology Review*, 96:5, July 1993, 38-46.

A wind power industry flourished briefly in the US after the energy crisis of the 1970s. But by the early 1980s, with the exception of California wind farms, US interest in wind as a large-scale electricity source almost disappeared. Despite initial problems, California's largest wind-farm developers improved their systems and their persistence paid off. "Wind-produced electricity is now less expensive than electricity produced by conventional fossil- or nuclear-powered generating plants in many parts of the world." But the US public's view of wind power remains distorted by its early failures. Many still wrongly believe that wind power does not have significant potential, that it is expensive and unreliable, that the technology is impractical for use by utilities, and that new and improved designs are needed (like the bicycle, the design is already well developed and likely to last for a long time). Recent studies of wind energy feasibility show 14 states, especially those in America's midsection, with a wind resource at least as great as that in California. If wind farms were built today on all available sites in North Dakota (excluding urban areas, parks, wetlands, and most farms), "that state alone could supply more than a third of all the electricity consumed in the continental US." A recent study in Minnesota showed that the wind resource at just one of the state's southwestern ridges could supply almost all of its electricity. Wind is now one of the least costly sources of new electric power in many locations, and the benefits of full-scale mass production have yet to be reached. Enthusiasm for wind power is much higher in Europe, which seeks to install more than 3,000 megawatts of new wind energy by 2000. Meanwhile, US utilities expect to add roughly 900 megawatts in the same period. The EC is also spending about 10 times more for wind-energy R&D than the US government. Proposes Federal government production incentives for wind, and regulatory reforms at state level. **(wind energy now practical)**

319

Renewable Energy from the Ocean: A Guide to OTEC. William H. Avery (Former Director, Ocean Energy Programs, Johns Hopkins U Applied Physics Laboratory) and Chih Wu (Prof of Mechanical Engineering, US Naval Academy). NY: Oxford U Press, March 1994/446p/$65.00.

The upper layers of the tropical oceans are a vast reservoir of warm water held at a temperature near 27°C (80°F). On an average day, the water near the surface absorbs more heat from the sun in one square mile of ocean area than could be produced by burning 7,000 barrels of oil. A practical technology for converting even 0.01% of the absorbed solar energy into electricity or fuel in a form suitable to consumers on land could have a profound impact on world energy availability and economics. Avery and Wu present the scientific and engineering fundamentals of ocean thermal energy conversion (OTEC), showing that the technology base is sufficiently well-established for large-scale demonstration plants to be built as forerunners to commercial plants and plantships that will be economically attractive and environmentally benign. Pioneering work in France and Japan is discussed, concluding that "OTEC-produced fuels can be an inexhaustible substitute for petroleum-based fuels and could ultimately be produced in quantities and at costs that would make it practical for this technology to supply a large part of the energy needs of the world."

Major subsystems of an OTEC system include a heat engine or power plant (operated by the temperature difference between the warm water on the surface and the cold water below the surface), a water ducting system, an energy transfer system, a position-control system, a platform to support the power plant (where deep water exists near the shore, OTEC plants may be onshore or shelf-mounted; 98 nations and territories have the prerequisites to use land-based or near-shore OTEC technology). Electrical energy generated on board an OTEC vessel can be stored as hydrogen or as ammonia (which is nine times as dense as liquid hydrogen and can be stored at ambient temperatures). The energy transfer method that offers the most promise for OTEC is production of methanol by combining hydrogen with carbon monoxide. Two scenarios for commercial development are proposed: ***1) Methanol Commercialization***: construction of 427 methanol plantships at about $500 million each with enough total capacity to replace the imported petroleum used in the US; if financial support is maintained, the program can be completed by 2020; ***2) Ammonia Commercialization***: construction of 1,681 ammonia plantships at about $450 million each to supply enough fuel to replace all gasoline used in the US in 1990. The principal differences between the two scenarios are that replacing gasoline with ammonia fuel would entirely eliminate carbon emissions, but would require a larger automobile adaptation cost. Ammonia could also be used in fuel cells for electricity generation. [**NOTE**: An attractive prospect accompanied with a wealth of technical detail.]

(ocean thermal energy conversion)

8. Transportation

*320

Car Trouble. Steve Nadis and James J. MacKenzie (WRI). World Resources Inst. Guides to the Environment. Boston: Beacon Press, Feb 1993/229p/$27.50;$12.00pb.

The automobile industry is the world's largest manufacturing activity, churning out some 50 million new vehicles a year. The US has 190 million motor vehicles, and Americans buy 15 million cars and trucks each year; in 1990, vehicle sales were more than $300 billion. One out of six jobs in the US is in auto or related industries. But our car-dominated transportation system has many environmental, economic, and safety problems with hidden costs. Chapters describe air pollution, roadway congestion, the search for more efficient engines, alternative fuels, high-tech toll collection systems, smart cars and highways, deterrents to solo driving (ride sharing, congestion pricing, parking freezes), the cost of parking as the single greatest determinant as to whether people will drive (free parking is available to about 90% of Americans who drive to work; $50 billion of parking spaces are provided free to US workers), driving bans and no-drive days, and alternative transport (light rail, buses, bikes, and walking). Several transitions are needed: **1) Phasing Out Oil Use:** aided by getting fuel prices right, high fuel efficiency standards for new cars, paying part of auto insurance at the pump, added fees on gas guzzlers and rebates on gas sippers, scrapping older cars; **2) Phasing in Clean Vehicles:** aided by regulatory changes as in California, feebate arrangements for electric and hydrogen-powered cars; **3) Shifting to a More Liveable Transport System:** aided by tech fixes to reduce congestion, integrating land use and transport planning, increasing motor vehicle fuel taxes, switching to low-impact transport. [**NOTE:** Good non-technical overview. **(reducing automobile problems)**

321

The Going Rate: What It Really Costs to Drive. James J. MacKenzie, Roger C. Dower, and Donald D.T. Chen (all WRI). Washington: World Resources Institute, June 1992/ 32p(8x11"/$9.95+$3.00p&h.

Heavy use of cars and trucks in the US is partly due to economic and political forces that mask the full costs of driving. Costs of driving are categorized as "market" or "external." Market costs are reflected in economic transactions such as buying a car or fuel. External or hidden costs are not reflected in these transactions. Market costs not paid by US drivers amount to $170 billion a year, including road construction and repair, highway services, and parking that is often supplied free. External costs, amounting to about $126 billion a year, include air pollution, the rising risks of climate change, security costs of importing oil, congestion, motor vehicle accidents, noise, vibration, and land loss. Together, these costs amount to almost $300 billion a year, more than 5% of GDP. Those who enjoy the benefits of motor vehicle use should pay for the costs directly, as close to the time and place where the cost is incurred as possible. Potential "polluter pays" policies include increased fuel taxes (perhaps including partial costs of auto insurance), an annual charge on trucks according to weight-per-axle and annual mileage, vanpooling incentives, tolls, time-of-day pricing of congested roadways, and reform of zoning and land use planning. [**NOTE**: Another WRI leading-edge exercise in more realistic economics. Full-costing of vehicle use might do more to conserve energy than any other measure!]

(full costs of driving: $300 billion/year)

322

The Keys to the Car: Electric and Hydrogen Vehicles for the 21st Century. James J. MacKenzie (Senior Associate, WRI). Washington: World Resources Institute, May 1994/128p/$14.95pb.

Growth in motor vehicle use has increased CO_2 emissions and US dependence on imported oil. In the 1970-1990 period, the number of cars on the road grew by 61%, the number of licensed drivers by 50%, and use of relatively inefficient trucks and vans by 278%. Recent projections indicate that US vehicle-fuel use and CO_2 emissions could increase 5-50% over 1987 levels by 2010. Worldwide CO_2 emissions from motor vehicles increasing 20-50% by 2010 are also plausible. Responding anew to the linked problems of urban air pollution, growing dependence on imported oil, and global warming, the Federal government and some states now encourage alternative fuels. High on their list are three carbon-based fuels (methanol, compressed natural gas, and ethanol blended with gasoline), plus electricity and hydrogen. The US would receive only marginal benefits from using the three carbon-based fuels as they are now produced. Electric vehicles or zero-emission vehicles (ZEVs), powered by some combination of batteries, hydrogen fuel-cells, flywheels, and ultracapacitors, are the most promising long-term measure. In the short term, few ZEVs can be put on the road, and US policy should stress new-vehicle fuel efficiency and reducing vehicle-miles traveled. ZEVs will have to be phased in—a transition that will take 3-4 decades.

Many benefits could result within 20 years if these new technologies were phased in aggressively. This will require an integrated national effort: 1) making gasoline prices more accurately reflect the full social costs of burning oil; 2) enhancing R&D to develop EV technology further (over the long run, hydrogen vehicles could surpass battery-powered EVs in range and refueling time); 3) developing a widespread system of user-friendly standardized recharging facilities; 4) encouraging fleet purchases to stimulate the market. California is leading the world in these efforts: by 1998, 2% of new cars and light trucks sold must be ZEVs, rising to 5% by 2001 and 10% by 2003.

(transition to ZEVs)

323

Steering a New Course: Transportation, Energy, and the Environment. Deborah Gordon (Sr. Policy Analyst, Union of Concerned Scientists). Washington: Island Press, Dec 1991/244p/$34.95;$19.95pb.

The US faces a deterioration of its transportation system across the board: cars, trucks, trains, and airplanes. Even if road capacity is increased by 20%, an unlikely achievement, road congestion in the US will triple in 15 years if present trends continue. As the infrastructure becomes increasingly unable to handle the growing volume of traffic, cracks that have already appeared will widen. Without corrective action, "a serious disintegration of service is likely by the end of the decade." Despite continuing gains in pollution control and efficiency improvement, overall emissions of pollutants are projected to increase by almost 40% by 2010, because we are driving more and under more congested conditions. Chapters discuss the history of US transportation in the 20th century, an overview of the US transportation sector, transportation in Europe and Japan (Americans use far more transportation energy than any other country), greenhouse gases and other air pollutants, alternative transport fuels, ultra-fuel-efficient vehicles, and innovative transportation strategies such as high-speed rail, maglev systems, personal rapid transit, intermodal freight transport, taxis and minivans, parking plans and ride-sharing, encouraging human-powered transport, and the limited improvements from advanced vehicle/highway systems.

(transportation deterioration in US)

324

Future Drive: Electric Vehicles and Sustainable Transportation. Daniel Sperling (Director, Institute of Transportation Studies, U of California-Davis). Washington: Island Press, Jan 1995/175p/$39.95;$17.95pb.

Small improvements in the environmental and economic performance of transport systems of past decades are being overwhelmed by rapid growth in people, cars, and travel. To remain satisfied with incremental improvements is to accept a slow deterioration in urban and environmental conditions. A more radical break is needed, pursuing a vision of an environment-friendly transport system that does not restrict freedom of movement. Electric propulsion is central to this more benign future, and we are not far from realizing this vision. The stage was set in 1990 when the California Air Resources Board issued its zero-emission vehicle mandate for a percentage of manufacturer's sales. "The ZEV mandate may be the single most important event in the history of transportation since Henry Ford began mass-producing cars 80 years ago." An optimistic vision views the possibility that, by 2020, most new vehicles in California are zero emission, with fuel cell vehicles accounting for half of all vehicle sales. ***1) Transport Options***: mass transit is playing a shrinking role, and alternative fuels and emission standards do not show anywhere near the promise of electric-drive technologies; **2) Electronic Propulsion**: electric vehicles outnumbered gas-powered vehicles in 1900, but were less than 2% of vehicles by 1915; between 1998 and 2000, the major automakers are likely to build at least one car and one small van or pickup truck (Peugeot-Citroen has already made a firm commitment to production); lead-acid batteries are likely to dominate the electric vehicle market at least through 2000; ***3) Neighborhood Electric Vehicles***: NEVs are larger than golf carts but smaller than subcompact cars; a principal force for accepting them is increasing multiple-car ownership so that every car no longer has to serve every purpose; ***4) Fuel Cells***: vehicles with fuel cells will be cleaner and more efficient than internal combustion vehicles, and cheaper and better than battery-powered vehicles; fuel cell buses could be commercially available by 2003, and fuel cell cars a few years later; early models will run on methanol or hydrogen from natural gas, but later models will run on hydrogen from water using solar energy; ***5) Hybrid Vehicles***: likely to be pushed aside by improvements in gasoline emissions and by the ZEV mandate; in the long-term, they will be squeezed by improving batteries and emerging fuel cells; ***6) Accelerating Regulatory Reform***: the current system is too inflexible to handle the growing diversity of technological options, very small vehicles, new propulsion systems, and new fuels. [**NOTE**: A strong argument, very clearly written.] **(transition to electric vehicles)**

325

Reinventing the Wheels, Amory B. Lovins and L. Hunter Lovins (Rocky Mountain Institute, Snowmass CO), *The Atlantic Monthly,* Jan 1995, 75-86.

In September 1993, the heads of the Big Three auto makers accepted President Clinton's challenge to collaborate in the Partnership for a New Generation of Vehicles. The PNGV's goal of tripling fuel efficiency in the next decade could become the core of a green industrial renaissance, instigating a profound change in what and how much we drive, and how our economy works. Yet this seemingly ambitious goal can be far surpassed. Modern cars are a sophisticated engineering achievement—the highest expression of the Iron Age. But they are obsolete, and the time for incrementalism is over Striking innovations have occurred in advanced materials, software, motors, microelectronics, electricity-storage devices, fuel cells, and computer-aided design and manufacturing. Artfully integrated, they could yield safe and affordable family cars getting hundreds of miles per gallon—several times the 80-odd mpg sought by the PNGV.

Achieving this will require a completely new car design—the ultralight hybrid, or "hypercar." The technologies already exist for a superefficient car, if we cut weight by 65-75% through use of advanced materials, cut aerodynamic drag by 60-80% through sleeker streamlining, and cut tire and road energy loss by 65-80% through better tires and lighter weight. Moldable composites are superior to steel because they are cheaper overall, do not require painting (the most polluting and costliest step in auto making), and do not dent, rust, or chip. Hypercars are also favorable to ultraclean alternative fuels such as compressed natural gas or hydrogen. Ultralight hybrids will probably be made and sold in completely new ways (made-to-order cars could be ordered directly from the local factory and delivered in a day or two, with maintenance by technicians who come to one's home or office with the aid of remote diagnostics). "The potential benefits of hypercars are enormous—in oil displacement, energy security, international stability, forgone military costs, balance of trade, climatic protection, clean air, health and safety, noise reduction, and quality of urban life." [**NOTE**: If Lovins can shake up the electricity industry, why not the auto industry too? ALSO SEE a companion essay (pp86-93) by Mark Derr of RMI on imaginative transport options such as low-speed "subcars" for local trips, aerodynamic tricycles that could accommodate a small motor, and a "car-bus" linked in trains.] **(ultralight hypercars)**

326

Improving Automobile Fuel Economy: New Standards, New Approaches. U.S. Congress, Office of Technology Assessment. Washington: USGPO, Oct 1991/$5.50 (S/N 052-003-01262-6). Summary free from OTA.

Higher fuel economy standards for new cars is a central issue in the US energy policy debate. Several bills seek to revise the 27.5 mpg standard because of growing oil import vulnerability, the trade imbalance, environmental concerns, recent leveling off of improvements in fuel economy, and the apparent success of the old standards which doubled new car fuel economy between 1973 and 1990. Without new policy measures, US new car fleet fuel economy is unlikely to improve significantly over the next decade, primarily because fuel costs continue to be a modest portion of total auto operating expense. OTA expects average new car fuel economy to be about 29 mpg by 1995 and perhaps as much as 33 mpg by 2001, but it could be much lower. New fuel economy standards could raise fleet averages to 35 mpg by 2001, though not without significant costs. With new technologies, fleet targets of 45-55 mpg could be practical by 2010. The form of the new standard is just as important: rather than the current uniform standard that places an unequal burden on automakers, standards that base fuel economy targets on the actual and changing attributes of each company fleet offer more flexibility to policymakers. [ALSO SEE: **Automotive Fuel Economy: How Far Should We Go?** (National Academy Press, Feb 1992/259p/$34.95pb) estimates "technically achievable" fuel economy levels for the new-car fleet at

31-33 mpg in 2001 and 34-37 mpg in 2006, and for the new light-truck fleet at 24-25 mpg in 2001 and 26-28 mpg in 2006.] **(fuel economy standards for 2001 and 2010)**

327

Back on Track: The Global Rail Revival. Marcia D. Lowe. Worldwatch Paper 118. Washington: Worldwatch Institute, April 1994/54p/$5.00.

Rail has many advantages over highway and air transport: greater energy efficiency (an intercity passenger train is three times as energy-efficient as commercial air, and six times as efficient as a car with one occupant), less air pollution, less dependence on oil, lower greenhouse emissions, less congestion, fewer injuries and deaths (in the US, the risk of dying in an auto accident is 18 times greater than in a train accident), less land paved over (two rail tracks can carry as many people/hour as 16 highway lanes), encouraging compact and efficient land use, and greater social equity (rail is a vital option for the young, the old, the disabled, and the poor). Since WWII, the role of rail declined in many countries (notably the US), but rail now seems poised to make a comeback in much of the world, encouraged by smog and traffic jams. Light rail systems are particularly popular. People can be encouraged to use rail systems by integrating them with other transport modes. "Bike-and-ride" is very popular in Japan and much of Europe. In Japan, people travel an average of 3,117 km/year by rail, in contrast to 84 km/year in the US and 72 km/year in Canada. This is reflected in the ratio of road kms to rail kms: Japan has a ratio of 2, France is 8, the UK is 14, Canada is 76 and the US is 228. [ALSO SEE: ***Railways Reap the Benefits of Trucking Woes*** (*Christian Science Monitor*, 12 July 1994, p8), reporting record levels of intermodal traffic for each of the past 12 years, a nationwide shortage of long-haul truck drivers, rapidly rising trucking costs, and railroads now hiring new employees.] **(rail revival worldwide)**

328

Breaking Away: Bikes Winning the Race (Cover Story), Ed Ayres (Senior Editor, *World Watch*), *World Watch*, 6:1, Jan-Feb 1993, 10-18.

The sprocket chain-driven bicycle and gas-fueled internal combustion engine-driven car were both invented in 1885, and have developed side by side. Sales of cars almost approached those of bikes in 1970, but bikes are now winning the race: between 1970 and 1990, annual car production worldwide increased by 14 million, while bike production grew by 60 million. In 1992, the number of new bikes produced exceeded that of new cars by almost 3 to 1. While cars still dominate personal transportation in the US and Europe, recent developments suggest that much of the world is not likely to adopt an American-style car culture. Large numbers of bicycles do not presage massive future conversions to cars because of: 1) **Affordability:** less than 10% of the world's people can afford cars, but about 80% can afford bicycles; moreover, the US-style automotive system requires not only a rich population but a rich public treasury to subsidize motor vehicle transportation; 2) **Functional Advantages:** for trips of just a few miles, bikes are increasingly competitive: they are more mobile and often quicker in congested areas; 3) **Lower Social Cost:** bicycles do not inflict fatal or disabling collisions, diseases caused by auto exhaust, disruption of urban life, and loss of productivity from chronic traffic delays and frustrations; 4) **Lack of Space**: poor countries cannot afford the large amounts of land required for car culture; 5) **Technology**: the bicycle shows more potential for technological improvement than the gasoline-power car, as shown by the surging popularity of "mountain bikes" that make it easier to pedal in hilly or unpaved areas. Concludes that "the era of almost complete domination by motor vehicles in the most developed cities may soon be over." Future transportation will involve a complex mix of walking, bicycles, non-polluting cars, and mass transit. [**NOTE:** A possible direction in new bike technology involves low-slung "recumbent bicycles" that are comfortable and efficient (***Laid-Back Bicycles***, *U.S. News & World Report*, 6 July 1992, 72-73).] **(bicycles outselling cars)**

329

Bicycle Transportation: A Handbook for Cycling Transportation Engineers (Second Edition). John Forester. Cambridge MA: MIT Press, Sept 1994/346p (8x11")/$25.00pb.

As either recreation or transportation, cycling is environmentally benign. Cycling is far better for the environment than motoring. There is no doubt and no discussion about that. But the general public believes that cycling on the roads with the threat of motor traffic from behind is just too dangerous for any rational person to undertake. This "cyclist-inferiority hypothesis" has an extremely strong hold on public and government opinion, but lacks any scientific support. The data support the principle that cyclists are most successful when they act like drivers of vehicles and society so treats them. No data support the orthodox opinion that cyclists need to act inferior. "General bikeway systems are at best useless and at worst are much more dangerous than properly designed roads." Most people, even children, can be trained to drive bicycles like vehicles. Cyclists who act like drivers of vehicles have far lower accident rates and car-bike collision rates than cyclists who do not act like drivers of vehicles. Decades of popular acceptance of cyclist inferiority programs have produced a negligible amount of cycling at an extremely high accident rate.

Chapters in this ostensible textbook on "cycling transportation engineering" discuss psychological foundations, the history and demography of cycling in Europe and the US (in Europe, motoring did not supersede cycling until the 1960s; in the US the transition to motoring occurred early and rapidly so that cycling and motoring never coexisted), the history of governmental actions regarding cycling ("the record of American governments with respect to cycling is appalling, a consistent record of discriminatory policy based on the cyclist-inferiority superstition"), cycling accidents and potential of various programs for reducing cyclist casualties, the effects of cyclists on traffic (under lane-sharing conditions, cyclists do not reduce highway capacity; conversely, conversion of some motorists to cycling could significantly reduce congestion), the complicating effect of bikeways, flow of cycle traffic, the difficulty of predicting cycling volume (although the knowledge we have does justify some actions), the importance of cycling proficiency (one cause of confusion is the low intellectual caliber of the authors of bike-safety programs), cycling organizations, the conflict between cyclists and environmentalists (whereby well-intentioned environmentalists fight against good cycling principles), nighttime protective equipment, the future of cycling (if transportation engineers produce useful bicycle programs, "it is quite reasonable to predict that 20-30% of urban trips for personal purposes will be made by bicycle"), a recommended cycling transportation program, the problem of police harassment, road design, traffic calming, bicycle facilities, integration with mass transit and long-distance carriers, and changing traffic law for cyclists. [**NOTE**: An

extraordinary book, by an avid cyclist, and son of British writer, C.S. Forester. Some data, charts, and formulas, but mostly a long and feisty essay that blows away various conventional wisdoms. This fresh and well-informed view of cycling could change transport policy, in turn relieving congestion and pollution. In any event, it's a good read, even for avid motorists.]
("cyclist-inferiority hypothesis" questioned)

9. Cities and Buildings

330
Mega-city Growth and the Future. Edited by Roland J. Fuchs (Vice Rector, UNU) *et al*. Tokyo: United Nations U Press, June 1994/428p/$43.00pb.

A major challenge for mankind is an informed response to unprecedented urban growth and the intelligent management of urban settlements. The 20th century began with perhaps a dozen cities of 1 million residents; it will end with several hundred cities of that size, including 28 urban areas with populations exceeding an arbitrary threshold of 8 million, thereby qualifying as "mega-cities." The vast majority of the mega-cities, and the most rapidly growing, are in the developing countries, where they serve as national and regional engines of economic growth, centers of creativity, homes for the poor, and sources of environmental pollution.

This book, based on an October 1990 conference in Tokyo, is the initial project of the UNU Programme on Mega-city Growth and Management. Topics include growth of the world's 28 megalopolises in the 1950-2000 period (London was the world's #2 city in 1950; by 2000, it will not rank in the top 28!), labor force change and mobility in Asia, the impact of the shifting techno-economic paradigm on the world city system, technology policy in the city of the future (by George Bugliarello), social and welfare impacts of mega-city development, impacts of mega-city growth on families and households (by Eleonora Barbieri Masini), mega-city management and innovation strategies (discussing land policy, housing, water supply, transport, environmental protection), financing infrastructure (property tax, motor vehicle tax, other local taxes, grants and transfers), severe land and shelter problems (arguing that more attention to rental housing is essential), strategic decisions for transportation, alleviating poverty, and deteriorating environmental conditions in mega-cities (compounded by myths that environmental management is an expensive luxury and that the poor—not the lack of urban services—cause pollution). A concluding chapter discusses priority urban management issues: 1) improving financial structure and management; 2) providing shelter, services, and infrastructure; 3) improving data collection and information systems; 4) strengthening urban informal sector roles (about 50% of the labor force in Third World cities is in the informal sector); 5) strengthening institutional capacities; 6) improving the urban environment (only 25-55% of waste generated in large LDC cities is collected and disposed of safely; about 40% of urban residents are not served by sanitation services; 25% do not have access to safe drinking water; air pollution in most large cities is getting worse). **(mega-cities and the environment)**

*331
The Gaia Atlas of Cities: New Directions for Sustainable Urban Living. Herbert Girardet (London UK). Foreword by Lester R. Brown. NY: Anchor Books/Doubleday, June 1993/191p/$16.00pb.

"If we are to continue to live in cities, indeed if we are to continue to flourish on this planet, we will have to find a viable relationship between cities and the living world—a relationship not parasitic but symbiotic, or mutually supportive." Humanity is becoming a predominantly urban species, and the limits of urban growth are increasingly seen as environmental. This book, on how modern cities might achieve sustainability, is in three parts: **1) The Ecology of Settlements**: city throughput, linear vs. circular metabolism, urban heat islands, "globalopolis" as a more accurate term for McLuhan's "global village," development of cities from settlements, urban planning visions, the growth of megacities; **2) Sick Cities, Sick World**: urbanization and overcrowding, the poverty crisis, cities as stressful places, the city as parasite, hidden costs of food, waste and pollution, the pervasive influence of automobiles, threats to cities from future sea level rise; **3) Healing the City**: making a city convivial, revival of city centers and neighborhoods, upgrading squatter settlements, healthy cities programs, the green cities movement and urban farming, combined heat and power stations, improving conditions for bicycles, traffic calming, recycling, cities learning from each other, the Megacities Project (run by Janice Perlman of NYU since 1987), safeguarding city hinterlands, China's eco-cities that are largely food self-sufficient, renovating waste water.

[**NOTE**: An excellent introduction and overview, as are the others in the UK-based Gaia Future Series (Norman Myers, General Editor): **The Gaia Atlas of Green Economics** by Paul Ekins *et al*. (#231), **The Gaia Atlas of Future Worlds** by Norman Myers (**FS Annual 1992** #10925), **The Gaia Atlas of First Peoples** by Julian Burger (**FS Annual 1992** #10970), **The Gaia Peace Atlas** by Frank Barnaby (**FS Annual 1990** #9028), etc.]
(sustainable cities)

332
Sustainable Cities. Graham Haughton (Leeds Metropolitan U) and Colin Hunter (U of Huddersfield). London: Jessica Kingsley Publishers and Regional Studies Assn, Sept 1994/357p/$24.95pb. (Order in US from JKP, 1900 Frost Rd, Ste 101, Bristol PA 19007.)

Interest in the urban challenges raised by the growing debates on sustainable development and global environmental change has increased rapidly in recent years. This textbook builds on the growing interest in the role of cities in the sustainable development process, examining both urban environmental degradation and the contribution cities can make to global sustainability. Chapters discuss population growth and urbanization, urban economic restructuring, urban sprawl, urban design and the natural environment, urban air pollution, fresh water resources and water pollution, guiding principles for urban development (minimize waste, prevention is better than cure, maximize use of renewable and recyclable materials, maintain requisite variety, encourage widespread public participation), policy instruments for improving the urban environment (impact assessment, strategic and land use planning policies, demand management, environmental charges and taxes), the ideal of "concentrated decentralization," and "the sustainable city manifesto" (eight general guidelines). [**NOTE**: A decent introduction for those who need it.] **(sustainable cities introduction)**

333
Sustainable Cities: Urbanization and the Environment in International Perspective. Edited by Richard Stren, Rodney White, and Joseph Whitney (all U of Toronto). Boulder CO: Westview Press, Feb 1992/365p/$29.85pb.

Papers from a June 1990 Toronto colloquium on "Human Settlements and Sustainable Development," addressing the question of the relationship of cities and the natural environment, and how cities can play a more constructive role in the sustainability issue faced by our planet today. The bulk of the book is devoted to nine chapters on population/environment aspects of cities in Canada, the US, Latin America, Western Europe, Eastern Europe, Africa, Southeast Asia, China and Hong Kong, and Japan. An overview chapter discusses carrying capacity and the metabolism of cities, pressure points at the local scale (job opportunities, food supplies, housing, the waste economy, energy use), implications of global warming and sea-level rise, and policy options for settlement design and management (the extended metropolis, dominant suburban centers, the socialist city, the self-help city, and cities sustained by international resources). Concludes that the trajectory toward sustainability is mixed but mostly bad: poorer cities are becoming less sustainable as their populations rise, while richer cities have a very limited consciousness of the global problems to which rich cities are major contributors. "The rich countries have become very used to having their cake and eating someone else's too." A bibliography of about 750 items is included. [**NOTE**: High-quality scholarly thought, but with no consideration of new technologies. On a more activist/idealist level, see the proceedings of the First International Ecocity Conference (Berkeley, 1990; **FS Annual 1992** #11119) which stimulated publication of *The Urban Ecologist* (PO Box 10144, Berkeley CA 94709), a 20-page quarterly newsletter on ecological rebuilding progress in the US and the world. The Second Ecocity Conference was held April 1992 in Adelaide, Australia.] **(sustainable cities)**

334

Sustainable Cities: Concepts and Strategies for Eco-City Development. Edited by Bob Walter (President, Eco-Home Network), Lois Arkin (Coordinator, L.A. Eco-Village), and Richard Crenshaw (Annapolis MD). Los Angeles: Eco-Home Media (4344 Russell Ave; 213/662-5207), Nov 1992/354p(8x11")/$20.00 + $3.50p&h.

Presentations from the First Los Angeles Ecological Cities Conference in 1991. Topics includes ***The Fate of Our Cities is the Fate of the Earth*** (by Ernest Callenbach, who states that "The challenge we face is to start rebuilding America yet again, in a way that is habitable by human beings instead of just cars, and that is vastly less ecologically destructive"), ecological planning principles for sustainable development, new strategies for suburban growth, retrofitting older neighborhoods, ecological community design (by Richard Register), low-toxic construction systems, a new architecture of sustainability, solar design, home and community water management, water harvesting (microcatchment elements to be integrated into the bioengineering process), urban forests, urban agriculture, integrated waste management, home and community composting, reducing traffic congestion, the role of bicycles, market approaches to reducing sprawl and congestion, changing bank lending policies, a Pedestrian Bill of Rights (sponsored by L.A. City Councilman Michael Woo), and a future scenario for ***Eco-L.A.*** by Sim Van der Ryn. **(L.A. ecocity conference)**

335

Toward Sustainable Communities: A Resource Book for Municipal and Local Governments. Mark Roseland (Centre for Human Settlements, U of British Columbia). Ottawa: National Round Table on the Environment and Economy, Dec 1992/340p/free (613/992-7189).

Seeks to enable local government officials and citizens to apply the concept of "sustainable development" in their communities, in that enlightened local decisions on environmental issues will be of global as well as local benefit. Chapters discuss the meaning of sustainable development, the city as eco-system, climate change and air quality, transport planning and traffic management, land use and growth management, energy conservation and efficiency, solid and hazardous waste reduction and recycling, water and sewage, edible planting and indigenous landscaping, economic development for a sustainable economy, healthy communities, environmental education, green investment and purchasing, leadership by example, environmental administration, and regional programs. Each chapter includes descriptions of key books, articles, and organizations. **(sustainable communities resource book)**

336

Shaping Cities: The Environmental and Human Dimensions. Marcia D. Lowe (Senior Researcher, Worldwatch Institute). Worldwatch Paper #105. Washington: Worldwatch Institute, Oct 1991/69p/$5.00.

Today's cities provide neither equilibrium with nature nor stability for citizens. And the growth of cities is an undeniable fact of the future. "At current rates of expansion, the world's population will double in 40 years, the urban population will double in 22, and the Third World urban population in 15—with momentous consequences for both human life and the environment." On a rapidly urbanizing planet, much is at stake in the way urban land is used. Many cities have long records of inadequate or ill-advised planning, or no planning at all. Good planning is essential, and various ideas are described as concerns: **1) Transport**: freeing cities from auto-centered problems, integrating homes with workplaces and other amenities so as to encourage walking and cycling, banning free parking downtown, encouraging compact growth (e.g., Portland, Oregon, has encircled itself with an Urban Growth Boundary to fend off sprawl); **2) Energy Efficiency**: reducing dependence on outside sources of energy by solar tanks, district heating, light-colored paints and building materials to reduce the heat island effect, careful landscaping (which can bring the temperature down by more than 8°C in high-density urban areas); **3) Protecting Water**: reducing pollution in aquifer recharge areas, identifying areas that are critical for water quality, adequate management of storm water; **4) Affordable Housing**: replacing exclusionary regulations with controls that promote a variety of housing types, land-sharing arrangements between illegal occupants and landowners, revising inflexible building codes and rent controls, creating land banks of tax-delinquent land, tax strategies to discourage speculation; **5) Humane Cities**: restraining auto traffic, devoting more space to trees and other plants, encouraging urban agriculture, liaison with neighborhood organizations. **6) Better Land Use**: "many cities have so much underused space that they could develop for decades to come, without bulldozing another square meter of undisturbed land." Concludes by advocating "compact growth" and carefully guided development to conserve energy, protect water resources, arrest destructive sprawl, and meet people's needs. To achieve these goals, three conditions are essential: better access to information on the community's population, greater regional cooperation, and stronger support from national governments. [**NOTE**: A useful global overview of practical ideas.]
(improving urban land use)

337

Garden Cities 21: Creating a Livable Urban Environment. John Ormsbee Simonds (EPD of Pittsburgh). NY: McGraw-Hill, Jan 1994/231p/$42.00.

Inspired by Sir Ebenezer Howard's **Garden Cities of Tomorrow** (1898; MIT Press, 1965), which found the industrial city to be unlivable, a leading US urban planner describes the livable city of tomorrow as one with a long range preservation, conservation, and development plan. Such cities will intensify the activity centers, preserve the integrity of the surrounding landscape, and engage in coordinated regional planning. Guidelines include strengthening the neighborhoods, incorporating the thinking of community leaders, preserving the best existing features, creating traffic-free activity centers of various types, using multimode rapid transit, preventing and eliminating all forms of pollution, planning integrated systems, creating an open space framework, planning the city and metro region together, and providing for flexibility so as to encourage creative individual expression. In the years ahead, the garden city envisioned by Ebenezer Howard may yet be realized, but in a different form. Within an overall parklike setting, there will be community gardens, hydroponic farms, orchards, vineyards, and solar-domed conservatories. City centers will be "hanging gardens," the main trafficways will be lineal parks, and local streets will be sheathed with canopies of foliage. As our cities are made to leaf and bloom again, gardening will be a favorite avocation and nature appreciation and outdoor recreation will become ever more a part of city living. [ALSO SEE: ***Designing the Future: A Cybernetic City for the Next Century*** by Jacque Fresco of Venus FL (*The Futurist*, May-June 1994, 29-33), on Fresco's "Venus Project," a circular city of free-form dome buildings surrounded by parks and gardens, designed to operate with minimal energy.] **(livable cities)**

338

Greenways: A Guide to Planning, Design, and Development. Edited by Loring LaB. Schwarz (Director of State Greenway Planning, The Conservation Fund, Arlington VA). Washington: Island Press, Nov 1993/351p(8x11")/ $45.00;$29.95pb.

Greenways are natural corridors crisscrossing a landscape that has been otherwise transformed by development—a generic term for a wide variety of linear open spaces that provide connections and foster movement of some sort, from neighborhood bicycle routes to wildland corridors that guide migrating wildlife. Interest has been renewed in this old concept because greenways or greenbelts can address many needs with minimal funding. Most greenways are community resources, and the lifeblood of most greenway initiatives is volunteer energy and landowner support. Charles Little's **Greenways for America** (Johns Hopkins UP, 1990; **FS Annual 1992** #11158) provided many examples of successful greenways. This book offers specific information on how to realize greenway visions, with chapters on developing a plan, building public support, funding, protection and ownership, promoting natural values, design of the trail and facilities, and management. **(greenway planning)**

339

Designing a Sustainable Future: Buildings, Neighborhoods, Cities. Edited by Robert Gilman. *In Context*, No. 35, Late Spring 1993/61p/$6.00 single issue.

Buildings account for more than 40% of all US energy use, and building placement dictates much of our need for transportation. There is no way to achieve a sustainable future without major changes in our built environment. Sustainable design has matured to the point where it can deliver a built environment with a much lower environmental impact while enhancing health, community, and quality of life—and saving money. The 16 brief articles describe restorative design, working with light, the National Audubon Society's new headquarters in a recycled Manhattan building (demonstrating whole-system success), institutional inefficiency (Amory Lovins on forces such as developers and lenders that frequently lead to inefficient buildings), new housing that collects energy and water, eco-community design in Bamberton BC (by Guy Dauncey), the Los Angeles Eco-Village concept, the Village Homes community in Davis CA, living machines to recycle human waste, the Winslow Cohousing group in Bainbridge Island WA, integrated and regenerative eco-cities (by John T. Lyle, author of **Regenerative Design for Sustainable Development**; Wiley, Nov 1993), the "pedestrian pockets" of Peter Calthorpe, and changing building patterns in Melbourne, Australia. **(sustainable design)**

340

A Building Revolution: How Ecology and Health Concerns Are Transforming Construction. David Malin Roodman and Nicholas Lenssen. Worldwatch Paper 124. Washington: Worldwatch Inst. March 1995/67p/$5.00.

Modern buildings represent an extraordinary achievement with a hidden cost. They make life easier for many today, but their construction and operation inflict grievous harm on the environment and on people (up to 30% of new and renovated buildings suffer from "sick building syndrome," subjecting occupants to unhealthy air). Buildings account for one-sixth of the world's freshwater withdrawals, one-quarter of its wood harvest, and two-fifths of its material and energy flows. Roodman and Lenssen discuss the rising complexity of construction and specialization of related roles, quality construction as a key to ecological design, the desirability of integrative design and adaptable buildings, the need to eliminate subsidies and reform tax policy so that true economic and social costs are felt, and new technologies for energy efficiency. No single measure will rectify the problems with modern buildings; a broad set of changes is needed to create a secure home for humanity. Some promising examples: 1) the Dutch National Environmental Policy Plan [also see #418] seeks to improve energy efficiency 25% by 2000 in new and renovated buildings, and to increase recycling of construction and demolition waste from 60% to 90% in the 1990-2000 period; 2) a Dutch NGO, the Association for Integral Biological Architecture, provides information and sells certified "green" products for builders and architects; 3) EPA's Green Lights program seeks to reduce lighting energy use by about half in at least 5% of US commercial floor space; 4) the Swedish "factory-crafting" system produces high-efficiency, high-quality homes; 5) Japan's largest construction company (Shimizu Corporation) adopted a Global Environmental Charter in 1991, pledging to incorporate environmental concerns into all of its work; 6) the Advanced Houses Program of the Canadian Energy Ministry is a showcase for state-of-the-art "green houses."

(buildings and the environment)

341

A Primer on Sustainable Buildings. Dianna Lopez Barnett (RMI Green Development Services). Snowmass CO: Rocky Mountain Inst, Spring 1995/135p/$16.95pb.

Many of today's buildings, construction practices, and land use patterns are not sustainable: they don't work now, and they won't work at all in the long run. Sustainable buildings should make appropriate use of land, use resources efficiently, enhance human health, strengthen local economies and communities, conserve species and habitats, be nice to live and work in, and be economical to build and operate. Green buildings cost about the same as conventional ones, but have many more benefits: 1) cheaper heating, cooling, and lighting (50% reduction in energy use is fairly easy to achieve, and 80-90% reductions are possible with good design); 2) increased worker productivity by 6-15% or more; 3) in several studies, when a company moved into a green building, absenteeism dropped 15-25%, and sick-leave use was significantly reduced; 4) higher initial sales prices, rents, and rental profit margins due to reduced water and energy costs. Chapters discuss green building design, site selection and development, land design, energy use, water-efficient landscaping, wastewater and alternative sewage, indoor air quality, and maintenance. **(green buildings)**

10. Health

342
Critical Condition: Human Health and the Environment. Edited by Eric Chivian (Harvard Medical School), Michael McCally (U of Chicago), Howard Hu (Harvard School of Public Health), and Andrew Haines (London Medical School). Cambridge MA: MIT Press, Sept 1993/ 244p/$29.95;$15.95pb.

A report by Physicians for Social Responsibility, which has led the effort to educate citizens about the medical consequences of nuclear war since 1961, and shared the 1985 Nobel Peace Prize with its international umbrella organization, International Physicians for the Prevention of Nuclear War. "The world now faces a similar threat to human health and survival from changes to the global environment... and there is a similar lack of understanding about the consequences." The problem of environmental degradation, however, is more complex by orders of magnitude than that of nuclear war, and its solution will demand much greater change in the way people will lead their lives. This report seeks to gather the best available medical information about the environment. Chapters discuss: **1) Risk Assessment and the Physician**: the training of physicians ought to shift from a bioengineering to an ecological model; all physicians should understand the relationship of environment to health; **2) Urban and Transboundary Air Pollution**: criticizes the current US approach to air pollution for inadequate control of ozone and lead, and for failing to address indoor air pollution; **3) Drinking Water Pollution**: "the number of different industrial and agricultural chemicals that threaten public and private water supplies is enormous;" nitrates, heavy metals, pesticides, and volatile organic compounds are of greatest concern; **4) Food Contamination**: the integrity of food is threatened by many man-made pollutants; most of these toxins are invisible and not easily detected by consumers ("moreover, the processing or cooking of food is generally not effective in neutralizing their impact"); **5) Occupational Exposures**: it is estimated that occupational disease is responsible each year for 50,000-70,000 deaths in the US, and for about 350,000 new cases of illness; **6) Radiation**: medical concern about nuclear power plants centers on four distinct public health problems: catastrophic accidents, nuclear waste containment, weapons proliferation, and hazards to workers and surrounding populations from routine operation; **7) War-Induced Environmental Damage**: the escalating numbers of weapons and the diverse technologies of destruction and delivery place the environment in greater jeopardy than ever before from aerial bombardment, land mines, toxic pollution, and despoliation; **8) Loss of Stratospheric Ozone**: increase in ultraviolet-B radiation reaching the earth is expected to have direct health effects on humans (skin cancer, aging skin, cataracts, sunburn of the cornea, impacts on the immune system) and to result in food shortages; **9) Possible Effects of Climate Change**: more summer deaths from high temperature, increased threats from communicable diseases, impacts on availability of food and water, impacts on mental health; **10) Population Growth**: physicians should have a major role in population control efforts, due to clear evidence of the threats to human health, the environment, and socio-economic development; **11) Species Extinction and Biodiversity Loss**: massive species extinctions forecloses the possibility of understanding and treating many diseases. [**NOTE**: Authoritative and readable overview. ALSO SEE: **Your Health, Our World: The Impact of Environmental Degradation on Human Wellbeing** by Diane Wiesner (Prism/Unity, dist in US by Avery, 1992/337p/ $14.95pb).] **(health and environmental problems)**

343
The Betrayal of Health: The Impact of Nutrition, Environment, and Lifestyle on Illness in America. Joseph D. Beasley, M.D. (Director, Institute of Health Policy and Practice, Bard College). NY: Times Books, Sept 1991/274p/$23.00.

Condensation of a 1989 report to the Kellogg Foundation, warning that "we are rendering ourselves toxic and malnourished while simultaneously destroying the ecological balance that supports life on earth." Modern medicine, for all its wondrous technology, has only succeeded in describing and treating the various symptoms of the crisis; it is unprepared to diagnose and treat the underlying causes. Chronic diseases, both social and medical, are really symptoms of a much more vast underlying problem: the culmination of years of inadequate nutrition, a toxic environment, sedentary lifestyles, familial and social disruptions, and dependence on artificial agents from cigarettes to cocaine. Chapters discuss: **1) Genetic Foundations**: man-made mutations are damaging the chemistry of our being; "induced mutations play a major role in the high 20th-century cancer rate and undoubtedly in a number of other chronic conditions"; **2) Nutrition**: our food supply has undergone vast changes in the past 100 years—some beneficial, many not; specific nutrient deficiencies continue to afflict most of the US population, and outright malnutrition is on the rise in many groups; **3) The Industrialization of Food**: the modern food-production system has greatly reduced ancient threats of contamination and spoilage while introducing two major new problems: man-made contaminants and reduced nutrient content; **4) Our Toxic World**: we are now exposed daily to chemicals in our food, air, water, homes, and workplaces; **5) Individual Vulnerability to Environmental Illness**: some of us are more resistant and others more susceptible due to genetic differences, but nutrients play a central role and pollutants destroy nutrients in the body; **6) Lifestyle**: some 23% of US children are obese for lack of exercise, and the fitness boom has completely bypassed many other groups; some 60 million Americans have sleeping problems each year; many Americans skip breakfast and snack on junk foods, the frequency and duration of stress is growing, and violence is increasing; **7) The Next Generation**: the deteriorating health of our children is a result of impaired development both inside and outside the womb: millions of children, from conception onward, grow up in a toxic environment and receive inadequate nourishment; **8) The Limits of Modern Medicine**: the modern approach can battle a host of diseases with heroic and specialized techniques, but is not nearly as effective in addressing the increas-ingly complex etiology of today's chronic diseases. Concludes by suggesting a new paradigm for medical care based on a broader education concerned with patient's needs, and lifestyle suggestions for individuals ("our best hope for a healthy life is not medical care but self-care"). [**NOTE**: A broadly integrative view, and the picture is not a pretty one.]

(nutrition, environment, and health)

344

Environmental Epidemiology, Vol 1. Public Health and Hazardous Wastes. National Research Council, Committee on Environmental Epidemiology (Anthony B. Miller of U of Toronto, Chair). Washington: National Academy Press, Aug 1991/282p/$29.95.

Part of our modern heritage is the increasing volume of waste created by all industrial societies (in the US, more than 6 billion tons per year). There is also unprecedented concern over potential consequences for public health and the environment from exposure to hazardous wastes. According to recent opinion polls, the American public believes that hazardous wastes are a serious threat. Many scientists and administrators in the field do not share this belief. "On the basis of its best efforts to evaluate the published literature relevant to this subject, the Committee cannot confirm or refute either view." A decade after implementation of Superfund, substantial public health concerns remain, and critical information on the distribution of exposures and health effects associated with hazardous waste sites is still lacking. Without doubt, toxic substances abound at these sites. Serious health effects cannot be ruled out, and have in fact occurred at some sites. A recent EPA study found that over 40 million people live within four miles of a Superfund site, and about 4 million people live within one mile. Residential proximity does not mean that air, water, and soil exposures are occurring, only that the potential is increased. In addition to the neglected task of assessing exposure and potential effects on public health, there is the critical question of whether Superfund and other hazardous-waste programs actually protect human health. "The Committee finds that the question cannot be answered." Until better evidence is developed, prudent public policy demands providing a margin of safety from exposures to hazardous-waste sites. We do no less in designing bridges and buildings, and in setting criteria for scientific credibility. [ALSO SEE: **Risk and Rationality** (#439), on rethinking risk assessment by laypersons and experts, and **Beyond Superfailure: America's Toxics Policy for the 1990s** (#161).]

(toxic waste and public health questions)

345

Chemical Deception: The Toxic Threat to Health and the Environment. Marc Lappé (Prof of Health Policy and Ethics, U of Illinois College of Medicine, Chicago). San Francisco: Sierra Club Books, Oct 1991/360p/$27.00.

We are in the midst of a chemical revolution. The chemical industry, and its allied fields of pharmaceutical and pesticide manufacture, are dominant forces that are shaping our world and insinuating themselves into our lives. "Nothing less than a science of planetary toxicology is needed to measure the results of chemical insults to the thin web of life on earth." But we have only studied this science on a small scale. "Our own and the planet's survival may depend on making accurate predictions about the composite impacts of new chemicals and electromagnetic radiation on all living systems." But we have thoroughly studied only a handful of the millions of known chemicals, and only a tiny portion of the 60,000 to 80,000 that are currently in commerce. Nearly 3,000 new chemicals are produced annually, far surpassing America's ability to test them for any cancer-producing capacity. As of 1990, only about 2,000 chemicals were tested in animals for carcinogenicity; about half were found to be positive. More than 100 billion pounds of hazardous waste are generated annually in the US, but only a few waste sites have been fully evaluated and cleaned up. More than 2 billion pounds of pesticides are used on our food crops each year, with some fraction contaminating the food chain.

Myths that perpetuate the toxics problem: 1) the problem is localized (the global reach of toxicity is made apparent by the ubiquity of air and water contamination, and occurrence of previously unrecorded cancers in fish and animals); 2) the body's defenses are adequate (the liver filters out potentially harmful substances, but is subject to damage and disease); 3) toxic effects not seen immediately will not occur (many substances leave their mark on body tissues for weeks, months, and even years after they have been excreted); 4) all effects of toxics disappear as doses diminish (this central dogma of toxicology is questionable); 5) the fetus develops out of reach of toxic danger (rather, it is the site where low doses are most likely to do harm); 6) "nonreactive" chemicals lack adverse effects (the case of silicone suggests that our understanding of toxicity must embrace immunologically provocative reactions); 7) the body's own chemicals are safe (estrogens and progestins, also in oral contraceptives, induce liver tumor development); 8) naturally occurring substances—especially dietary factors—cause most cancer (describes the flaws of the argument by Bruce Ames of U of California-Berkeley); 9) tap water is safe to drink (groundwater—about half of all drinking water—is the primary vehicle for human exposure to chemical pollutants; chlorination of drinking water itself may be mutagenic); 10) the environment is resilient (this flawed belief has blocked essential solutions to control toxic substances). Concludes that "the ability of the planet and persons to cope with toxic substances has been clearly overestimated," and that we must reduce emissions, eliminate certain chemicals, return to sustainable agriculture, and foster regenerative ecosystems.

(toxic chemical threat to health)

346

Evolutionary Medicine: Rethinking the Origins of Disease. Marc Lappé (Director, Center for Ethics and Toxic Substances, Gualala CA; formerly UI-Chicago). San Francisco: Sierra Club Books, Oct 1994/255p/$30.00.

"Denial of evolutionary realities has been the bane of modern medicine." Our present medical woes are not self-induced; rather, they result from a blindness to the natural forces that have shaped human disease. All human illnesses have their roots in the forces that shaped evolution. Emergence of new infectious diseases is a function of the extent to which pathogens can cross species lines to infect new hosts, and the evolutionary resistance of those hosts. Chapters discuss ecosystem disruption and disease, natural selection, the evolutionary view of cancer (that powerful selective forces operate on a cellular level), antibiotic resistance, the global resurgence of TB, possible new strategies for AIDS, evolutionary processes and the immune system, vulnerability to disease, evolution and asthma, and controlling the malaria parasite.

A crucial message of the evolutionary perspective is that we remain at risk for disease despite medical ad-

vances. The interconnectedness of living things on earth, poetically embraced by the environmental movement, is also the basis for our continuing peril. Our most vigorous defense against one disease often sets the stage for another. We arrive in the 21st century with some 4 million years of primate evolutionary baggage such as vulnerability to viruses, chemical hypersensitivity, parasitic infections, and autoimmune diseases. Modern medicine reinforces these vulnerabilities by using chemical controls such as antibiotics, when it could be strengthening our innate resistance mechanisms. The evolution of new diseases has been encouraged by rapid movement and dispersion of human populations, new patterns of land use and economic development, and the breakdown of public health control measures. To regain control, we need a dramatic strengthening of vaccination and surveillance programs, better control of disease vectors, and special education programs. "Without a new coevolutionary ethic, medicine will be trapped in its present reactive mode, with little or no chance for the dramatic reintegration of human and nonhuman biology that a safe world demands."
(evolution and disease)

347
The Coming Plague: Newly Emerging Diseases in a World Out of Balance. Laurie Garrett (NYC; *Newsday* Science Writer). Preface by Jonathan M. Mann (Harvard AIDS Institute). NY: Farrar, Straus and Giroux, Oct 1994/750p/$25.00.

The shock of the AIDS epidemic prompted many virus experts to ponder the possibility that something new was happening and that something worse might emerge. As stated by Mann in the Preface, the history of our time will be marked by recurrent eruptions of newly discovered diseases, epidemics of diseases migrating to new areas, diseases that become important through human technologies (e.g., superabsorbant menstrual tampons favored toxic shock syndrome), and diseases that spring from insects and animals to humans through man-made disruptions in local habitats. To some extent, each of these processes has been occurring through history. What is new is the increased potential that at least some of these diseases will generate large-scale, even worldwide epidemics. AIDS "may well be just the first of the modern, large-scale epidemics of infectious disease." The driving force behind the "globalization of disease" is the dramatic increase in worldwide movement of people and goods.

Garrett devotes chapters to the "age of optimism" in the 1950s and 1960s (when it was thought that disease could be eradicated), Bolivian hemorrhagic fever, the Marburg virus, Lassa fever, efforts to eradicate yellow fever, Ebola virus in Zaire, the Swine Flu affair and Legionnaire's Disease in the US, toxic shock syndrome, hantaviruses in the US, the growing evidence that links DNA viruses to cancer, cities as "microbe magnets" or disease centers, sexually transmitted diseases, drug-resistance (in bacteria, viruses, and parasites), "thirdworldization" (the interactions of poverty, poor housing, and social despair with disease), and the need for preparedness. [**NOTE**: Well-paced journalistic overview of many scary cases, but no good summary or recommendations. ALSO SEE: the academic version of the same argument, **Emerging Viruses** (Oxford UP, 1993); **The Hot Zone** by *New Yorker* writer Richard Preston (Random House, Oct 1994/400p/$23.00) on the Ebola Reston scare; ***Revenge of the Killer Microbes*** (*Time* Cover Story, 12 Sept 1994, 62-69); and **Plagues, Products, and Politics: Emergent Public Health Hazards and National Policymaking** by Christopher H. Foreman Jr (Brookings Institution, Oct 1994/210p/$34.95;$14.95pb), on the need for enhanced surveillance and effective resilience, and avoiding both underreaction (e.g., AIDS) and overreaction (e.g., the Swine Flu fiasco).]
(emerging diseases)

348
Man and Microbes: Disease and Plagues in History and Modern Times. Arno Karlen (NYC). NY: Jeremy P. Tarcher/Putnam, April 1995/266p/$24.95.

"An alarming tide of new and resurgent diseases has been rising around the world for decades. Now it advances faster than ever. This signals a crisis in the history of the human species." Until recently, few people seemed aware of the "epidemic of epidemics," and that we live in "a new biocultural era." Much has been written about AIDS, but far less about other new diseases that have shattered the dream of a sanitary utopia. Karlen lists 28 new diseases (Lassa fever, Lyme disease, Ebola fever, toxic shock syndrome) and 14 diseases that have revived or become widespread (chlamydia, cholera, malaria, TB). In the landmark study by the National Institute of Medicine (**Emerging Infections: Microbial Threats to Health in the US**, National Academy Press, 1992; **FS Annual 1994** #12581), six major, synergistic causes of disease emergence were emphasized: breakdown of public health measures, economic development and changing land use, international travel and trade, new technology and industry, changes in human demographics and behavior, and microbial adaptation and change. In the *best-case global scenario*, population growth will slow, and development will proceed in an atmosphere of social and political stability. The *worst-case scenario* includes a massive global die-off caused by pandemics (the Black Death killed one-quarter or more of humanity; "it is conceivable that pandemics could do so again.") But history offers grounds for cautious optimism. Our relationship with pathogens has always been one of sporadic crises and adaptations. The pace of change is faster than ever, but our tools for responding are better. [**NOTE**: Despite the lurid but quite plausible thesis, the text is remarkably restrained.] **(epidemic of epidemics)**

349
Climate Change: The Threat to Human Health, George F. Sanderson (International Trade Canada, Ottawa), *The Futurist*, 26:2, March-April 1992, 34-38.

Former deputy director of information for UNEP describes several negative consequences from the widely-anticipated greenhouse effect: 1) worsened air pollution, especially in large urban areas, by altering the composition, concentration, and duration of chemical pollutants in the atmosphere; 2) a likely proliferation of heat-related illness; 3) a probable shift in the range of conditions favoring certain pests and diseases such as malaria, hepatitis, cholera, etc.; 4) sea level rise could spread infectious disease by flooding sewage and sanitation systems; 5) the increased number of "environmental refugees" will lead to overcrowding, social stress, and instability.
(health effects of climate change)

350
Should We Fear a Global Plague? Yes—Disease Is the Deadliest Threat of Rising Temperatures, Paul R. Epstein (New Disease Group, Harvard School of Public Health) and Ross Gelbspan (*Boston Globe*), *The Washington Post,* Sunday, 19 March 1995, C1.

We are all familiar with future-horror stories about global warming, but "a much more imminent—and deadly—threat from climate change is already upon us and could be felt in North America as early as this summer."

This threat is a global spread of infections due to worldwide redistribution of disease vectors—the animals, insects, microorganisms, and plants that transmit disease to humans. Most reports on the redistribution of infections focus on growing resistance to antibiotics, ecological changes, or the economic inequities that increase vulnerability to disease. The accounts largely ignore the critical—and growing—role of global climate change in worldwide movement by many disease-carriers. Weeds, rodents, insects, bacteria, and viruses rapidly reproduce and colonize disturbed environments. Predatory birds and animals are superior competitors under stable conditions, but submit to the opportunists when their habitat is fragmented, polluted, or altered by rapid climate change. Fossil records show that insects have proliferated and extended their range faster than plants. One sign of the current spread of climate-related disease is the appearance of disease-carrying mosquitoes at higher elevations in Latin America and Central Africa. **(plague and global warming)**

11. Education

351

Envisioning a Sustainable Society: Learning Our Way Out. Lester W. Milbrath (Director, Program in Environment and Society, SUNY-Buffalo). Albany NY: State U of New York Press, Dec 1989/403p/$57.50;$18.95pb.

Author of **Environmentalists: Vanguard for a New Society** (SUNY Press, 1984; **FS Annual 1985** #6706) argues that we have built a civilization that cannot sustain itself. Our environmental predicament has its origin in the very nature of modern society, and we can only extricate ourselves by transforming society. Envisioning a new and sustainable society, and improving our social learning, can make the transition easier. Chapters discuss: 1) why our direction is not sustainable (countering four arguments that deny the problem: humans are clever, we will develop unlimited energy, the market will take care of it, we can maximize productivity from renewable resources); 2) transforming the male dominator society; 3) values for a sustainable society (new ways of thinking whereby we are conscious of our ways of knowing, take control of technology, think holistically and integratively, live lightly on the earth, etc.); 4) elements of a learning society (better ways to disseminate information, probabilistic thinking, emphasis on values as well as facts, consciously anticipatory, employing systemic and futures-thinking modes); 5) competing belief paradigms (the dominant social paradigm vs. the new environmental paradigm); 6) left-right vs. ecological thinking (ecological ideology does not fit on the standard left-right spectrum; we should avoid overemphasis on cooperation or competition, on equality or freedom); 7) ecosystem viability as a top societal value; 8) regenerative agriculture; 9) fulfilling work and enjoying life without material indulgence; 10) directing science and technology to serve a sustainable society; 11) a governance structure that helps society to learn (proposes a Council for Long-Range Societal Guidance, staffed by some 100-200 trained generalists, as an integral part of the government; the Council "would provide a basic institutional capability for bringing thoughtful consideration of future probabilities into present policymaking;" such a Council is probably the most effective thing we can do to program our government to learn); 12) sustainable development for a fragmented world; 13) strategies for intentional efforts to induce social change (empirical-rational approaches are seldom persuasive; power-coercive approaches are widely used and commonly believed to be most effective; elite change can bring about social change quickly; also social inventions and movements, and normative reeducative approaches); 14) a plausible scenario of paradigm shift despite barriers to social learning. **(learning for a sustainable society)**

352

Earth in Mind: On Education, Environment, and the Human Prospect. David W. Orr (Prof of Environmental Studies, Oberlin College). Washington: Island Press, Oct 1994/213p/$29.95;$16.95pb.

Essays from the 1990-1993 period, joined by the belief that the environmental crisis originates with the inability to think about ecological patterns, systems of causation, and the long-term effects of human actions. "Ultimately, the ecological crisis concerns how we think and the institutions that purport to shape and refine the capacity to think." Topics include rethinking education for human survival, the dangers of formal schooling (rendering students narrow technicians who are morally sterile, and deadening their sense of wonder), the marriage between the academy and the worlds of power and commerce, educating minds capable of building a sustainable world order, characteristics of real intelligence, how oil and the automobile have undermined our intelligence (by increasing the speed at which we fly over the landscape, by leading us to think that urban sprawl and ugliness are normal and necessary), how the language of politics has become confused and convoluted, the need to establish national goals for ecological literacy, rating colleges on their ability to move in more sustainable directions, the failures of economists, the problem of academic disciplines and the discipline of problems, how professionalism has Balkanized the intellectual landscape, the need to support disciplinary boundary crossings, ecological design competence (requiring the ability to comprehend patterns that connect), campus architecture as pedagogy, the need to include agriculture as part of a complete liberal arts education, long-term education for a global constituency ("authentic citizenship" contrasted to the mindset of "cheap citizenship" that requires little from us), a world that takes its environment and sacred places seriously, the true costs of the US food system, why urban areas have grown beyond our ability to manage them, reinhabiting rural America sustainably ("twenty-first century agrarianism"), and a proposed nationwide effort to build sustainable rural community. [**NOTE**: Original and insightful essays, sort of a cross between Wendell Berry (cited) and Ivan Illich (not cited). Clearly the leading edge in thinking about education for sustainability. No one else comes close.] **(education for sustainability)**

353

Ecological Literacy: Education and the Transition to a Postmodern World. David W. Orr (Prof of Environmental Studies, Oberlin College). SUNY Series in Constructive Postmodern Thought. Albany NY: State U of New York Press, Jan 1992/210p/$14.95pb.

We are in a long, difficult, and risky transition to a postmodern world that is sustainable, at peace, just, and democratic. Education in the modern world, designed to further the conquest of nature and industrialization of the planet, has produced unbalanced, underdimensioned people to fit the modern economy. Postmodern education must be life-centered: designed to heal, connect, liberate, empower, create, and celebrate. These essays, written in the 1984-1989 period, address such themes as: **1) The Crisis of Sustainability**: seen as a product of social traps in the way we use fossil energy, land, water, forests, minerals, and biodiversity; honest bookkeeping would deter entry into social traps; **2) Two Meanings of Sustainability**: "technological sustainability" believes that there are technoanswers and/or market solutions; "ecological sustainability" seeks alternatives to the practices that got us in trouble, and is often rooted in past practices and traditions, with nature as a model for design; **3) Four Types of Change**: the simplistic New Age view that change is inevitable, relying on markets and economic self-interest, relying on public policy and government regulation, and changing values through education (civic renewal is a prerequisite for sustainability); **4) Ecological Literacy**: implies a broad understanding of how people and societies relate to each other and to natural systems, the dynamics of the modern world, and of environmental problems and their solutions; **5) A Critique of Allen Bloom**: the bestseller on **The Closing of the American Mind** (Simon & Schuster, 1987; **FS Annual 1988-89** #9458), which extols the not-so-Great Books as antidote to fragmentation, "is indicative of the closure of the purely academic mind to ecological issues"; it seeks to exit the 20th century backwards, with a mummified curriculum; **6) A Syllabus for Ecological Literacy**: a finely categorized bibliography of about 400 books and articles; **7) Place and Pedagogy**: the integration of place into education requires the combination of intellect with experience; Thoreau's *Walden* as a model of possible unity between personhood, pedagogy, and place; **8) What is Education For**: a discussion of paradigms, values, scope, definition of knowledge, and the challenge of interrelatedness arising from the environmental perspective; **9) Environmental Education as Oxymoron**: education may be part of the problem—it has made us more clever but not ecologically wiser; conservation education must be woven through the entire curriculum and through all operations of the institution. [**NOTE**: Eloquent and radical thinking about education; as important to today's greening sensibilities as Ivan Illich's **Deschooling Society** was two decades ago.]

(education for sustainability)

354

Schools for the Twenty-First Century, David Orr (Prof of Education, Oberlin College), *Resurgence*, No 160, Sept-Oct 1993, 16-19. (Extracts from an Oct 1992 Schumacher Lecture; full text from Schumacher Society, Box 76, RD3, Gt. Barrington MA 01230.)

Most people in industrialized societies believe that their educational systems need radical reform. Although divided about how to go about it, there is wide agreement on the need to make society more competitive in the global economy. But there are better reasons to rethink education related to issues of human survival. The generation now being educated will have to do what we have been unable or unwilling to do: stabilize world population, reduce greenhouse gases, protect biological diversity, conserve soils, use energy and materials with great efficiency, etc. For the most part, however, we are still educating the young as if there were no planetary emergency. "Resolution of the great ecological challenges of the next century will require us to reconsider the substance, process, and purposes of education at all levels." Necessary tasks include: 1) developing more comprehensive and ecologically solvent standards for truth—a broader science and a more inclusive rationality; 2) challenging the hubris in the hidden curriculum that economic growth is good and all knowledge is equally valuable; 3) addressing the flaw in the modern curriculum that teaches a great deal about individualism and rights, but little about citizenship and responsibilities; 4) questioning the widespread assumption that our future is one of constantly evolving technology and that this is desirable; 5) encouraging ecological education, which requires the re-integration of experience into education, a new curriculum organized around the "ecological design arts," spreading ecological intelligence throughout the curriculum, thinking how we might reconnect things, and ecological management of institutions. [**NOTE**: This long-term futures perspective is notably absent from other educational critiques. ALSO SEE: **Education, Cultural Myths, and the Ecological Crisis** by C.A. Bowers (SUNY Press, 1993; **FS Annual 1994** #12613), and **Critical Essays on Education, Modernity, and the Recovery of the Ecological Imperative** by C.A. Bowers (NY: Teachers College Press, June 1993/ 220p/$45.00;$18.95pb).]

(ecological education for long-term survival)

355

Educating for the Future: A Practical Classroom Guide. David Hicks (Director, Global Futures Project, Bath College of Higher Education). Godalming, Surrey, GU7 1XR: World Wide Fund for Nature (Panda House, Catteshall Lane), 1994/106p(8x11")/£10.95 + £1.64 P&H.

Provides practical support for teachers who wish to make a concern for the future more explicit in their classroom practice, based on the premise that developing skills of forward-looking thinking is vital for our time of rapid change and an essential pre-requisite for responsible citizenship. The book is primarily intended for use with pupils in the 9 to 14 age range, although ideas herein can be adapted for both younger and older age groups. Hicks discusses the concept of alternative futures, "sustainable society" as the most essential element of any preferable future, educational rationale for a futures dimension, the insufficiency of the UK's National Curriculum, teacher support, and classroom activities in four areas: ***1) Thinking About the Future***: continuity and change, images of the future, exploring consequences, the 200-year present; ***2) Envisioning the Future***: preferable futures, rights of future generations, goals for a better world, letters to the future, an imaginary Council of All Beings; ***3) Choosing the Future***: more of the same, technological fix, sustainable development; ***4) A Sustainable Future***: projects for a better world, energy options, transport choices, global warming. **(futures-thinking in education)**

356

Learning Peace: The Promise of Ecological and Cooperative Education. Edited by Betty Reardon (Director, Peace Education Program, Teachers College, Columbia U) and Eva Nordland (Prof Emeritus, Institute of Pedagogical Research, U of Oslo). Albany NY: State U of New York Press, May 1994/234p/$49.50;$16.95.

Essays from the Project on Ecological and Cooperative Education (PEACE), launched in 1988 by US, Russian, and Norwegian scholars. Nordland asserts familiar arguments that ecology and economy are increasingly interlinked, the widening rich/poor gap is dangerous, sustainable lifestyles are needed, and education must be to learn for life and for ecological responsibility. Reardon advocates authentic security for all people, cooperation as a basic component of all sustainable social systems, cooperative learning as an intentional teaching mode, universal human rights in a democratic world society, a shift from domination to mutuality, and learning to care. Other

essays consider new thinking and universally shared values, the "big ideas" of ecology (systems, cooperation, cycles, risk analysis, balance, change, food chains), social responsibility and ecological culture, and educational planning for an ecological future. [ALSO SEE: **Ecology in Education** edited by Monica Hale (Cambridge U Press, Nov 1993/191p/$54.95), on practices in the US, Britain, Australia, Japan, Indonesia, and Africa.]
(cooperative and ecological education)

357
Green Guide: A User's Guide to Sustainable Development for Canadian Colleges. Ottawa: National Round Table on the Environment and the Economy, Nov 1992/190p/free (613/992-7189).

An overview of steps involved in creating an environmentally sustainable college, prepared in cooperation with the Association of Canadian Community Colleges. The **Guide** begins with a six-page "green vision" of mythical Gaia College in Enviroville, Canada, 20 years hence. It features a new generation of non-disposable buildings powered by solar-generated electricity converted to hydrogen gas, living decorative elements, recycling as a standard part of college life, widespread vegetarianism, campus design for pedestrians, participation of aboriginal people in instruction ("the aboriginalization of the Canadian national consciousness helped us see life in community over many generations and connected to many species"), environmental issues as part of nearly all courses ("the most pervasive transformation of curriculum since the introduction of systematic instruction"), and the college as a more active partner in community development. The bulk of the **Guide** covers case studies such as the Task Force for Sustainable Development at Red River Community College in Winnipeg, evaluating education and training for sustainable development at Assiniboine CC in Brandon (Manitoba), energy conserving measures at Fleming College, and brief sketches of programs at various colleges. [ALSO SEE: **Education for the Earth: A Guide to Top Environmental Studies Programs** (Princeton NJ: Peterson's Guides, Jan 1993/175p/$10.95pb), which outlines undergraduate programs at some 100 US colleges and universities, selected from over 400 programs.]
(green colleges in Canada?)

358
The Global Environment: Securing a Sustainable Future. Penelope ReVelle (Essex Community College and JHU) and Charles ReVelle (Johns Hopkins U). Boston: Jones & Bartlett Publishers (1 Exeter Plaza), March 1992/480p(8x10")/$45.00.

A textbook designed for a one-semester course, with instructor's resource manual, student study guide, transparency masters, videos, and a test bank available from the publisher. Chapters discuss ecosystems and evolution, global habitats, the flow of material and energy (ecosystem dynamics), population growth curves, human population growth, limiting human population, erosion and desertification, pesticides, increasing the world's food supply, biodiversity, conventional and renewable energy resources, solid and hazardous wastes, urban air pollution, acid rain, CO_2 and global climate change, water resources, and pollution of drinking water. A concluding chapter discusses eight components of a sustainable society: renewable energy, sustainable use of the atmosphere, maintenance of soil fertility and food supplies, sustainable use of water, recycling of scarce mineral resources, maintenance of biodiversity, sustainable use of biological resources, and limitation of population to the carrying capacity of the environment. [**NOTE**: Valuable for high school and freshman levels. **The Global Ecology Handbook** appears suited for a higher college level, and includes chapters on social aspects: foresight capability, development, global security, ethics, economics, and suggestions for action.]
(sustainable society textbook)

359
Sustaining the Earth: An Integrated Approach. G. Tyler Miller Jr. (Adjunct Prof of Human Ecology, St. Andrews Presbyterian College, Laurinburg NC). Belmont CA: Wadsworth Publishing Co, 1994/360p(8x11").

An elementary textbook with chapters on environmental problems in "an exponential age" and their causes, matter and energy connections, population growth and urbanization, energy resources and improving energy efficiency, evaluating energy alternatives (a two-page chart assessing 25 energy resources as to short-term and long-term availability, project cost, and environmental impact), biodiversity and conservation, wildlife and fishery management, sustaining soil and producing food, solutions to world food problems, the case for and against pesticides, types of chemical and biological hazards, risk and human health, risk assessment, weather and climate, possible effects of a warmer world, indoor and outdoor air pollution, water resource problems, reducing and reusing solid waste, and hazardous waste. A final chapter on ***Economics, Politics, and Worldviews*** considers the flaws in GNP measures, improved indicators, full-cost pricing, regulation vs. market forces, ways to reduce poverty, the transition to a sustainable-earth economy, the 1992 Rio earth summit, and human-centered vs. life-centered worldviews. Concludes with an extensive glossary and a list of about 500 further readings. [**NOTE**: An excellent, broad-ranging, and fair-minded introduction, with many charts, diagrams, and boxes. Related items include an instructor's manual, transparency acetates, test items on electronic disk, a workbook (**Green Lives, Green Campuses**) to help students apply environmental concepts to personal and campus life, and a special version of STELLA II software for developing critical thinking. Especially with these supplements, it compares well with **The Global Ecology Handbook** by Walter Corson (Beacon, 1990; **FS Annual 1991** #10453).] **(sustainability textbook)**

III. POLITICS AND PLANNING

1. The Global Context

360
Report on the World Social Situation 1993. UN Dept of Economic and Social Development. NY: United Nations, Aug 1993/226p(8x11")/$24.95pb. (Sales # E.93.IV.2)

The 13th in a series of reports prepared every four years. **Part One: Social Conditions**. ***1) Population and Urbanization***: expected world population of 8.5 billion in 2025, changes in immigration flows, growth of refugees, growth of youthful populations without employment; ***2) Hunger and Food Supplies***: chronic hunger presently afflicts some 550 million people worldwide; ***3) Health***: emerging public health concerns (spread of cholera, malaria, AIDS, TB) in the midst of remarkable advances in medical technology; declining infant mortality rates and rising life expectancy largely due to strong government intervention; ***4) Education and Literacy***: despite gains in school enrollment there were 948 million adults aged 15 and over who were illiterate in 1990; the number of book titles published worldwide increased from some 521,000 in 1970 to 834,500 in 1988; educational reforms attracted considerable attention in all countries; ***5) Housing and Sanitation***: housing conditions in most countries have deteriorated in recent years, especially in the indebted poor countries; ***6) Unemployment and Low-Productivity Employment***: a major problem in all parts of the world, and by far the most important cause of poverty; all countries are affected by the rapid advent of labor-saving technical progress; ***7) Income Distribution and Poverty***: increasing—but not uniform—polarization in living conditions; "in both developed and developing countries, women have fallen into poverty at a significant rate"; ***8) Government Expenditures on Social Services***: slow economic growth almost everywhere has imposed tightening fiscal constraints; one of the most significant trends has been privatization of social service delivery in all countries; ***9) Quality of Life***: notes a "paradoxical combination of material progress and social deterioration."

Part Two: Major Issues and Dilemmas. ***10) Major Changes in Economic and Social Institutions***: the State is being divested of means of production, mechanisms for social security are being redesigned, laws are establishing private ownership; ***11) Financial Crisis***: social security policies face severe challenges in all countries due to aging populations and slow economic growth; ***12) Ethnic Conflicts and National Disintegration***: a "great escalation" of violent conflict in the world since the end of the Cold War, destroying conditions necessary for economic and social development; ***13) The Environment***: protecting the environment from wasteful patterns of production and consumption remains a problem of "momentous importance"; ***14) Social Consequences of Advances in Technology***: spread of infotech and biotech; new technologies generally support sustainable development; ***15) Drug/Tobacco/Alcohol Use***: data on consumption patterns and trends. [**NOTE**: An excellent integration of global data and sophisticated analysis, serving as general background to the World Summit for Social Development.]

(UN on world social trends)

361
An Agenda for Development. Boutros Boutros-Ghali (Secy-General of the UN). NY: UN General Assembly, Agenda Item 91 (A/48/935), May 1994/50p.

This report is "intended as a first contribution to the search for a revitalized vision of development." Development as a common cause is in danger of fading from the forefront of our agenda. The poorest nations fall further behind, and many of the poor are dispirited. Development must be seen as a fundamental human right, and the most secure basis for peace. It has five major and interlinked dimensions: ***1) Peace as Foundation:*** development cannot proceed easily where military concerns are at or near the center of life; traditional approaches to development presuppose that it takes place under conditions of peace, but this is rarely the case; ***2) Economic Growth as the Engine of Development:*** without economic growth there can be no increase in public capital, health, welfare, or security, but this growth must be sustained and sustainable, and requires a supportive national and international climate; ***3) Environment as Basis for Sustainability:*** the environment permeates all aspects of development, and one cannot be addressed without reference to the other; "sustainability must be strengthened as a guiding principle of development"; ***4) Justice as Pillar of Society:*** a vigorous civil society is indispensable to lasting and successful social development, and popular participation at all levels of society is vital; ***5) Democracy as Good Governance:*** democracy must be regarded as a process that grows and must be sustained over time; it is the only long-term means to regulate the tensions that threaten to tear societies apart; without democracy, "development will remain fragile and be perpetually at risk."

Addressing the global development challenge requires building a common awareness of the many dimensions of development, and a better appreciation of the importance of the various actors in development. Each of the five dimensions of development is vital to the success of all others, as well as the core concept of human-centered progress. "A culture of development, in which every major dimension of life is considered as an aspect of development, is emerging." It can encompass the goals of peace, justice, and progress in a single, comprehensive vision and framework for action. Development has to be oriented not only to each person in the world, but also generations yet to come ("the welfare of future generations must not be compromised"). "Signs of a global era of development can be observed... (yet), progress is not inherent in the human condition; retrogression is not inconceivable." [**NOTE**: Important background for the 1995 World Summit for Social Development; Boutros-Ghali hopes that it "could be a synthesizing event of world significance."]

(development: a broad and revitalized vision)

362
Overcoming Indifference: Ten Key Challenges in Today's Changing World. Edited by Klaus Schwab (Prof of Bus. Policy, U of Geneva; President, World Economic Forum). NY: New York U Press, Jan 1995/368p/$50.00.

The 95 very brief contributions are arrayed in ten chapters: ***1) Disintegration of Value Systems***: death of the modern, family breakdown, decline of the US as inspiration to emerging nations, the positive values of East Asia (strong families, work ethic, communitarianism), pirates in modern society, an ethic for the 21st century, the need for standards of conduct in government and business, the need to overcome indifference (by Elie Wiesel); ***2) Maintaining Global Security***: a sustainable strategy for stabilization (promoting "stability projection" to replace "power projection"), the transformation of war, the explosion of crime, further expansion of democracy (by Samuel P. Huntington); ***3) The New Inequalities***: the new North-South conflict, growing inequality, increasing exclusion of young people (fostering a parallel economy based on crime), the new decoupled elite; ***4) Ensuring Sustainability in an Overpopulated World***: the evolving concept of sustainability, population growth, insufficient grain and food resources (by Lester R. Brown), insufficient water resources (by Sandra Postel), overcoming the "cowboy economy" (by Robert U. Ayres), international environmental policy; ***5) Living in the New Information Society***: the infotech superindustry, transition to the digital age, impact of information glut (by Neil Postman, who sees a growing sense of impotence), the universal information infrastructure; ***6) A Globalizing Economy***: liberalization of trade, emerging regionalism, the global corporation and the decline of nation-states, global shifts in manufacturing and services; ***7) Integrating Asia***: the resurgence of Asian confidence, Asia as the new center of the world economy, China's emergence; ***8) Creating Sufficient Employment***: the increased labor supply in the world economy, long-term structural unemployment (or "jobless growth"), permanent education as a must; ***9) National Policy-Making in a Global World***: why domestic economic policies still count, the end of the inflationary era (by Roger C. Altman), reconstructing the welfare state by expanding the low-wage sector, making the state efficient and consumption levels sustainable, expansion of privatization in development, the need to overcome short-term problem solving; ***10) Reengineering the Corporation***: the shift in global economic leadership away from the G-7, transition to lean processes, rightsizing global companies.

Schwab concludes that there are four interrelated revolutions underway: the revolution of information, the spread of the market economy, the reduced role of values, and the reduced input of labor in production. He also identifies three general megatrends: decreasing social and ecological sustainability of humanity's existence, the compression of time and acceleration of change (leading to discontinuous and fragmented lives; "paradoxically, the future also counts less"), and the increasingly complex set of issues on the international level. [**NOTE**: A valuable primer on global issues by an impressive list of international thinkers and leaders. The essays lack footnotes or cross-references; most follow an effective four-category format: Description of Change, Reasons for Change, Probable Consequences, and Proposed Actions.]

(ten global changes)

*363

Preparing for the Twenty-First Century. Paul Kennedy (Prof of History, Yale U). NY: Random House, March 1993/428p/$25.00.

Author of **The Rise and Fall of the Great Powers** (Random House, 1987; **FS Annual 1988-89** #8862) shifts his focus to consider humanity's encounter with technology, economic change, and population growth. "The greatest test for human society as it confronts the 21st century is how to use the power of technology to meet the demands thrown up by the power of population; that is, how to find effective global solutions in order to free the poorer three-quarters of humankind from the growing Malthusian trap of malnutrition, starvation, resource depletion, unrest, enforced migration, and armed conflict—developments that will also endanger the richer nations, if less directly." Chapters in **Part 1** discuss general trends: the demographic explosion, the communications revolution and the rise of multinational corporations (global transmission of information may not necessarily lead to universal enthusiasm for the Western way of life), world agriculture and the biotechnology revolution (biotech could very well worsen the indebtedness and trading position of poor countries), robotics and automation (longer-term implications threaten to exacerbate the global dilemma by eliminating many jobs), dangers to our natural environment, the declining usefulness of the nation-state (but no adequate substitute has emerged to replace it). In sum, the population explosion in the poor nations will lead to increased environmental damage and social stress, while globalization and new technologies in the rich nations may undermine traditional methods and locations of agriculture, manufacturing, and business in general.

Chapters in **Part 2** consider the preparedness of the world's regions and nations: **1) Japan**: probably the country least likely to be hurt by overpopulation, mass migration, environmental disasters, and globalized production; Japan has a purposeful and utilitarian education system, and is engaged in the most enormous industrial retooling for growth that the world has ever seen (but it faces considerable dangers such as nuclear proliferation in Asia and a surge in Chinese power); **2) India and China**: both nations are in a race against time to keep their population growth from eroding gains in agriculture and manufacturing; gains in the standard of living threaten the local environment and the earth's atmosphere; **3) The Developing World**: the developed economies appear to have all the trump cards in their hands and their advantages are growing as technology erodes the value of labor and materials (the chief assets of LDCs); a small but growing number of countries is moving away from "have-not" status, but many more remain behind; **4) The Erstwhile USSR and its Crumbled Empire**: the region's many problems could have serious impacts elsewhere, but the nations of Eastern and Central Europe do have lots of talented and ambitious people, and "at least they are out of jail"; **5) Europe**: the EC nations appear relatively comfortable, but face major challenges in creating a thriving unified body which will assume a responsible world role; **6) The US**: it may not be a "loser" in the face of global change, but it could be less than a clear "winner"; the US will probably continue to muddle through (as Britain did a century ago when faced with a similar erosion of power), with steady and relative decline in the long term.

Concludes by emphasizing three key elements in any general effort to prepare for the 21st century: enhancing the role of education, improving the status of women, and intelligent political leadership. "If my analysis is roughly correct, the forces for change facing the world could be so far-reaching, complex, and interactive that they call for nothing less than the reeducation of humankind." [**NOTE**: Outstanding survey of global trends and sobering regional prospects. Amply deserves its Spring 1993 best-seller status.] **(population/technology impacts)**

364

Global Trends: The World Almanac of Development and Peace. Edited by Ingomar Hauchler (German Parliament, Bonn) and Paul M. Kennedy (Chair, International Security Studies, Yale U). NY: Continuum, Dec 1994/416p/$39.50;$19.95pb. (First published in 1993 by Development and Peace Foundation, Bonn.)

The world is tending toward a single whole, but globalization and interdependence are taking place in a highly asymmetrical, asynchronous, and conflict-laden manner. The seven parts of this Almanac, with illustrative entries and ideas, include: **1) World Development:** a multiplicity of often countervailing trends, growing rich-poor gap, reform and strengthening of the UN (generally agreed to be urgent, but has made no headway in the last decade), the paradigms of world development that underlie the ineffectiveness of the international system (the Western model of development and its transferability, international division of labor), reorganizing the world economy, development as a preventive peace policy; **2) World Order:** concepts of order (the "realistic" vs. "idealistic" model), growth of global players (with a shift from a world of states to a world of societies), causes of ethnonationalism, principles for strengthening the UN; **3) World Society:** the UNDP Human Development Index, poverty trends in world regions, main dimensions of human development, regional trends in political human rights, population trends, migratory movements; **4) World Peace:** downward trend in worldwide military spending (projections through 2000 assume an annual 3% decline), trends in the worldwide arms trade, the UN "Agenda for Peace"; **5) World Economy:** growth of the world economy, trend toward a world service society, uncoupling of finance markets from the real economy, structures of trade, growing unemployment everywhere; **6) World Ecology:** dangers of climate change, soil and water, global destruction of forests, energy supply risks, renewable energy; **7) World Culture:** main currents in religion, the World Ethos Project of Hans Küng, communication of the world cultures, the proliferation of TV sets, new technologies, speeding up of technology development. [**NOTE**: Unique format blends German, American, and UN sources with a variety of interesting charts, tables, and sketchy ideas. ALSO SEE **World Futures and the United Nations: An Annotated Guide to 250 Recent Books and Reports** (Bethesda: WFS, Apr 1995/116p/$25).] **(global trends almanac)**

365

The Road to 2015: Profiles of the Future. John L. Peterson (President, The Arlington Institute, Arlington VA). Corte Madera CA: Waite Group Press (200 Tamal Plaza), Sept 1994/372p/$18.95pb. (Dist by Publishers Group West, Emeryville CA; 800/788-3123.)

Offers an integrated framework to look systematically at the forces that drive change in the major segments of human life, in response to the need for systems thinking for our new era of infotech and mounting environmental problems. Chapters consider 11 areas of life: 1) new ideas in science (complexity/chaos theory, inventing life, consciousness research); 2) extraordinary new technology (the age of the microprocessor, globe-girdling networks, AI, virtual reality, biotechnology, and emerging technologies such as molecular nanotechnology, cold fusion research in Japan, and condensed charge technology); 3) environmental problems (including overlooked issues such as desertification, more hurricanes, and growing natural disaster destruction); 4) population (the global explosion, immigration pressures, changing US demographics and families); 5) energy shifts (including possible breakthroughs in room-temperature superconductivity and zero-point energy that captures energy out of electromagnetic fluctuations in a vacuum); 6) new directions for transportation (maglev trains, high speed trains, magnetohydrodynamic ship propulsion, electric automobiles); 7) commercial uses of space (communications, power generation, remote sensing); 8) health threats and gains (life extension, AIDS, virulent diseases revived); 9) changing social values with changing generations; 10) economic shifts and uncertainties (trade imbalance, inflation, national debt, insurance companies weakened by a record series of catastrophes, a potential pension crisis); 11) major political trends (democratization, multilateralism, fragmentation, diffusion of power, weapons proliferation). Chapter 13, 50 pages on wild cards and crosscuts of major trends (each explained, with implications discussed) alone is worth the price.] **(trends and wild cards overview)**

366

Global 2000 Revisited: What Shall We Do? Gerald O. Barney (Executive Director, Millennium Inst). Arlington VA: Millennium Institute, 1993/105p(8x11")/$20.00pb; 16p summary, $8.00. (Order from Public Interest Pubs, PO Box 229, Arlington VA 22210; 800/537-9359.)

A brief, updated version of **The Global 2000 Report to the President** (1980; **FS Annual 1980-81** #1620), prepared for the 1993 Parliament of the World's Religions held in Chicago. [Barney was the Study Director of the first G2000 report, which sold 1.5 million copies in eight languages.] The overall conclusion: "If present beliefs and policies continue, the world in the 21st century will be more crowded, more polluted, less stable economically and ecologically, and more vulnerable to violent disruption than the world we live in now. Serious stresses involving inter-religious relations, the economy, population, resources, environment, and security loom ahead. Overall, Earth's people will be poorer in many ways than they are today." Critical issues highlighted: our growing numbers and basic needs, food and the loss of arable land, the need to double agricultural yields, loss of biodiversity, global energy for sustainable development, greenhouse gases and ozone loss, and poverty and violence.

To avoid a tragic wasteland and create a just, sustainable future, North and South must build a new world order, including: 1) new participatory mechanisms to replace those of the Cold War era such as the UN Security Council; 2) make development peace and focus on successful transfers; 3) a strong, non-discriminatory body of international law; 4) a continuing forum for global discussions of the whole human mega-problem (the agenda must cease to be limited to a few narrow topics); 5) replace the outmoded and misleading UN System of National Accounts with a new set of national indicators that measure the degree to which a nation is living sustainably; 6) establish everywhere social conditions that actively resist corruption and favoritism; 7) accelerate transfer of up-to-date, socially and environmentally beneficial technologies. The North should make its primary contribution to global sustainability by stabilizing its resource consumption and using "defense" funds to invest in the South; the South can make its primary contribution by achieving population stability.

Questions are also addressed to spiritual leaders about the adequacy of their traditions (Barney doubts "that there is a faith tradition on Earth today that can provide the moral foundation needed for the 21st century"). Over the next five years, humanity "must prepare to die to 20th century ways of thinking and being," while preparing to see the opportunities in our new condition. For our entry into the 21st century, we need a new symbol, such as crossing a stream or river to a new place and a new way of being. [**NOTE**: Of special interest are 20 charts showing long-term trends from 1600 to 2200, highlighting expected changes in the 75-year lifetime of a child born today.]

(need to build new world order)

367

The Choice: Evolution or Extinction? A Thinking Person's Guide to Global Issues. Ervin Laszlo (Montescudaio, Italy). Afterword by Federico Mayor (Director-General, UNESCO). NY: Jeremy P. Tarcher/Putnam, Aug 1994/215p/$17.95.

"Our generation is called upon to make the choice that will decide our ultimate destiny... We are forced to choose, for the processes we have initiated in our lifetime cannot continue in the lifetime of our children." The four parts discuss: ***1) The Threat to Survival:*** drivers of the grand transition (informatization, and globalization of government and business), the five shock waves of this century (communism, fascism, decolonialization, glasnost, and *global stress* caused by poverty, overpopulation, militarization, and environmental degradation), obsolete beliefs and misguided practices (the "Neolithic illusion" that nature is infinite, economic blind spots, policy specialization and fragmentation); ***2) The Imperatives of Perception:*** the urgent need for evolutionary literacy, preserving the integrity of the world's cultures within the "culture of interexistence," catalyzing social creativity (the role of science, art, religion, and education); ***3) The Imperatives of Action:*** three basic rules (think globally, act morally, live responsibly), priorities of human development so as to escape the dinosaur syndrome (providing the necessary minimum of education, free and unbiased channels of communication, relevant and reliable information), strengthening UNESCO (which is active in all of the domains where priority action is called for), creating a functional world order ("a global peacekeeping system, an ecologic authority, and a financial regulatory body would serve every nation's interest; they would not impair the autonomy of national and local decision-making in other policy domains"), thorough reform of the UN system (better coordination of agencies, and opening membership to regional organizations, multinational business and NGOs); ***4) The Outlook:*** signs of hope (value change, green politics, the shift in corporate culture, the shift toward a systems paradigm in science, articulation of a new theology), the hope for an "evolutionary future" where some societies develop highly advanced environmental technology while others evolve in different ways guided more by social and spiritual goals ("a world of this kind would be diverse, but not inequitable and unjust"). [**NOTE**: Concludes with a selected list of 26 books by Laszlo on evolution, systems thinking, and world problems. Laszlo is a member of The Club of Rome, President of The Budapest Club, and Director of Planetary Citizens.]

(action for an evolutionary future)

368

The First Global Revolution: A Report by the Council of The Club of Rome. Alexander King (former President, CoR) and Bertrand Schneider (Secretary-General, CoR). Foreword by Ricardo Diez-Hochleitner (President, CoR). NY: Pantheon Books, Sept 1991/259p/$15.00pb.

We are in the early stages of forming a new type of world society. The magnitude of changes in the population explosion, world climate, global food security, energy availability, and the geopolitical situation amounts to a major revolution on a worldwide scale. The global revolution has no ideological basis; humanity must grope its way toward an understanding of the new world. Since its founding in 1968, the Club of Rome, comprising 100 independent individuals from 53 countries, has sought a deeper understanding of "the world problematique"—the massive and untidy mix of problems that form the predicament of humanity. This is described in chapters on the whirlwind of change, areas of acute concern, mismanagement of the world economy, evaporation of generalized religious faith and respect for the political process, and the worldwide human malaise.

A corresponding term, "the world resolutique," is introduced here to connote a coherent, comprehensive, and simultaneous attack on many of the diverse elements of the problematique. Three zones of the problematique that demand immediate attention: 1) reconversion from a military to a civil economy (all governments should institute active policies for their declining arms industries, with freed resources devoted to improving the social structure); 2) rethinking development strategies (with much greater priority to be given to the needs of the marginalized and forgotten millions of the rural poor, and to small-scale projects); 3) creation of an environment for survival (requiring reduction of global CO_2 emissions, reforestation, alternative forms of energy, and greater energy conservation). Suggestions for action on the environment include a Worldwide Campaign of Energy Conservation and Efficiency working with National Energy Efficiency Councils, a competent high-level body to consider in depth and over a long time frame the implications of macropollution, a UN Environmental Security Council parallel to the existing Security Council for military matters, Global Development Rounds (regular meetings of industrial leaders and governments similar to the Tariff Rounds of GATT), creation of National Centers for Clean Technology, a UN scientific meeting to plan a comprehensive World Alternative Energy Project, and concepts of global development integrated into UNESCO educational programs.

Unfortunately, the deficiencies of government are at the root of many strands of the problematique. Some of the more obvious inadequacies of national governments and intergovernmental bodies include retaining the principle of sovereignty in the face of interdependence, priority given to short-term solutions, organization in the form of sectoral ministries with a confused mass of specialized units and vested interests, and swollen bureaucracies. New qualities of governance must include a strategic vision and a global approach, a capacity for learning and innovation, and an ethical perspective. Other "agents of the resolutique" must be the challenge of continuous learning in a transitional society ("Introduction to the Great World Problems" should be a compulsory course for all students), science and technology for development within a human framework, and the mass media (the Club of Rome seeks to engage in a broad debate with journalists and top media leaders). These ideas "are offered as a basis for learning our way into the future." [**NOTE**: Some important ideas, but with no index, no bibliography, and only a few very abbreviated footnotes, mostly referring to other CoR reports. There is a hint of insular hauteur to this report, rather than reaching out in the spirit of intellectual interconnectedness demanded by our times.]

("world problematique" and "world resolutique")

369

International Futures: Choices in the Creation of a New World Order. Barry B. Hughes (Prof of International Relations, U of Denver). Boulder CO: Westview Press, Aug 1993/205p/$49.95 (includes software disk).

A textbook on potential global developments, combined with a computer simulation model called International Futures (IFs) that simulates population, food, energy, environmental, economic, and political developments from 1990 to 2035 in 10 geographic regions of the world. Chapters discuss action in the face of uncertainty, values and the future (material progress vs. sustainability, economic growth vs. equality, individual security vs. collective security), global change (demography, energy, environment, technology, economy), using a model to investigate change, pursuing security and peace, pursuing economic well-being and equality, modernism vs. ecoholism, and preferred futures (IFs allows selection of up to ten "goal variables"). [**NOTE**: Does not advocate any specific vision, but seeks to help the student develop one. Hughes is also the author of **World Futures** (Johns Hopkins, 1985; **FS Annual 1985** #6465) and **Continuity and Change in World Politics** (Prentice-Hall, 2nd Ed, Aug 1993), which he recommends as a companion volume to **IF**.]

(International Futures computer model)

2. Environmentalist Critiques

370

A Moment on the Earth: The Coming Age of Environmental Optimism. Gregg Easterbrook (Arlington VA). NY: Viking, April 1995/745p/$27.95.

"This volume . . . is not an attack on environmentalism." Ecological consciousness is a leading force for good in world affairs; without the unstinting pressure on government and industry from modern environmentalism, the Western world today might actually be in the kind of ecological difficulties conventional wisdom assumes it to be in. "Instead, the Western world today is on the verge of the greatest ecological renewal that humankind has known; perhaps the greatest that the Earth has known. Environmentalists deserve the credit for this remarkable turn of events." But, although environmentalists are surely on the right side of history, they are "increasingly on the wrong side of the present, risking their credibility by proclaiming emergencies that do not exist." Ecorealism will be the *next wave* of environmental thinking, based on principles of logic and accurate understanding. "In the Western world pollution will end within our lifetimes, with society almost painlessly adapting a zero-emissions philosophy;" indeed several categories of pollution have already ended. Nearly all technical trends are toward new devices and modes of production that use fewer resources, produce less waste, and cause less ecological disruption. Even a growing human population can take a constructive place in the natural order. "Fashionable pessimism about the environment could not be more wrong."

Chapters discuss thinking like nature (arguing that factory-based economies may ultimately be the sustainable ones), understanding the strength and resilience of nature (absent nuclear war, no environmental threat humanity causes or is likely to mount is unprecedented), nonlinear evolution, why Rachel Carson was wrong, the technological life ("done properly, a high-consumption technological existence actually may be less taxing to the environment than a subsistence lifestyle"), acid rain, air pollution, the spotted owl (not an endangered species, but a subspecies), chemicals, the clean-tech vision, global cooling and warming, energy, the romanticized past of enviros, farms and forestry, genetic engineering (which may be able to reverse extinctions and increase the pool of resources), land use issues, natural and artificial radiation, species extinctions (it is better to protect general habitats rather than individual species), the Third World, toxic wastes, water clean-up in the US, what the New Nature may be like (possible high-tech alterations may create "a gentle dawn"), and the possibility of greening Mars (extending the size and number of biospheres could enable a large expansion of life). Concludes that our moment on the Earth could be the juncture at which a profound positive development begins: the moment when people, machines, and nature negotiate terms of truce. [**NOTE**: Wordy and provocative; often takes high-tech optimism to an extreme.]

("ecorealism" as next wave?)

371

Green Delusions: An Environmentalist Critique of Radical Environmentalism. Martin W. Lewis (Asst Prof of Geography, GWU). Durham NC: Duke U Press, Oct 1992/288p/$24.95.

The global environment faces two profound ideological threats: the anti-environmentalists (Julian Simon, Dixy Lee Ray, Ben Wattenberg) who would have us believe that the ecological crisis is a mirage, and eco-radicalism. The dominant version of eco-radicalism rests on the postulates that primal peoples exemplify how we can live in harmony with nature, thoroughgoing decentralization is necessary, technological advance is inherently harmful and dehumanizing, the market system is inescapably destructive and wasteful, and that economic growth is by definition unsustainable. Five main variants of this eco-extremism are identified: antihumanist anarchism (a scft variant of deep ecology, advocating decentralization and reintegration with nature), primitivism (the harsh variant of deep ecology, advocating the active destruction of civilization), humanist eco-anarchism (dominated by Murray Bookchin and his school of "social ecology"), eco-Marxism (strictly an academic movement, viewing the root of the problem in sociopolitical structures), and radical eco-feminism (which blames the male mode of apprehending reality for our ills).

In almost all respects, eco-radical prescriptions are in the end just as harmful as the disease originally diagnosed: 1) eco-extremists threaten the environment by fueling the anti-environmental countermovement; 2) some radicals oppose necessary environmental reform, arguing that it is worse than useless because it postpones thorough transformation; 3) embrace of "appropriate technology," often little more than well-engineered medieval tools, would undercut foundations of all technological R&D; 4) reverting to self-sustaining rural communities would only recreate the conditions that have historically led to steady population gains; 5) primal economies have rarely been harmonized with nature as they are depicted; many have actually been highly destructive; 6) small-scale structures can be just as violent and ecologically wasteful as large ones: "small is sometimes ugly and big is occasionally beautiful"; 7) anti-growth views are biased against even nonconsumptive forms of economic growth (while the global economy cannot grow indefinitely in volume, it can continue to expand in value by producing better goods and services more efficiently—a type of economic growth that is absolutely essential); 8) the fear of compromise seriously diminishes the possibility of creating a broader coalition for environmental action; 9) rather than seeking to reconnect humanity with nature, human society should strive to separate itself as much as possible from the natural world (the greatest hope for virtually complete decoupling may lie in nanotechnology).

Concludes that radical environmentalism's ecology is outdated and distorted, based on "erroneous ideas fabricated from questionable scholarship." Most eco-radicals show an ignorance of history and a willful dismissal of economics. Their thought is mired in idealism, which inspires eloquent jeremiads but is less conducive to designing concrete strategies for effective social and economic change. In contrast to "Arcadian Environmentalism" and "naive globalism," Lewis advocates a "Promethean Environmentalism" that harmonizes technology to a new environmental vision and favors "guided capitalism" as the only way to realize general prosperity, social equity, and environmental stability. A 4-page tabular appendix compares these two worldviews as to general orientation, ecological views, human scale, technology, economics, and global development. [**NOTE**: A strong moderate/reformist argument: sophisticated, smoothly-written, and amply-documented.] **(eco-radicalism questioned)**

372

No Turning Back: Dismantling the Fantasies of Environmental Thinking. Wallace Kaufman (North Carolina). NY: BasicBooks, Aug 1994/212p/$25.00.

An environmentalist who lives in the woods "farther from neighbor or passersby than Thoreau was at Walden Pond" worries that "the environmental movement has almost lost touch with reality" and that "it has started to exercise power in ways that may do more harm to nature than good." Instead of seeing the natural world as manageable, it sees a fragile, unmendable lace of life. Instead of individualized responsibility, it wants centralized government. Instead of recognizing that economic growth goes hand in hand with stewardship, it has declared growth the enemy. Chapters discuss differences between the conservation and environmental movements, false prophecies of doom, Malthusian views (in contrast, humankind uses technology to increase the earth's carrying capacity faster than it has increased its numbers), today's most fashionable idea of "sustainable development," the "distorted" historical views of Wendell Berry, the 1854 speech of Chief Seattle widely cited by environmentalists (it was invented in 1971), mistaken views of the wisdom of primitive peoples, environmental journalists who "sell the myths of the movement," the biased analysis of the Worldwatch Institute and reverent coverage of its reports, environmental education as teaching obedience to uncriticized authority, the property rights movement, the questionable cost-effectiveness of energy conservation programs, the "determined antitechnology and antibusiness crusade of environmentalists," and the real choice we face: not between simple living and the apocalypse, but among forms of high-tech, sophisticated dominion. "In wildness is not the preservation of the world, but in civilization is the preservation of wildness and the rest of nature."

(environmentalist thinking questioned)

373

Life on a Modern Planet: A Manifesto for Progress. Richard D. North (Herefordshire UK). Manchester UK: Manchester U Press (dist in US by St. Martin's), June 1995/326p/$29.95.

An environmental writer for 20 years seeks to reconcile his preoccupation about human relations with nature, and about the well-being of the human species. The century has seemed to accelerate human exploitation of nature and a countervailing green thinking. But the Green's contribution is "often spiritually deficient and usually practically redundant... the core message of many Greens, that Western industrial civilization is dangerous by its very nature, will, I think, be proved wrong. Indeed, I think purist Green thinking is itself the greater danger." Though human numbers have grown dramatically and will soon double, "this species and planet have rather good prospects", if we see the well-being of humankind depending on technical sophistication. Chapters discuss the expected 21st century world of ten billion people, feeding the future billions through the bio-industrial revolution, fuelling the future billions through solar and nuclear power (newer generations are likely to have a more relaxed attitude to nuclear power than their parents), the myth of ecological disaster (despite hype by the media and environmental campaigners, there is "little evidence that the technosphere is out of control"), the chlorine controversy, recycling and eco-labelling, the idea of paradise, wilderness and the "Manscape" (the whole planet has more or less become a Manscape, where nature is subject to arrested development), the idea of wilderness (far more potent in the US than the UK), Green economic thinking (books such as **For the Common Good** by Daly/Cobb are "the preserve of the virtuous rather than the realistic"), seeking a balanced approach ("the time for cool cost and benefit assessment has come"), and humanizing development. [**NOTE**: A clearly-written critique of purist Greens from a more-or-less centrist position; the "manifesto for progress" is not developed, but only implicit.]

(Green thinking questioned)

*374

The True State of the Planet. Edited by Ronald Bailey. A Project of the Competitive Enterprise Institute. NY: The Free Press, April 1995/472p/$15.00pb.

The back cover proclaims that "ten premier scholars shatter the myths of overpopulation, food, global warming, and pesticides, while redirecting environmentalists' concerns to the far more urgent problems of fisheries, fresh water, and third-world pollution—and the political causes behind them." 1) **Nicholas Eberstadt** (Harvard Center for Population Studies) notes that the global population boom has coincided with an explosion of health and of productivity, and that reversals and catastrophes that have afflicted large populations in low-income countries can be traced directly to the perverse policies of governments; 2) **Dennis Avery** (Hudson Institute) argues that food is more abundant and cheaper today than at any time in history, and that 10 billion people could be fed a minimal diet with high-yield technology and global free trade in food [see **Saving the Planet with Pesticides and Plastic**, #384]; 3) **Robert C. Balling Jr** (Arizona State U) questions the greenhouse scare, arguing that the earth's atmosphere has actually cooled by 0.13°C since 1979, and that corrective policies would have only a minuscule impact on climate; 4) **Stephen Moore** (Cato Institute) faults "the prophets of doom (as) wrong in virtually every prediction they made in the 1960s and 1970s" and calls the **Global 2000** report "a case study in the abuse of forecasting as science"; 5) **Bruce N. Ames** and **Lois Swirsky Gold** (both UC-Berkeley) question the "epidemic of human cancer" caused by synthetic industrial chemicals such as DDT; "pollution appears to account for less than 1% of human cancer" (tobacco is the most important global cause, contributing to about one-third of US cancer); 6) **Roger A. Sedjo** (Resources for the Future) provides a global overview of forests and the positive role of plantation forests (commercial logging is not a major cause of deforestation; expanding agriculture is); 7) **Stephen R. Edwards** (IUCN-The World Conservation Union) discusses sustainable conservation of biodiversity and questions dire predictions of massive extinctions (sustainable use appears to be the only approach that appears to be grounded in objective methodologies); 8) **Terry L. Anderson** (Montana State U) considers global water options, advocating full-cost water pricing, water marketing, and private water rights to provide incentives to use water efficiently; 9) **Kent Jeffreys** (National Center for Policy Analysis) worries that many ocean resources have been seriously damaged or depleted by open access to the commons, and argues for property rights to coastal zones and individual transferable fish catch quotas that may be sold or exchanged; 10) **Indur M. Goklany** (US Dept of the Interior) questions Paul Ehrlich's I=PAT formula, arguing that affluence and economic growth encourages environmental cleanup, and that air quality is worst in the developing world.

In the Prologue, ***Environmentalism for the 21st Century***, Ronald Bailey, author of **Eco-Scam: The False Prophets of Ecological Apocalypse** (1993), argues that the *first wave* of the modern environmental movement has scored some major successes, but it has been "spectacularly wrong" about some things because it operated under the false assumption of Malthusianism and advocated top-down imposition of laws and regulations. It is time for a *second wave* environmentalism that views private ownership as the best defense against environmental degradation. In the Epilogue, **Fred Smith** (President, Competitive Enterprise Institute) assesses both "doomsayers" and "cornucopians," noting that the two groups are talking past each other in many respects and also advocating a "second wave" of free market environmentalism focusing on improved institutional arrangements. A 60-page Appendix presents ***Benchmarks*** indicators such as US pollution control expenditures (up nearly fivefold since 1972), world population projections to 2050, US wetland gains and losses since 1987 (since 1991, the gains have been greater than losses), rising world cereal production, world agricultural production per capita, world aquaculture production, world commodity index, and declining consumption of energy per unit of GDP. [**NOTE**: A generally tough challenge to the "doomsters" and an alternative set of "Vital Signs" (contrast to #015).]

("second wave" environmentalism)

375
Seven Doomsday Myths About the Environment, Ronald Bailey (New River Media, Washington), *The Futurist*, 29:1, Jan-Feb 1995, 14-18.

Author of **Eco-Scam: The False Prophets of Ecological Apocalypse** (St. Martin's, 1993) and editor of **The True State of the Planet** (Free Press, May 1995) socks it to "the likes of Paul Ehrlich, Lester Brown, Al Gore, Stephen Schneider, and Carl Sagan. There is no scientific evidence to support the often heard claim that there is a global ecological crisis threatening life on the entire Planet Earth." There are local problems, but no global threats, and "there is a record of enormous environmental progress and much to be optimistic about." The "doomsters" are set straight by "seven false doomsday predictions—many of which are still being peddled by unscrupulous activists" along with "a hard look at what actually happened": ***1) Global Famine***: whereas Paul Ehrlich predicted in 1968 hundreds of millions of people starving to death in the 1970s, food production has tripled since WWII, the real price of wheat and corn dropped by 60%, and worldwide life expectancy has risen by over 16 years since 1950; ***2) Exhaustion of Nonrenewable Resources***: the 1972 "Limits to Growth" study predicted that the world would run out of various raw materials, but humanity hasn't come close to running out of any mineral resource; ***3) Skyrocketing Pollution***: in contrast to forecasts of pollution increasing exponentially, air pollutants in the US have dramatically declined despite growth of population and the economy, and water pollution is abating; ***4) The Coming Ice Age***: Nigel Calder feared the threat of a new ice age as equal to that of nuclear war in 1975; now we are supposed to fear global warming; ***5) The Antarctic Ozone Hole***: a program manager at the National Science Foundation declared in 1989 that if ozone holes keep growing "they'll eventually eat the world"; ozone holes are now seen as a "purely localized phenomenon"; ***6) Ozone Hole Over America***: in 1992, NASA declared that an ozone hole could open up over the US, but then admitted that none had opened up; ***7) Global Warming***: the "Mother of all environmental scares" is not supported by recent data showing that the planet has actually cooled, and that glaciers in both Antarctica and Greenland are accumulating ice, not melting. People are afraid of environmental problems "because so many interest groups have a stake in making them afraid." [**NOTE**: Many other interest groups have a stake in *not* making people afraid. Bailey's exaggerations are arguably worse than those that he seeks to undo. For an apolitical assessment of climate trends, see #053.]

(eco-doomsters exposed!)

376
Why Do We Hear Prophecies of Doom From Every Side? Julian L. Simon (Prof of Management, U of Maryland), *The Futurist*, 29:1, Jan-Feb 1995, 19-23.

Environmental apocalypses are less dramatic than their religious counterparts, but environmental breakdown is more widely believed than the doom of religious visionaries—by more than half of the US public, and by many leaders. Many have said that environmentalism is socialism under another label. Environmentalism does not call for government ownership of resources or distributive justice, but "perhaps the issues become confused because many environmentalists in the US are *also* socialists." Both environmental and religious prophecy differ from worldly thinking in their imperviousness to counter-evidence. The environmentalists at least pay lip service to the evidence, but "shut their eyes to long-run trend evidence, all of which contradicts their prophecies." Many people hold these beliefs against all available evidence so that they can elicit research funding, or because they lack an understanding of the process of building wealth and resources. The "enviro-prophets" refuse to engage in true scientific discussion of the validity of their forecasts. "In fact, the current crop of prophets will not even bet with me (winnings to charity) that their forecasts will come about." [**NOTE**: See *New York Times Magazine*, 2 Dec 1990 (**FS Annual 1992** #11099) on a widely-publicized bet made by Paul Ehrlich and Julian Simon on the prices of five metals over the 1980s. Prices on all five declined and Ehrlich lost; a cautionary tale of betting on the wrong indicators. ALSO SEE: **Scarcity or Abundance? A Debate on the Environment** by Norman Myers and Julian L. Simon (W.W. Norton, Aug 1994/245p/$21.00), and **The State of Humanity** edited by Julian Simon (Blackwell, late 1995).]

(enviro-prophets challenged)

377
Apocalypse Not: Science, Economics, and Environmentalism. Ben Bolch (Prof of Economics, Rhodes College, Memphis TN) and Harold Lyons (Prof of Chemistry Emeritus, Rhodes College). Washington: Cato Institute, April 1993/139p/$19.95;$10.95pb.

Views the environmental movement as a religion and a utopian crusade. "No doubt this book errs on the side of presenting the case for restraining people whose stock in trade is environmental hysteria. Our basic thesis is that the world is not about to come to an end and that by nearly any measure the people who reside in the market-oriented economies of the West enjoy a cleaner and safer environment than ever experienced in modern history." The authors discuss rationality and morality (fearing that "a world run by radical Greens would result in a mass extermination of human beings"), a revisionist view of the population problem (global food production has more than kept pace with population growth; "Malthus was wrong 200 years ago, and the odds are great that neo-Malthusians are wrong today...many people who worry about population clearly aspire to impose controls on others—and make a living doing so"), chemophobia in America (the Alar scare, the electromagnetic scare, unnecessary asbestos removal, the mandate to reduce dioxin emissions to undetectable levels), the radon scare, questionable theories of global warming ("there is a greater probability that the future will be made worse by controls" to prevent it), the missing-ozone controversy (agreements to cut use of CFCs "were made before scientific research could establish how much the compounds contributed to what might be a preponderantly natural event"), the lack of evidence that acid rain is doing any significant harm to forests, and nuclear power as "the only viable alternative to fossil fuels given present technology" (the waste problem can be managed by dissipating the radioactivity of rods in water, after which remaining uranium and plutonium is reprocessed).

(apocalyptic environmentalists questioned)

378

Environmental Gore: A Constructive Response to *Earth in the Balance*. Edited by John A. Baden (Dept of Economics, U of Washington: Chairman, Foundation for Research on Economics and the Environment). San Francisco: Pacific Research Institute for Public Policy, Sept 1994/270p/$21.95.

In contrast to Al Gore's book (Houghton Mifflin, Jan 1992; #117), most of the writers in this volume endorse "adaptive economics," which stresses liberty, individual responsibility, the mischief of concentrated power, private property rights, market incentives, voluntary action, and entrepreneurial creativity. The essays are in four parts: ***1) Civilization in the Balance***: **Earth in the Balance** as "an uneasy mixture of apocalyptic environmentalism and political realism" (with persistent reliance on misplaced metaphors and absurd analogies), Gore's Global Marshall Plan questioned, Gore's visionary approach to sustainable development (a blend of religion, revolutionary zeal, and homespun economics), Gore's "intolerance for those who disagree with him," a critique of Gore's philosophical core from an Ayn Rand perspective; ***2) Climate Issues***: the Gore vision vs. climate reality (by Robert C. Balling Jr), the origins and nature of alleged scientific consensus on global warming, arguments against the ozone apocalyptics; ***3) Resource Issues***: tropical rain forest myths and facts, rethinking the waste problem, electricity from nuclear power (by Dixy Lee Ray), promoting sustainable development through free trade; ***4) Toward Real Reform***: property rights as a central organizing principle, self-organizing sustainable communities, and the longer time horizon that nourishes a more optimistic view. [**NOTE**: A serious challenge, despite the quirky title.]

(Al Gore questioned)

379

Taking the Environment Seriously. Edited by Roger E. Meiners and Bruce Yandle (both Clemson U). The Political Economy Forum. Lanham MD: Rowman & Littlefield, Oct 1993/265p/$42.50.

After two decades of high-cost, low-output Federal efforts to protect and improve US environmental quality, it is time to take the environment seriously and consider market-oriented solutions. "The American environment is much cleaner than many other places around the world," but these successes, created by the substantial wealth of the market system, have come at a high cost. "Our resources devoted to the environment are too large and too valuable to continue to squander... new policies must take economic considerations and relative real risks to the American public explicitly into account. They must reflect the common sense that has been missing thus far in the environmental field." These essays, the third volume from the Political Economy Forum (sponsored by the Political Economy Research Center in Bozeman MT, which supports New Resource Economics or free market environmentalism), discuss the importance of property rights, environmental harms from government (e.g., draining of wetlands by the Army Corps of Engineers), water pollution control through the market approach and use of the common law (allowing users to sue polluters), Superfund as a program that does not work, acid rain control, property rights structures to solve the fishery commons problem, marketable permits to control agricultural nonpoint source pollution, problems of expert risk assessment, and understanding environmentalism as an exercise in theology (viewing it as "environmental Calvinism" that makes strong moral judgments while uninformed of any basic economics). [**NOTE**: Although not explicitly a "conservative" worldview, the recurring signs are all here: 1) vigorously extolling the merits of the private sector, the wisdom of economists, and the sins of big government; 2) looking only at US environmental issues and not at any global issues; 3) ignoring arguments for sustainability, rather than countering them (the "S-word" is absent from the index); 4) advocating more science and common sense while ignoring a vast scientific literature on US and global environmental problems; 5) attacking environmentalist simplifications and assumptions while employing many other simplifications and unexamined assumptions; 6) seeking the verbal high ground by labeling these essays as "taking the environment seriously." Such chutzpah!]

(economists question environmentalists)

380

All the Trouble in the World: The Lighter Side of Overpopulation, Famine, Ecological Disaster, Ethnic Hatred, Plague, and Poverty. P. J. O'Rourke (March Hare Farm, Sharon NH). NY: The Atlantic Monthly Press, Oct 1994/340p/$22.00.

Author of **Parliament of Whores** and **Give War a Chance**, writer for *Rolling Stone*, and "H.L. Mencken Research Fellow" at the Cato Institute describes his travels to Bangladesh, Somalia, an eco-tour of the Amazon, the UNCED conference at Rio, the former Yugoslavia, Haiti, and Vietnam, making pungent comments along the way: 1) "this is a moment of hope in history... the bad political ideas that have menaced our century—fascism, communism, Ted Kennedy for President—are in retreat"; 2) right now is the best moment of all time, and right here in the US is the best place to be; yet, "I hear America whining, crybaby to the world"; 3) professional worriers put our fears to use; Masters of Sanctimony have an agenda, working toward a political goal; "politics is the business of getting power and privilege without possessing merit"; 4) "worry itself is fairly pointless. Worrying is a futuristic matter... To worry is an act of sublime ignorance"; 5) "some kind of central planning seems to be the object of most environmental activists... the grave worries facing the world today mostly don't have solutions. That is, they don't have solutions outside ourselves"; 6) "property rights, rule of law, responsible government, and universal education: That's all we need... (the worriers) are haters of liberty and loathers of individuals. They wish to politicize everything"; 7) "there happens to be no empirical evidence to support the Malthus theory"; 8) "fretting about overpopulation is a perfectly guilt-free—indeed, sanctimonious—way for 'progressives' to be racists"; 9) "the people who believe that, as a result of industrial development, life is about to become a hell, or may be one already, are guilty, at least, of sloppy pronouncements"

[further comments aim at Al Gore and **50 Simple Things You Can Do to Save the Earth**, "which leaves you wondering whether you're going to die from environmental disaster or intellectual annoyance"]; 10) "the US government has been a terrible steward of the environment... while the rest of America was hugging trees, the US Forest Service was selling them at throwaway prices"; 11) "government improves the environment about as well as it does most other things. Government usually doesn't work ... (it) doesn't work because of the problem of concentrated benefits and diffuse costs"; 12) pollution control is not free; the environment turns out to be a luxury good; "our government persists in acting as though an immaculate earth were a God-given right instead of a taxpayer-purchased benefit"; 13) "Ecopolis: City of the Future [a scale model shown at the Global Forum in Rio] was pyramidical in cross-section—as I believe all futuristic city designs are required to be by some United Futurists trade-union work rule"; 14) rich is what we must be if we are going to fix the world; a rich Bangladesh wouldn't be an overcrowded hell hole: it would be a suburb, albeit one with bad basement flooding problems; 15) "money is preferable to politics... to rid ourselves of all the trouble in the world we need to make money. And to make money we need to be free." [**NOTE**: These quotes suggest the libertarian political essence, but don't do justice to the engaging humor. O'Rourke is funny. He's a good writer. His anti-government views are now politically correct. And he was reviewed (favorably) by *The New York Times Book Review* (16 Oct 1994). *Time*, however, calls it "the *Wall Street Journal* editorial page with a laughtrack" (31 Oct 1994).]
(world's troubles questioned)

381
Environmental Overkill: Whatever Happened to Common Sense? Dixy Lee Ray (former Governor, State of Washington, and Chairman, Atomic Energy Commission), with Lou Guzzo (KIRO-TV/Radio, Seattle). Washington: Regnery Gateway, April 1993/260p/$19.95.

Authors of **Trashing the Planet** (Regnery, 1990; **FS Annual 1992** #11115) continue their attack on government regulators and "the anti-human, anti-Western civilization attitude" of environmentalists, with chapters on: 1) the UNCED Earth Summit ("all of the actions taken there were without the benefit of competent scientific input"; the objective of UNCED is to change the present system of independent nations with a future of world government and central planning by the UN); 2) global warming (we should clear away some of the real uncertainties before devoting significant resources to a phenomenon that may exist only in computer simulations; 3) the ozone "hole" (the theory has not been proven; ultraviolet radiation levels at the earth's surface are going down, not up); 4) urban air pollution (many careful studies do not find a relationship between air pollution and public health; if all automobiles and industry were to be removed from the Los Angeles Basin, "it would still fail to meet EPA ozone standards on most days, due entirely to natural conditions and weather"); 5) food and population (the Malthusian attack on fertilizers and pesticides is ill-informed); 6) disappearing species (such hypotheses are based on guesswork and unwarranted assumptions, and lead to spendthrift environmentalism); 7) wetlands regulation (a war on private property and probably on public health, in that preserving wetlands has led to an enormous increase in mosquitoes); 8) deforestation (most projections are based on government owned and managed lands, and ignore "unofficial" forests on private land); 9) education (plays a large role "in spreading the propaganda of the environmental extremists"); 10) the media ("the primary culprits in misleading the public"); 11) the Gospel according to Al Gore ("an extremist on environmental issues"; his proposals for a Global Marshall Plan would cost the US billions).
(environmentalists questioned)

382
Science Under Siege: Balancing Technology and the Environment. Michael Fumento (Los Angeles CA). NY: William Morrow & Co, Feb 1993/448p/$27.50.

Author of **The Myth of Heterosexual AIDS** (Basic Books, 1990; **FS Annual 1991** #10759) questions the accuracy of activists who warn of environmental health problems of today as a scourge to be likened to infectious diseases of earlier ages. Chapters discuss the false alarm over spraying Alar on apples (the Great Apple Scare of 1989 was "the result of one of the slickest, most cynical fear campaigns in recent American history"), the politics of cancer testing (questioning extrapolation of animal tests to predict for cancer in humans, and the goal of zero exposure to allegedly carcinogenic pesticides), the pitfalls of am ateur epidemiology in determining cancers, the dioxin scare and overreaction at Love Canal, claims of Vietnam veterans about Agent Orange, the opponents of food irradiation (generally, the same people "who have a horror of any recent chemical or process devised by man"), the coercive utopians ("brownshirts dressed in green" who insist that we must all eat nonirradiated food), the facts about electrical and magnetic fields (questioning the books of journalist Paul Brodeur), risk evaluation fallacies ("many of the health hysterias in recent years have had their roots in the risk-free society"—the endless effort to reduce all levels of risk to zero, no matter what the trade-offs), health problems attributed to video display terminals (again questioning Paul Brodeur), the myth of environmentally benign ethanol, businesses that dishonestly capitalize on fears and anxieties, agrarian utopians and technophobes (e.g., Jeremy Rifkin), and environmental organizations as "the crisis industry." Concludes that we are in the midst of an environmental revolution that "correlates with the Reign of Terror in the French Revolution"; in the name of ideals, we are expending amazing amounts of energy to accomplish amazingly little. We need to restore sanity and sound principles to the revolution, shaping policies based on science, and rejecting the ideas of fanatics and faddists. It is time for the pendulum to swing back to center from the two extremes of "I don't care" and "React to every scare." [**NOTE**: Backed by 60 pages of footnotes. Consider this: "Eating vegetables and fruits treated with pesticides... is incomparably safer than downing the delicious ice cream produced by the environmentally and politically correct company Ben and Jerry's, chock-full of saturated fat."] **(environmental extremists questioned)**

383
Higher Superstition: The Academic Left and Its Quarrels with Science. Paul R. Gross (Director, Center for Advanced Studies, U of Virginia) and Norman Levitt (Prof of Mathematics, Rutgers U). Baltimore: The Johns Hopkins U Press, March 1994/314p/$25.95.

"Muddleheadedness has always been the sovereign force in human affairs." It blunts our wisdom, misdirects our compassion, and is the "chief artisan of the unintended consequences that constitute human history." The authors "take on a species of muddleheadedness" that abounds in the American academic community, described here for convenience as "the academic left." The academic left dislikes science, and is openly hostile toward the content of scientific knowledge and toward the assumption that such knowledge is reasonably reliable and rests on sound methodology. Much of this critique is informed or inspired by "post-modern" thought and its concomitant value system—the assumption that the ideological system sustaining the practices of Western European civilization is bankrupt and on the point of collapse. The most perplexing difficulty in responding to the newly fashionable critique of science is the absence of a central body of doctrine that can constitute the quintessence of the post-modern view. Each criticism goes off on its own tack, showing little correlation with others.

Gross and Levitt devote chapters to the history and politics of science, the "cultural construction of cultural constructivism," postmodernism and literary theory, feminist criticism, racial justice and the AIDS epidemic, and environmental alarmism. They charge "ecoradicals" with being willing and even eager to accept work with alarmist and apocalyptic implications, unwilling to accept work that modifies alarmist theories, depicting "science" as a tool of repression and multinational polluters, and making a long sequence of unrealized predictions of disaster that will desensitize the public when real environmental problems arise. They have become inclined to champion "the wisdom of gurus and shamans—Deep Ecologists, Earth Firsters, James Lovelock, Rupert Sheldrake, and the like—whose views are open to serious question." The postmodern left is ready to "scorn the canons of logic, evidence, objectivity, and coherence on which most of them used to depend for a living." **("academic left" lambasted)**

384

Saving the Planet with Pesticides and Plastic: The Environmental Triumph of High-Yield Farming. Dennis T. Avery (Director, Hudson Institute Center for Global Food Issues). Indianapolis IN: Hudson Institute (317/545-1000), Jan 1995/432p/$12.95pb.

The environmental movement is valid and important. Environmentalists force us to recognize that we are increasingly capable of—and responsible for—saving natural resources. But, although environmentalists recognize the problems, they do not always see the solutions; in the case of agriculture, "they have come to exactly the wrong policy solutions." Land is increasingly in demand for human recreation, and every bit of wildlife habitat is important. Agriculture and forestry are the only sectors where we can "create" more land without sacrificing the environment; high yields are the only way to do it. "The world cannot save its wildlife without high-yield agriculture and the careful use of farm chemicals." High-yield plantation forestry is necessary to preserve truly wild forests from logging pressures. If we aggressively pursue yield-enhancing agricultural and forestry research, "we should be able to feed, house, and clothe 10 billion people on less land than we use for farming and forestry today."

Some other provocative observations: 1) integrated pest management, though useful, is not a way to replace pesticides but only to make them more effective; 2) composting and organic fertilizer can only add marginally to plant nutrient supply, unless huge tracts of land are cleared for green manure crops, thus sacrificing wildlife habitat; 3) the low yields of organic farming threaten wildlife far more than the occasional losses to pesticides; 4) according to David Seckler of the Winrock Foundation, the most likely projection of world population is that it "will peak at about 8 billion people in 2030, and then trend downward" 5) "eco-zealots" demand that we prove zero risk in pesticides or ban them, yet "the natural chemicals in our foods themselves are 10,000 times more toxic than the pesticide residues"; 6) less than 3% of all cancer deaths are caused by environmental contamination and pollution; despite extensive research, modern pesticides have not been implicated in causing human cancer; 7) health problems definitely linked with properly used pesticides on US farms are relatively minor; even in the Third World, pesticide health risks are rarely severe; 8) Rachel Carson implied that DDT was a serious threat to humans, but "no danger to humans has yet been identified"; 9) Sub-Saharan Africa gets 90 million new malaria cases per year; many of these people could be saved if we used more DDT; 10) neither organic farming nor chemically intensive farming as we know it deserves to dominate: what we need is the "*radical middle ground*," where farmers use the most effective inputs to get high yields at the lowest economic and environmental costs; 11) we truly need to abandon the old farm subsidies and international farm trade barriers; 12) modern farming systems are radically cutting soil erosion around the world by raising yields so that less land is exposed to erosion and by using herbicides and reduced tillage to kill weeds; 13) most of the world has ample fresh water if we stop wasting it (the world's farmers use 70% of water consumed for irrigation; the vast majority of this irrigation water is wasted); 14) Californians could double the water available for non-farm use at virtually no public cost by giving farmers who own water rights actual title to the water; farmers would sell their surplus water to cities and use the income to pay for more efficient irrigation systems.

Plastic is a vital element in high-yield farming that can make room on the globe for both people and wildlife. The biggest use of agricultural plastics is for polyethylene sheeting as mulch, which helps warm the soil, cuts water evaporation and soil erosion, and prevents weeds from competing with crops. Plastic sheeting to cover greenhouses, which grow many of our fruits and vegetables, is far less expensive than glass greenhouses. Plastic sheeting is also used to protect silage, obviating the need to build silos. An important potential use of plastic is for drip irrigation, which uses permanently installed plastic tubes to feed water directly to plant roots, thus ending evaporation and runoff. Plastic is also valuable for chemical containers, nearly permanent fence posts, buckets, watering tanks, and other farm implements. [**NOTE**: Each chapter begins with quotes from "mythmakers" such as Lester Brown, Al Gore, Paul Ehrlich, David Pimentel, and Jeremy Rifkin ("mythmaker extraordinaire"), followed by quotes of what "realists say." This highly combative stance detracts from a bold and interesting argument, especially concerning use of plastics. The strategy of capturing the "radical middle ground," while pushing opponents outward as extremists and anti-scientific, aptly demonstrates the art of political sumo wrestling pioneered by Herman Kahn.] **(high-yield farming as necessary)**

385

Seizing the Future: How the Coming Revolution in Science, Technology, and Industry Will Expand the Frontiers of Human Potential and Reshape the Planet. Michael G. Zey (Prof of Management, Montclair State College; Executive Director, Expansionary Institute). NY: Simon & Schuster, 1994/414p/$24.00.

"The human species is about to burst the boundaries of nature and unleash the power of its technology and human ingenuity, hurtling itself to the next stage of evolution." Genetic engineering will revolutionize agriculture, initiating a new era of abundance and eliminating starvation. Medical breakthroughs will increase the human life span on average to 100 years. Cold fusion should extend cheap energy and national prosperity to all countries. We are living through the onset of the Macroindustrial Era, an exciting new stage of hyperprogress and societal development. The Imperative of Growth states that the human race must grow, materially and intellectually. The answer to pollution ought to be more economic growth. Advanced biotechnology can eliminate toxic waste, bacteria can "eat" oil spills and other contaminants, and impure air can be scrubbed or processed by laser beams.

The human species should summarily reject "the siren call of sustainable development," visions of zero growth, and returning to preindustrial society. Positive comments are made about maglev supertrains, the next generation of nuclear power plants (advanced light water reactors), nanotechnology, space exploration, the Human Genome Project, superdrugs, the capability of feeding 40 billion people (citing a 1984 estimate by Roger Revelle), the advanced materials revolution, the case for encouraging population growth, and creating the expansionary culture we need. But, in the coming battle for humanity's future, "the zero-growth, antiprogress faction seems to exert a strong influence in the media, the social sciences, and some parts of the political spectrum. They even have a prominent voice in the home of futurology, the World Future Society itself." Once humanity comprehends the true nature and scope of its abilities, it will demand actualizing. "History beckons, the cosmos awaits. Our destiny demands that we seize the future." [**NOTE**: An unabashed view of soaring techno-fix progress.]

(zero-growth policies deflated)

386

Sound and Fury: The Science and Politics of Global Warming. Patrick J. Michaels (Associate Prof of Environmental Sci., U of Virginia; Sr Fellow, Cato Inst). Washington: Cato Institute, Oct 1992/196p/$21.95;$11.95pb.

The most popular new religion to come along since Marxism is apocalyptic environmentalism. The Popular Vision of global warming is wrong. Michaels points to many signs that the opposite is occurring: we are creating a world in which the winters warm and the summers do not, in which the nights warm and the days do not (thus enhancing crop yields), in which the growing season lengthens and the great ice fields of Greenland and Antarctica change little (they may even be enlarging), and in which the CO_2 we are emitting to the atmosphere helps plants grow better. In 20 years, we will more fully appreciate the effects of modest warming and a greener planet. The report of the Intergovernmental Panel on Climate Change (#052) is often cited as the consensus of climatologists, but there is a "profound lack of consensus that exists among even the community of scientists and reviewers who worked on the IPCC document." Three recent surveys document that the true consensus among scientists does not coincide with the Popular Vision of climate apocalypse. In fact, through most of the past one billion years, CO_2 concentration of the atmosphere has been greater than it has today: "from the perspective of both geological and evolutionary history, the atmosphere is currently impoverished in CO_2." We have had some warming over the past 100 years, probably between one-quarter and one-half of a degree (C). The temperature record will warm because the number of people surrounding weather stations will continue to increase. The greenhouse enhancement is likely to contribute to warming, too, even though there are competing inhibitors, such as sulfate aerosol and other cloud enhancers. The combination of El Niño events, along with artificial urban warming, will create the illusion of some brief periods of rapid warming. This is likely in the mid-1990s, after Mt. Pinatubo's ash disperses and a strong El Niño roars in. The environmental lobby will press for carbon-emissions reduction; "the cost will be several trillion dollars to a nation already deeply in debt" (based on estimates by the Electric Power Research Institute). The lobby is faulted for defining the problem as apocalyptic, presenting the vision as mainstream, and supporting profound legislation before completing the necessary science. [**NOTE**: The opposing argument that the IPCC report is understated is offered by **Global Warming: The Greenpeace Report** (Oxford UP, Dec 1990; **FS Annual 1992** #11130).]

(climate change as gradual and benign)

387

But Is It True? A Citizen's Guide to Environmental Health and Safety Issues. Aaron Wildavsky (UC-Berkeley; died, Sept 1993). Cambridge MA: Harvard U Press, May 1995/574p/$35.00.

Co-author of **Risk and Culture** (U of California, 1982) and author of **Searching for Safety** (Transaction, 1988) and many other books on public policy seeks to inform citizenship in a sci/tech age so that people can understand the scientific bases for rival claims, engage in informed discussion, and make reasoned judgments. "If more citizens when confronted with claims of harm begin to ask 'But is it true?' I shall have achieved my purpose." [**NOTE**: One presumes that the good citizen should also address the same question to claims of safety.] Chapters discuss the cranberry scare, **Silent Spring** and dieldrin, the saccharin debate, regulations governing PCB residues, the DDT debate, Dioxin and Agent Orange, the costly exodus from Love Canal, Superfund's abandoned hazardous waste sites, the asbestos and alar scares, arsenic in drinking water, animal cancer testing, the effects of acid rain (less than 0.1% of our forest resources have experienced damage from acid rain, and few will be damaged in the next several decades even at present emissions levels), CFCs and ozone depletion (there is no clear evidence of global ozone depletion), disagreement on global warming, reporting environmental science, and detecting errors in environmental and safety studies. Concludes that we should reject the "precautionary principle" of worst-case scenarios. Conservatives used worst-case rules for Cold War defense policy; now liberals are using these rules for environmental and safety policies. Too often, small gains in health come at the expense of other valued qualities; overall, "there are no health benefits from regulation of small, intermittent exposures to chemicals."

("precautionary principle" rejected)

388

Green Eyeshades: Environmentalism Encounters Economic Reality, Rick Henderson (Washington Editor, *Reason*), *The American Enterprise: A National Magazine of Politics, Business, and Culture* (American Enterprise Institute), 6:1, Jan-Feb 1995, 83-85.

"Quite suddenly and unexpectedly, the political movement for ever-more-extensive environmental controls is on the ropes." Major items on the US environmentalist agenda have stalled. The Sierra Club is "teetering toward bankruptcy." Activists are in retreat. The bills are coming due. "After more than two decades of basically unrestrained growth, a new movement is afoot to control environmental costs, to balance them more scientifically against benefits, and to make sure that particular individuals and localities don't get unfairly stuck paying for other people's environmental niceties. The hard reality of the green eyeshade is descending on Green land." Risk analysts, economists, and even some environmentalists agree that the estimated $150 billion Americans pay each year for environmental protection isn't well spent. The notion that Americans can afford to live risk-free is being strongly questioned. Environmental advocates rarely acknowledge the necessity of setting cleanup priorities or leaving "the last little bit" of contamination in place. In these days of exaggerated environmental fears and perpetual $200 billion Federal deficits, the attitude that all things that are "bad" in any way must be attacked—full throttle and all at the same time—is a recipe for gridlock. [ALSO SEE: ***A New Environment: Braking the Green Machine*** by Rick Henderson (*Reason*, Feb 1995, 36-37) on prospective regulatory changes in 1995 that "could help bring environmental extremism under control."]

(risk-free environmentalism questioned)

389

The Greenpeace Guide to Anti-Environmental Organizations. Carl Deal (Asst Editor, *Covert Action Quarterly*, Washington). Berkeley CA 94707: Odonian Press (PO Box 7776), March 1993/110p/$5.00pb.

Environmentalism has now become a mainstream cause; predictably, this success has brought a backlash. With support from polluting industries and the far right, anti-environmental groups have declared war on every issue environmentalists support. Under the banner of free enterprise, democracy and economic growth, they advocate nuclear power, fossil-fuel development, and repeal of crucial environmental legislation. They have developed deceptive jargon like *wise use, integrated resource management, sustainable development, multiple use*, and *balance*. The groups fall in six categories: ***1) public relations firms*** (E. Bruce Harrison Co. of Washington is foremost in the "greenwashing" field); ***2) corporate front groups*** (the B.C. Forest Alliance was created by the Canadian timber industry via the NYC public relations firm, Burson-Marstellar); ***3) think tanks*** such as the Heritage Foundation, Cato Institute, Citizens for the Environment, and the Science and Environmental Policy Project (a Moonie-funded organization directed by Fred Singer); ***4) "public-interest" legal foundations*** (the founding president of the Coors-funded Mountain States Legal Foundation was James Watt); ***5) endowments and charities*** (in 1991, the Lilly Endowment, Carthage Foundation, John Olin Foundation, and Sarah Scaife Foundation collectively disbursed in excess of $150 million to right-wing groups); ***6) the Wise Use movement*** in the US and the Share movement in Canada (local organizations based mainly in the West). Profiles are provided of 54 groups; in addition to those mentioned above, they include The Abundant Wildlife Society, Accuracy in Media, Alliance for America (claiming over 500 member organizations), Alliance for a Responsible CFC Policy (an industry group that seeks to slow down the phase-out of CFCs), Business Council for Sustainable Development (seeking economic growth and easy access to raw materials), California Desert Coalition, Center for the Defense of Free Enterprise (the premier think tank and training center for the Wise Use movement), Committee for a Constructive Tomorrow, Consumer Alert, The Evergreen Foundation (a timber industry front group), Foundation for Research on Economics and the Environment (seeking privatization of Federal lands—[see #128]), The Global Climate Coalition (a front group of 46 corporations and trade associations advancing the view that global warming is a myth), Information Council for the Environment (also argues that global warming is a myth), The Institute for Justice, Keep America Beautiful (a "greenwashing operation" funded by some 200 companies—[see #151]), National Wetlands Coalition, Public Lands Council, Putting People First, The Reason Foundation (largest right-wing think tank outside Washington), Scientists and Engineers for Secure Energy (a think tank funded by the nuclear energy lobby), US Council for Energy Awareness, and Wilderness Impact Research Foundation. Each profile lists funders and officers, interconnections with other groups, and quotes from the principals. [**NOTE**: A small book (4x7"), but a large eye-opener. Are these groups waxing or waning? Are there any equivalents outside North America?] **(54 anti-environmental groups)**

3. Environmentalist Groups

390

World Directory of Environmental Organizations. Fourth Edition. Edited by Thaddeus C. Trzyna (President, CIPA) and Roberta Childers. Sacramento CA: California Institute of Public Affairs, Aug 1992/231p(8x11")/$45.00 + $2.00 p&h. [Fifth edition, 1996/$50.00.]

Published earlier in 1973, 1976, and 1989 [**FS Annual 1991** #10455], this handbook sponsored by CIPA, the Sierra Club, and IUCN (International Union for Conservation of Nature) describes over 2,600 organizations and programs in over 200 countries concerned with protecting the earth's resources. **1) Introduction**: includes a glossary of common abbreviations and acronyms, and a listing of landmark events (founding of IUCN in 1948 through UNCED in 1992); **2) Who's Doing What**: a guide to organizations concerned with specific problems in 52 areas such as appropriate technology, biotechnology, future studies, land use, mountains, noise, nuclear energy, polar zones, soil, tourism impacts, water, etc.; **3) World Regions**: reviews organizations concerned with the major regions of the world; **4) The UN System**: description and addresses of various programs and agencies with activities related to environmental protection and natural resource management; **5) Other Intergovernmental Organizations**: describes about 80 organizations such as ASEAN, EC, International Whaling Commission, NATO, OECD, OAS; **6) International Non-Governmental Organizations**: describes about 400 NGOs such as Alp Action, Business Council for Sustainable Development, Centre for Our Common Future, Ecotourism Society, Greenpeace, IFIAS, IIASA, IIED, WFSF, World Sustainable Agriculture Association; **7) Country and Area Listings**: addresses and phone numbers of government agencies and NGOs within various countries. An appendix lists 55 other directories and databases.

In the Foreword, Trzyna notes that three generations of organizations are described here: **Conservation Organizations** founded in the late 19th and early 20th centuries to protect wildlife and natural areas; **The Environmental Movement** that began around 1970 and looks at the human environment as a system with a knot of interrelated problems; and **The Sustainability Movement** that started around 1980, revolving around the idea of living within the limits of Earth's resources. The next wave, the fourth, may link sustainability to reforms in governance (referring not only to government but to the whole set of institutions that steer society). [**NOTE**: A very useful and intelligently constructed "who's who".]

(guide to 2,600+ environmental organizations)

391

Environmental Profiles: A Global Guide to Projects and People. Linda Sobel Katz, Sarah Orrick, and Robert Honig. Foreword by Vice President Al Gore. NY: Garland Publishing (717 Fifth Ave, Ste 2500), Sept 1993/1,083p (8x11")/$125.00.

Profiles of environmental organizations at global and regional levels in 126 countries. Each profile provides address, phone, contact person, brief history and goal statement, description of projects and people, and listing of information resources and publications. A 50-page Appendix lists organizations, projects, and individuals under the following headings: air quality/emission control, biodiversity/species preservation, deforestation, energy, global warming, health, population planning, recycling/waste management, sustainable development (some 600 listings!), transportation, and water quality. The 135-page index lists organizations, subjects, individuals, and projects such as Adopt-a-Forest, Belize Zoo, Canada's Green Plan, Data Network for Conservation, Earth Aid Society, Future in Our Hands, Global Action Network, Global 500 Honour Role, Global Legislators for a Balanced Environment (GLOBE), Institute for 21st Century Studies, etc. In all, nearly 3,000 people are mentioned, and nearly 7,000 projects, programs, and campaigns. [ALSO SEE: **Who Is Who At the Earth Summit** (Visionlink Education Foundation and Terra Christa Communications, 130 Biodome Drive, Waynesville NC 28786: 1992/481p/$30.00pb), which lists some 18,000 individuals and 1,400 organizations at UNCED.] **(environmental organizations/people)**

392

Environmental NGOs in World Politics: Linking the Local and the Global. Thomas Princen (U of Michigan) and Matthias Finger (IHEAP, Lausanne). London & NY: Routledge, Dec 1994/262p/$16.95pb.

Environmental non-governmental organizations have grown tremendously in size and number, and there is growing awareness that these NGOs are integral to world environmental politics. Almost all environmental NGOs, networks, and coalitions were started in the 1980s; the fourth edition of the **World Directory of Environmental Organizations** (CIPA, 1992; **FS Annual 1994** #12305) lists 365 international environmental NGOs. Foreign affiliates of Greenpeace grew from 5 in 1979 to 24 in 1992; affiliates of Friends of the Earth grew from 25 in 1981 to 51 in 1992. NGOs appear to be key actors in moving societies away from current trends in environmental degradation and toward sustainable economies. NGO politics are a crucial counterweight to dominant trends in the global political economy. They are critical because the biophysical and social conditions necessary for sustainability must be translated into a politics that is both local and global, and both economic and moral. This translation is not being made by the dominant actors—states and corporations. NGOs perform key roles as independent bargainers and as agents of social learning (or learning our way out of the vicious circles we have created). They are most effective at the transnational level to the extent that they exploit transnational opportunities and transnational linkages, and inject scientific and earth-centered concerns into political and economic situations. Four case studies show NGO roles in the Great Lakes water quality agreement, the ivory trade ban, the Antarctic environmental protocol, and the UNCED process.

(international environmental NGOs)

393

The Green Rainbow: Environmental Groups in Western Europe. Russell J. Dalton (Prof of Poli Sci, U of Cal-Irvine). New Haven CT: Yale U P, Dec 1994/350p/$35.00.

Chronicles the evolution of environmental interest groups in Europe, from their first mobilization in the late 19th century (e.g., a society for the protection of seabirds was formed in Britain in 1867) to the present (based on in-depth interviews with leaders of 69 environmental organizations). A recurring theme in this study is the organizational diversity of the environmental movement, notably in ideological variation or "the green rainbow." The green movement is an important new development for Western political systems, as warranted by its size alone. But radical and anarchic elements within the movement are scarce, even if visible in the media. Rather, most environmentalists are products of Western affluence and strong advocates for democracy, especially in pressing for more direct citizen participation in policy making. Their beliefs and behaviors represent a reformist orientation, which delays or modifies transformational goals. There is a movement for social change, but no clear utopian vision of what the future should look like. "Given this evidence, we think it unlikely that ecologists will be agents for rapid and radical social change. Instead, we shall probably witness continuation of the incremental progress of the past two decades. Rather than supplanting the dominant paradigm, ecologists will modify it." Whether this produces the changes some deem necessary for the health of the planet remains to be seen. **(European environmental groups)**

394

The Greens and the Politics of Transformation. John Rensenbrink (Research Prof in Political Science and Ecology, Bowdoin College). Foreword by Jay Walljasper (*Utne Reader*). San Pedro CA: R. & E. Miles, March 1992/269p/$22.95;$12.95pb.

A founder of the Maine Green Party in 1984 and a leading coordinator of the SPAKA process to create a "Green Program USA" [**FS Annual 1991** #10588] describes development of a green-centered social movement in the US to shift public priorities and leadership into transformative channels. Chapters discuss the dynamics of denial among those in positions of power, ferment from below in the US and elsewhere, the abyss of mistrust that separates today's leaders from the misled, the US political oligarchy that caters to the very few with money and power, the contributions of feminism and Native American traditions to the Green vision, the penchant for moralizing among activists (as well as the lure of failure and fear of success), the Greens as harbingers of new politics unfettered by the baggage of Left and Right ideologies, the quest for community as the basis of the Green movement, development of Green parties in the US, sustainability as a guiding green philosophy, the drafting of the US Green Program, the many rooms of "The House of Green" (a large, rambling, but elegant structure with a common center), the Green idea of leadership as partnership, and the fourth national gathering of US Greens in West Virginia in August 1991 (they now boast over 300 local action groups). [ALSO SEE: ***Green Party Politics Around the World*** by Wolfgang Rüdig of the U of Strathclyde in Glascow (*Environment*, 33:8, Oct 1991, 6-9ff), who notes the very marginal position and general lack of Green success; major obstacles are lack of unity among different Green groups and refusal to join coalitions with established parties.]

(Green politics in US and world)

395

The Fading of the Greens: The Decline of Environmental Politics in the West. Anna Bramwell (OECD-Paris; formerly Trinity College, Oxford). New Haven CT: Yale U Press, Dec 1994/224p/$25.00.

Author of **Ecology in the 20th Century** (1989) discusses the decline of the Green movement—the end of the brief era of dedicated Green national politics—which is not to imply that environmentalism is finished. Indeed, the environmentalist constituency is now far wider than any party. But the gap between support for environmental issues and a vote for Green parties is wide everywhere in Europe. The parties fell because the most successful Green party was the German one, which was untypically urban-left oriented and locked into stereotyped reactions. Using them as a model hampered development of Green parties elsewhere. Also, established political parties and international agencies took on those environmental programs that could be incorporated into political life. The importance of environmentalism has forced the main political parties to take account of it. Industry is beginning to adopt green policies; if clean technology can replace pollution control, then many demands of deep ecology will lose even more support. The cultural criticism of Green ideology is not likely to go away, but it is likely to be further marginalized. "It may be that national Green parties have outlived their usefulness." **(decline of Green parties)**

396

The United States Goes Green (Special Report), Brad Knickerbocker, *The Christian Science Monitor*, Tues, 12 Jan 1993, 8-13.

"The environmental movement in the US today stands poised to have greater influence than ever before." There are as many as 10,000 environmental organizations in the US today, ranging from local groups that may exist for only a few months to established institutions with dozens of lawyers, scientists, and economists on their staff. Environmental groups in the past few years have seen steep increases in contributions, largely due to threats from the Reagan and Bush administrations. The 20 largest groups together can now raise $1 billion a year. The National Wildlife Federation (the largest of the so-called "Big 10" environmental groups, with 5.3 million members and supporters) has an annual budget larger than the UN Environment Programme. Other articles in this report include an interview with Al Gore Jr. ("who may be characterized as the nation's senior environmentalist"), and profiles of the greening of business, grass-roots activists, and the "wise use" counter movement (much smaller than the environmentalists, but it has won some significant victories). Ron Tipton of the Wilderness Society predicts that "One of the results of the election is that the wise-use movement is going to be bigger and meaner and better financed, now that they have a bogeyman in the White House." **(growth of environmentalism)**

397

American Environmentalism: The US Environmental Movement, 1970-1990. Edited by Riley E. Dunlap and Angela G. Mertig (both Dept of Sociology, Washington State U). Philadelphia: Taylor & Francis, April 1992/121p/$12.95pb.

Essays on trends among rapidly-growing national environmental organizations, parallel growth of grassroots activism in local communities (many local groups are moving beyond a NIMBY stance to one of NIABY—Not in Anyone's Back Yard, and will increasingly be drawn into the national policy arena), African-American groups that have begun to challenge the legitimacy of environment-jobs trade-offs and environmental inequities, deep ecology and trends in radical environmentalism (by Bill Devall), the growth of global environmentalism (by Lynton K. Caldwell), a critique of 20 years of change in the movement (by Michael McClosky of the Sierra Club, who views a split into three camps: radical, mainstream, and those seeking accommodation with industry), and trends in public opinion (by Riley Dunlap, who reports that public support for environmental protection is extremely high). The editors conclude that environmentalism has become a much more diverse movement since 1970. When judged against the typical social movement, it has been "a resounding success." Although the movement has largely failed in its goal of protecting the environment, the situation would be far worse had the movement not been around. **(environmental movement trends)**

398

The Green Revolution: The Environmental Movement 1962-1992. Kirkpatrick Sale (NYC). NY: Hill & Wang, July 1993/124p/$22.95;$7.95pb.

A brief history, from Rachel Carson and the "sixties seedtime" to the doomsday decade of the 1970s and the Reagan reaction of the 1980s. Considerable achievements are obvious: within a single generation, environmentalism has become embedded in American life, law and custom, text and image, classroom and workplace, practice and consciousness, politics and culture. But to understand prospects, one must consider the shortcomings of environmentalism: ***1) Structural***: people in the important mainstream organizations are very largely white, well-off, and male; their concerns tend to mirror those of suburban well-to-do constituencies (moreover, there is also the practice in some parts of the movement of working with and accepting money from corporate America, which may compromise ideals); ***2) Institutional***: a function of success, the large organizations tend to solidify into hierarchy, centralization, and dullness; ***3) Ideological***: both the mainstream majors and the grass-roots NIMBYs tend to see environmental problems as isolated aberrations within a functioning system; this condemns the movement to a kind of perpetual tinkerism and finger-in-the-dike-ism (radical environmentalism sees the need for systemic change, but suffers from "the factionalism endemic to the faithful"). In the future, the movement might continue to operate largely as a reformist citizen's lobby, seeking technofixes, reformed tax policies, and more environmental R&D. Or the movement might deepen and darken its analyses and criticisms, and try to work for structural changes. [ALSO SEE: **A Fierce Green Fire: The American Environmental Movement** by *Greenwire* publisher Philip Shabecoff (Hill & Wang, Jan 1993/352p/$25.00), who concludes that "to make the political breakthrough necessary to achieve their goals, the environmentalists *must* make common cause with other sectors of our society that have a stake in changing the political and economic status quo." A similar call for broadening is made by Robert Gottlieb (UCLA Urban Planning) in **Forcing the Spring: The Transformation of the American Environmental Movement** (Island Press, Oct 1993/413p/$27.50).] **(environmental movement histories)**

399

Losing Ground: American Environmentalism at the Close of the Twentieth Century. Mark Dowie (former publisher and editor, *Mother Jones*). Cambridge MA: MIT Press, April 1995/317p/$25.00.

American environmental history is seen in three waves: 1) the *first wave* began with the conservationist/preservationist impulse of the late 19th/early 20th centuries and coincided with the closing of the frontier; 2) the *second wave* came in the brief era of environmental legislation that began in the mid-1960s and was abruptly halted by the Reagan Administration in the 1980s; 3) the *third wave* is "a relatively fruitless and hopefully brief attempt to find a harmonious conciliation between conservative environmentalists and corporate polluters." The American environmental movement has become a wealthy complex of organizations dominated by a dozen or so large national groups centered or headquartered in Washington and dominated by lawyers and MBAs. Their strategy—a legislative/litigative initiative focused largely on the Federal government—is central to the environmental effort. But this keeps environmentalism narrowly defined. Meanwhile, the antagonistic Wise Use movement "is expanding while the environmental movement is shrinking." (An Appendix reprints the 25 goals of the Wise Use Agenda.) The dismissive arrogance and general elitism of many environmentalists has been one of the best organizing tools of the Wise Use activists. As the third wave washes back into history, a *fourth wave* of new environmental passion is gathering force, and "should crest sometime early in the 21st century." It is part wilderness preservation, part toxic abatement, part ecological economics, part civil rights, part human rights, part secular, part religious, and parts of many ecologies, fueled by anger and commitment to democratic processes. The central sentiment that will define this next generation is a sense of justice, almost completely absent from the US environmental imagination. It will combine splinters from the mainstream movement, the "environmental ad hocracy" (some 10,000 NIMBYs that share similar tactics), the people-of-color network for environmental justice, and the new conservation movement. Its common ground will be biocentrism, challenging all economic orthodoxies, sustainability, clean air/water and environmental justice as basic human rights, and a land ethic taught in schools. [**NOTE**: An interesting vision, despite reading just like an informed but feisty editorial from *Mother Jones*.]

("fourth wave" of environmentalism ahead?)

400

Environmentalism and the Future of Progressive Politics. Robert C. Paehlke (Prof of Political Science and Environmental Studies, Trent U, Ontario). New Haven CT: Yale U Press, April 1989/325p/$25.00.

"Environmentalism... has the potential to become the first original ideological perspective to develop since the middle of the nineteenth century." It is now at a stage of development comparable to that of socialism a century ago. It can be developed into an ideology as coherent as any of the three classical ideologies of liberalism, conservatism, and socialism. And it may be more effective than either of the contemporary Western ideologies, neoconservatism and progressivism. It provides a useful base from which to make individual life choices, from which to take collective political action, and from which to decide a broad range of public policy issues. It is an ideology distinctive in its unwillingness to maximize economic advantages for its adherents. It is the first ideology to be deeply rooted in the natural sciences (Marxist claims notwithstanding). And it may help to develop a "third way" able to defuse the ideological duality of the contemporary world. It could be seen as a third wave of progressivism, developing in response to the atrophication of liberalism in the West and socialism in the East. In this respect, perhaps in defiance of all historical experience, environmentalism "is an ideology whose time has come."

But environmentalism still lacks a mass following, and is by no means certain to attract such a following. To do so, environmentalists must develop clear and consistent positions on the full range of political and social issues, which becomes possible only when environmentalism is seen as neither "left" nor "right." Core priorities of a contemporary environmental progressivism might include: 1) long-term international agreements to allow a gradual real increase in oil prices; 2) a long-term emphasis on productivity-enhancing technologies coupled with a commitment to full employment; 3) materials policies to promote recycling and durable/repairable products; 4) strengthened environmental protection policies; 5) restoration and enhancement of urban cores and urban transportation systems; 6) commitment to deficit reduction by reduced military spending and reduced subsidies and tax expenditures for various industries; 7) expanding aid to LDCs by matching-fund incentives focused on environmental protection; 8) promoting sustainable agriculture and forestry.

(environmentalism as potential major ideology)

401

Choosing Our Future: A Practical Politics of the Environment. Ann Taylor (British Shadow Minister for Environmental Protection). London and NY: Routledge, June 1992/235p/$59.95;$17.95pb.

The environment is not and never will be a non-political issue. In a sense, all environmental problems are fundamentally political, because they result from the way our society is organized, the way that resources are distributed and put to use, and the kinds of economic and social goals we pursue. Chapters discuss the holism of socialist vision and its belief in democratic planning for the kind of world we want, how the present structure of the world economy is inimical to sustainable development, the need for government to create frameworks for individuals to play a greater role in planning and policy decisions, the government's responsibility for clearly presented environmental information (so as to stimulate environmental awareness and increase public trust in government), the greening of industry, the problem of public alienation from environmental issues because the debate is felt to be too technical, and the role of government in coordinating science and technology.

Concluding observations: 1) the no-growth lobby has retreated to the Green Party; the new focal point for discussion is not how to halt growth, but how to regulate it; 2) a regulatory approach is absolutely essential to environmental protection, as to all safeguards of collective welfare; 3) rather than revise GNP calculations (the *bête noire* of many environmentalists), it may be better to retain it as a yardstick but to clarify what it measures, while regularly publishing an "environmental quality index" of parallel indicators; 4) it is simply not feasible nor desirable to freeze all consumption and economic activity; in their anxiety to save the earth, some environmentalists disregard claims of the poor and oppressed and "obscure all judgement as to what is practical"; 5) sustainability is the approach we must adopt, but it is not clear what this involves; "the real debate is between rival understandings of just what sustainability implies, and how it is to be achieved"; 6) in practical decision-making, one of the frequently perceived limiting factors of environmental enhancement is the cost of necessary measures; much less often mentioned is the cost of neglect; 7) environmental efficiency is a key to the approach we must adopt, with energy as an obvious example; 8) an energy tax has an ideological appeal to non-socialists because it is market-based, but such taxes hit the poor hardest; a more promising approach is efficiency standards; 9) we have a responsibility to safeguard the future on the basis of "preventive expenditure" which will not only greatly enhance quality of life, but also avert much greater costs if we fail to act now; 10) inaction is a policy itself and an extremely dangerous one; far better that we "seize the opportunity, in a constructive way, of choosing our future." [**NOTE**: Thoughtful, plain-spoken, and relentlessly practical. ALSO SEE:

Environmentalism and Political Theory: Toward an Ecocentric Approach by Robyn Eckersley of U of Tasmania (Albany NY: State U of New York Press, June 1992/274p/$44.50;$14.95pb), arguing that "the Green movement will ultimately stand or fall on its ability to generate *practical* alternatives to the advanced industrial way of life."] **(practical environmental politics)**

402
Environmental Politics and Policy: Theories and Evidence (Second Edition). Edited by James P. Lester (Prof of Poli Sci, Colorado State U). Durham NC: Duke U Press, July 1995/386p/$59.95;$22.95pb.

In the last 30 years, environmental protection has moved from being a nonissue to being one of the most significant issues of our time. Until the late 1960s, there was no clear or explicit formulation of a public responsibility for the state of the environment. Since the National Environmental Policy Act of 1969, numerous environmental laws have been passed. These essays highlight the work of political scientists that bears on environmental politics and policy. Topics include the evolution of environmentalism, state environmental policy and innovations, public opinion, interest groups, political parties, Congress, the bureaucracy, the courts, policy elites, evaluating environmental policy outcomes, and a vision of the politics of hope based on an ecologically sound paradigm. In ***Alternative Views of the Environmental Problematic***, John S. Dryzek and James P. Lester present a six-fold taxonomy of environmental worldviews: Hobbesians and Structural Reformers proposing central governance, those who espouse a Technocratic Guardianship, Reform Ecologists who seek a higher voice for ecological values in the present system, Free-Market Conservatives (including regulatory reformers, new resource economists, and cornucopians), Social Ecologists who seek radically decentralized structures, and Deep Ecologists with limited interest in human communities. [**NOTE**: The taxonomy has some value, but misses the pervasive movement toward sustainability that falls between these poles.] **(six environmental worldviews)**

403
Environmental Politics and Policy (Third Edition). Walter A. Rosenbaum (Prof of Poli Sci, U of Florida). Washington: CQ Press/Congressional Quarterly Inc, Jan 1995/372p/$22.95pb.

When the first edition of this book was written 20 years ago, many "experts" then dismissed environmentalism as another trendy and transient issue. But it has endured, although this decade will be decisive for the movement. Chapters discuss: 1) the politics of regulatory discontent (Americans are still reluctant to accept many of the costs and inconveniences that are essential for environmental protection); 2) the state of the environment ("progress has been uneven at best," and many problems have been created or exacerbated by new technology); 3) new structural and legal arrangements for the environment (environmental management has become a permanent new policy domain within Federal and state governments); 4) institutional and economic obstacles to implementing environmental policy effectively (congressional overcontrol of agencies, legislative reluctance to create clear mandates and priorities, excessive fragmentation, resistance of subgovernments to reform); 5) the complex problems of risk assessment (limitations arise from the frequent absence of essential data, ambiguity of available data, or disagreement among experts); 6) the unfinished agenda for air and water (despite spending more than a half trillion dollars over the past 20 years, "the nation's air and water remain seriously polluted"; a very long time is needed to regain the healthful air and water we once had); 7) the regulatory thicket for toxic and hazardous substances (progress here is slower than in other major areas of environmental policy); 8) the politics of energy (enabling continuing coal use and surface mining, and perpetuating the nuclear dream); 9) the battle for public lands (the emerging "wise use" movement seeks economic development and dislikes environmentalists). Some necessary directions: cultivation and reinforcement of public environmental values, defining environmental protection as national security, a congressional declaration of environmental priority, a greater role for ecological forecasting, and more attention to risk assessment in technology development. **(US environmental politics)**

4. Global Action

404
International Environmental Policy: Emergence and Dimensions. Second Edition. Lynton Keith Caldwell (Prof of Public and Environmental Affairs, Indiana U). Durham NC: Duke U Press, Dec 1990/460p/$35.00;$18.95pb.

First published in 1984, this update surveys the worldwide movement for protecting the human environment, with emphasis on intergovernmental agreements and institutional arrangements. The unifying concept is the transformation of institutions through social learning in response to the findings of science and the perceived impacts of environmental change. "The movement belongs to a larger transformation in human social thought, which may be likened to a second Copernican revolution. The first revolution removed the earth from the center of the universe; the second removes humanity from the center of the biosphere." The so-called environmental crisis marks the end of a half-millennium of exuberant, exploitive expansion of (chiefly) western society that we have called modern times. The "crisis" is not in the environment, but in human behavior as it affects the environment.

Chapters discuss critical issues of environmental policy, the road to the 1972 UN Conference on the Human Environment in 1972, the legacy of the Stockholm Conference, evolving structures of environmental law/policy/cooperation, bilateral and multilateral regional arrangements, the growing acceptance of the "sustainable development" concept in the 1980s, enhancing quality of life by protecting nature and preserving culture (with emphasis on the leading role played by UNESCO), protecting the international commons (the atmosphere, outer space, oceans, and Antarctica), and the changing international order. Emphasis is placed on three international programs that illustrate strategies to reshape the human dimension of environmental attitudes and values: 1) UNESCO's Man and the Biosphere, an interdisciplinary research program that began operations in 1976; 2) the UNESCO/UNEP Inter-national Program for Environmental Education, launched in 1975; 3) the World Conservation Strategy, started in 1980 by IUCN with sponsorship of UNEP and financing from the World Wildlife Fund (during the first decade of the Strategy, 45 countries developed national plans). Concludes that the emergent politics of planet earth differ from earlier geopolitical strategies, in that nations seek to cooperate rather than compete. And "it seems plausible that by the 21st century the environment will have become the primary subject of international policy." [ALSO SEE: **International Law and Pollution**, edited by Daniel Barstow Magraw (U of Pennsylvania Press, April 1991/368p/$37.95), with special emphasis on pollution from nuclear accidents and from acid deposition.] **(global environmental movement)**

405
International Governance: Protecting the Environment in a Stateless Society. Oran R. Young (Director, Institute on International Environmental Governance, Dartmouth College). Ithaca NY: Cornell U Press, Oct 1994/221p/$32.50;$13.50pb.

Governance systems are social institutions or sets of rules guiding the behavior of those engaged in identifiable social practices. *Governments* are organizations or material entities to administer the provisions of governance systems. Simply establishing a government offers no assurance that the function of governance will be fulfilled. Under a variety of conditions, "the operations of governments are not only insufficient to ensure that growing demands for governance are met, but also may be unnecessary for the provision of governance." This realization is liberating. It does not necessarily reject arguments about the importance of governments, but it does allow us to think seriously about the idea of governance without government. And it allows a critical assessment of organizations that may prove useful in administering various governance systems. In addressing questions of governance, international environmental problems can be grouped into four broad clusters: 1) the international commons (systems such as Antarctica, the high seas, the geostationary orbit); 2) shared natural resources (fish stocks, migratory birds, complex ecosystems); 3) transboundary externalities (when activities within one state, such as loss of biodiversity, produce effects in other jurisdictions); 4) linked issues (efforts to protect the environment and to promote trade). Chapters discuss creating international governance systems, bargaining leverage, governance systems and international organizations, international legal regimes, and case studies of negotiating a global climate regime and managing the Arctic's shared natural resources. [**NOTE**: Important, but very dense. ALSO SEE: **Governance Without Government** (Cambridge UP, 1992; **FS Annual 1994** #12206) and **The Global Commons** by Harlan Cleveland (UP of A, 1990; #408).]

(environmental governance without government)

406
Global Accord: Environmental Challenges and International Responses. Edited by Nazli Choucri (Prof of Political Science, MIT). Foreword by Maurice F. Strong. Global Environmental Accords Series. Cambridge MA: MIT Press, June 1993/562p/$50.00.

"Effective management of the global environment may well become the most significant institutional challenge for the 21st century." Development of global accord on environmental management will be facilitated to the extent that sound theoretical orientation is coupled with realistic strategies. These essays discuss the main processes leading to global change, impacts from the lack of a coordinated intervention policy, inherent stresses between growth and sustainability, the rise of global eco-politics, perspectives on eco-development, multinational corporations as agents of environmental degradation, technology as problem and solution, an international legal framework for intergenerational equity, intertemporal time comparisons, the evolution of international environmental law, institutional bargaining, and practical approaches to regime building for complex global problems. Choucri concludes that meeting imperatives for the 21st century requires framing global norms (essential principles to guide strategy). Five strategic norms are suggested in this book: legitimacy, equity, efficacy, volition (responses must be voluntary), and universality (all actors must participate). Together, these norms lead to a critical policy precept: maintenance of adjustable target priorities in pursuit of ever more energy-efficient technologies and resources. **(global environmental accord)**

407
Institutions for the Earth: Sources of Effective International Environmental Protection. Edited by Peter M. Haas (U of Mass-Amherst), Robert O. Keohane (Harvard U), and Marc A. Levy (Princeton U). Global Environmental Accords Series. Cambridge MA: MIT Press, May 1993/448p/$37.50;$17.95pb.

Global environmental threats have attained a prominent position on the international agenda. These transnational problems cannot be managed by one country acting alone; effective institutions are needed to facilitate cooperation across borders and organize political energies for environmental policy change. Since the 1972 UN Conference on the Human Environment, over 60 multilateral environmental treaties have been signed. Essays discuss the effectiveness of international environmental institutions (in boosting concern, building capacity, and/or facilitating agreement), protecting the ozone layer, controlling acid rain in Europe, protecting the Baltic and North Seas, preventing oil pollution from tankers, fisheries management, managing pesticide use in developing countries, and international population control institutions. These international institutions are not always successful, nor do they provide a magic formula for success. Some advice to architects of these institutions: nurture coalitions with like-minded governments and supportive NGOs, foster open-ended knowledge creation, create a dynamic process that moves from norms to rules, and recognize the often central importance of building capacity within the state and in civil society. [ALSO SEE: **Forging International Agreement: Strengthening Intergovernmental Institutions for Environment and Development** by Lee A. Kimball (World Resources Institute, April 1992/84p/$9.95pb), which calls for integrating environmental protection with economic and social goals.]

(international environmental institutions)

408
The Global Commons, Harlan Cleveland (President, World Academy of Art and Science; former Dean, Humphrey Institute, U of Minnesota), *The Futurist*, 27:3, May-June 1993, 9-13.

The acts of individuals are producing a pervasive threat to the global environment and creating a new class of problems that are not only global, but behavioral. They are behavioral because they require literally hundreds of millions of people to do something or stop doing something. We need to treat the global environment not as a series of separable puzzles, but as parts of an integrated Global Commons, with hundreds of millions of people [not billions??] responsible for its health. Four enormous environments, still mostly unexplored, are already treated in international law and custom as parts of a Global Commons: outer space, the atmosphere, the oceans, and Antarctica. The problem is not "managing the Global Commons," but managing human behavior in the Commons by stabilizing population, pricing products at their full social cost, and reducing wasteful consumption (and thus pollution). An organization to protect and supervise the "common heritage of mankind" is needed. It could be a revitalized version of the UN Trusteeship Council, which the success of decolonization has now left without a function. The now-dormant Council could readily form a Global Commons Trusteeship Commission to negotiate norms, standards, and guidelines for exploring and exploiting the Global Commons. The Commission would keep the health of the Commons environments under open and continuous review, and make room in its work for the views of major NGOs. It might be empowered to ask governments, NGOs, and individuals to serve as "centers of initiative" on various problems. The new genre of problems depends on wide participation: the Global Commons will be governed pluralistically or not at all. [**NOTE**: Further thoughts based on Cleveland's two recent books: **Birth of a New World** (Jossey-Bass, April 1993; **FS Annual** 1994 #12207) and **The Global Commons: Policy for the Planet** (Aspen Institute/UPA, 1990; **FS Annual 1992** #11104).]

(a Global Commons Trusteeship Commission?)

409

The Gnat Is Older Than Man: Global Environment and Human Agenda. Christopher D. Stone (Prof of Law, USC). Princeton NJ: Princeton U Press, April 1993/ 341p/$21.95.

The earth faces many perils, immediate and remote, well-defined and vague. How do we best prepare for them? What sacrifices are merited? What legal and diplomatic structure can best handle global environmental problems? Chapters consider the condition of the earth from the legal perspective, transboundary pollution, managing the global commons, the pros and cons of treaties, the economist's prescriptions of taxes and tradable permits, prevention vs. cure, forms of calamity insurance, and moral dimensions. Offers two proposals: 1) a system of guardians who would be legal representatives for the natural environment (the success of such a regime would depend on significant changes in the substantive law that the guardian would be empowered to invoke); 2) a Global Commons Trust Fund, with revenues from ocean use (a tax of 0.5% of the fish catch would raise $250 million; the same rate on offshore oil and gas would add $375 million), a CO_2 tax of ten cents a ton (raising $2.2 billion per year), selling the rights to park satellites in the geostationary orbit, fees on minerals from the seabed, etc. There will of course be questions on how to allocate the fund: to national governments or the UN, based on population or need, and how to measure need. [**NOTE**: The reference to gnats in the title is obscure and poorly developed. It was suggested by the papers of Stone's journalist father, the late I. F. Stone. Also see proposal for an Environmental Security Trust Fund in Al Gore's **Earth in the Balance** (#177).]

(Global Commons Trust Fund proposed)

410

Partnership for the Planet: An Environmental Agenda for the United Nations. Hilary F. French (Sr. Researcher, Worldwatch Institute). Worldwatch Paper 126. Washington: Worldwatch Institute, July 1995/71p/ $5.00.

When the UN was created 50 years ago, environmental degradation was not considered much of a national threat, let alone a pressing global problem (the UN Charter does not mention the word "environment"). Today, environmental security has emerged alongside economic and military security as both a major concern of national decision-makers and a third pillar of international relations. The UN has helped to develop hundreds of international agreements and programs, albeit with some duplication and inefficiency in the resulting patchwork quilt. If global ecological decline is to be reversed, this disorganized system must be streamlined. Various proposals have called for a new or newly-strengthened environmental institution within the UN, an environmental security council, a Trusteeship Council to manage the global commons, and a Global Environmental Organization to develop basic environmental principles. French argues for transforming UNEP into an operational UN Environment Agency that would gather and provide information (as UNEP already does on a small scale), execute projects, elaborate some common minimum international environmental standards, and serve as an umbrella organization for the current scattered collection of treaty bodies.

(UN Environment Agency needed)

411

Global Environmental Politics. Gareth Porter (Environmental and Energy Study Institute, Washington) and Janet Welsh Brown (World Resources Institute). Boulder CO: Westview Press, Dec 1991/208p/$43.50;$10.95pb.

Less than ten years ago, global environmental politics were regarded as low politics—a set of minor issues to be relegated to technicians. But the withering of superpower competition and appearance of a new set of global environmental issues have given environmental politics a new status. Chapters discuss various issues (world population growth, the ozone layer, global warming, endangered resources, environmental accounting), actors (nation-states, international organizations, NGOs, and corporations), the formation of environmental regimes (on ozone depletion, whaling, acid rain, the ivory trade, international toxic waste trade, Antarctic minerals, global warming, tropical forest destruction), the increasing linkage between environmental problems and national security, and global economic inequality and natural resources. The new agenda of problems will demand international attention in the 1990s and beyond. Three broad alternative strategies have been suggested: **1) Incremental Change**: a continuation of incremental changes is most likely, assuming that progress can be made within the parameters of existing global political institutions, diplomatic practice, and socioeconomic realities; but this would do little to bind developing countries to global agreements for action; **2) Global Partnership**: a "global bargain" strategy of concerted effort by rich and poor countries to collaborate widely on sustainable development; **3) Global Governance**: some UN authority that could take effective action even if unanimous agreement has not been achieved (an idea from a 1989 conference in The Hague, now supported by over 30 nations but opposed by the US, USSR, Britain, China, and Japan), making the "common heritage of mankind" and "intergenerational equity" legal principles applicable to all global environmental issues. Concludes that something beyond traditional power politics is clearly at work; the issue is not whether nation-states will move toward more effective cooperation on global environmental threats, but whether they will do so rapidly enough.

(global environmental regime ahead?)

412

Global Ecology: A New Arena of Political Conflict. Edited by Wolfgang Sachs (Institute for Cultural Studies, Essen, Germany). London & Atlantic Highlands NJ: Zed Books; Halifax NS: Fernwood Publishing, Dec 1993/262p/ $59.95;$25.00pb.

Nearly everybody has now turned environmentalist, and the conflicts of the future will center on who stands for what kind of environmentalism. These 17 essays probe critically the new language put forth by the rising breed of environmental professionals who push for a worldwide rationalization of life-styles, provide survival strategies for the powers that be, and promote the aspirations "of a rising eco-cracy to manage nature and regulate people world-wide." **1) In the Wake of Rio**: Wolfgang Sachs on the dangers of linking environmentalism to development thinking (sustainability is thus conceived of as a challenge for global management), Nicholas Hildyard on putting the corporate foxes in charge of the chickens, Matthias Finger on the failed politics of the UNCED process, Tariq Banuri on North-South conflicts at UNCED, Klaus Meyer-Abich on the growing North-South gap due to climate change; **2) Confusion Over Sustainability**: Paul Ekins on how to genuinely attain sustainable development, Christine von Weizsäcker on competing notions of biodiversity, Donald Worster on deep flaws of the sustainable development ideal; **3) Against Global Environmental Management**: Vandana Shiva on the greening of the global reach or "green imperialism," Larry Lohmann on resisting green globalism, Eduardo Gudynas on the fallacy of eco-messianism (critiquing the scientific superiority complex, the North's pedagogic tendency, and the new ecocratic discourse), Guy Beney on Gaia as "the globalitarian temptation"; **4) Ecology from Below**: Frédérique Apffel Marglin on regenerating the land and the community, a plea for cultural and ecological pluralism in India, Anil Agarwal and Sunita Narain on green villages. [**NOTE**: Crypto-anarchist critiques similar to those in **The Development Dictionary**, also edited by Sachs (Zed Books, 1992; **FS Annual 1994** #12225).]

(global environmentalism questioned)

413

The Earth Brokers: Power, Politics, and World Development. Pratap Chatterjee (Inter Press Service, Washington) and Matthias Finger (IHEAP, Lausanne; formerly Columbia U). London & NY: Routledge, Nov 1994/191p/$69.95;$16.95pb.

A critical overview of the UNCED/Agenda 21 process, discussing the development paradigm, the green critique of development, the Bruntland Commission, the South Commission and its focus on economic growth, the 1992 Earth Summit in Rio, the evolution of the green movement in the North (toward global environmental management and New Age environmentalism), NGO participation in UNCED, promoting big business at Rio (with critical comments aimed at Stephan Schmidheiny and the Business Council for Sustainable Development), and the "impossible" task of the Commission on Sustainable Development, formed to monitor the implementation of Agenda 21. Concludes that "as a result of the UNCED process the planet and most of its inhabitants will be worse and not better off." Even the most modest hopes have been disappointed. The institutional and financial outcomes bear no relationship to the expectations raised, let alone the needs of the global environmental crisis. The philosophy of sustainable development on which UNCED was based, and the set-up of its process, were flawed from their inception. Old thinking about economic growth prevails, old institutions promoting such growth persist, and the old development establishment has repackaged itself in green. The monitoring of what is believed to be the solution to the global crisis was handed over to a powerless commission, and implementation of projects has been appropriated by a subsidiary of the World Bank (the Global Environmental Facility). The major lesson: the global approach is at best a useful tool for awareness raising. But the word "global" is a mirage—an illusion created by the traditional agents and major stakeholders. We must focus on the local, its people, and its communities, and collectively un-learn the paradigm of modern society. [**NOTE**: An overly harsh and premature indictment?] **(UNCED questioned)**

414

Environmental Diplomacy: Negotiating More Effective Global Agreements. Lawrence E. Susskind (Prof of Urban and Environmental Planning, MIT; Director, MIT-Harvard Public Disputes Program). NY: Oxford U Press, Feb 1994/201p/$45.00;$14.95pb.

Processes used to negotiate global agreements are as important as the technical capabilities and scientific understanding that negotiators bring to the bargaining table. Based on a close look at 15 environmental treaty-making efforts, including those culminating in the Rio Earth Summit, Susskind identifies four procedural shortcomings that account for most negotiation failures: 1) representation and voting procedures do not guarantee that all countries and interests are treated fairly; 2) scientific and political considerations are not balanced to insure the wisest possible agreements; 3) linkages among environmental concerns and other policy issues are rarely explored or crafted adequately; 4) effective monitoring and enforcement arrangements are not implemented. Global environmental negotiations are not likely to produce adequate results until these shortcomings are overcome. Susskind is "especially enthusiastic" about a new system of *sequenced negotiation*, a multi-step process that synchronizes worldwide expectations and moves systematically from Level I treaties (spelling out principles, definitions, timetables, targets, and responsibilities), to Level II treaties (requiring commitments to minimal levels of performance, in exchange for explicit benefits), to Level III treaties (offering maximum benefits for maximum effort, based on what can be learned from shared efforts). Many other comments are offered concerning roles for scientific advisors, the precautionary principle, the value of contingent agreements (which sidestep the need for consensus on what the future holds), pros and cons of issue linkage, the need for a "Green Amnesty International" to monitor noncompliance, increasing compliance by posting bonds to be paid back upon performance, and the Salzburg Initiative to reform the UN-sponsored system of environmental treaty-making (advocating prenegotiation assistance to countries, new approaches to treaty-drafting, different standards of responsibility for different countries, and removing penalties for constructive unilateral action). [**NOTE**: A valuable book on a neglected topic.]

(improving environmental treaty-making)

415

Environmental Change and International Law: New Challenges and Dimensions. Edited by Edith Brown Weiss (Prof of International and Environmental Law, Georgetown U). Tokyo: United Nations U Press, Dec 1992/493p/$60.00;$40.00pb.

International law has been based on the relationship between independent states that exercise exclusive national sovereignty over their territories. Global change is altering this vision by causing states to realize that they are locked together in sharing the use of a common global environment. "This has led to a growing awareness that the interests of humankind must constrain the interests of individual states." Moreover, actors other than states have become essential to managing global environmental change. These developments are leading to a fundamental shift in the paradigm of international environmental law. Modern law in this area dates to about 1972 (the time of the UN Stockholm Conference). Since then, the scope of international agreements has expanded significantly, and there is not a single example in which the provisions of earlier conventions have been weakened. There is also a growing realization in the international community that the time has come not only to monitor and research environmental risks, but also to reduce them. Thus we have moved toward agreements that require reductions in pollutant emissions and changes in control technology. In the next decade, more international agreements and other legal instruments will be concluded, and they are likely to be ever more demanding in the actions they ask of countries. Further, there is likely to be increased emphasis on monitoring compliance with the conventions, and on providing means to facilitate implementation and compliance. The role of NGOs in negotiating and implementing agreements will likely continue to grow, and new attention will be paid to trade implications of agreements. At the same time there is a countervailing theme, in that some countries are becoming overtaxed by the resources needed to engage in negotiations and to effectively implement agreements already concluded. [**NOTE**: An important element in the ongoing erosion of national sovereignty. Weiss provides a chapter on alternative approaches to intergenerational equity, derived from her book, **In Fairness to Future Generations** (Transnational Publishers, 1989; **FS Annual 1991** #10472).]

(growth of international environmental law)

416

International Environmental Law. Alexandre Kiss (President, European Council for Environmental Law) and Dinah Shelton (Prof of Law, Santa Clara U). Foreword by Maurice F. Strong. Ardsley-on-Hudson NY: Transnational Publishers, 1991/541p/$95.00.

Concepts about nature changed fundamentally in the late 1960s, when governments began to demonstrate concern over the state of the environment, under growing pressure from international public opinion. Developments in international environmental law paralleled this evolution within states. There are now several hundred multilateral and bilateral agreements in the field, and new global initiatives are constantly adding to the areas codified. Chapters describe the necessity for interdisciplinary approaches and the difficulty in delimiting the field, questions about how much risk to assume now and for the future, the evolution of international environmental law (there remain divisive issues over implementation of norms, but the split over the need for these norms has substantially decreased), institutional cooperation (on the role of various UN agencies and NGOs), the sources of environmental law, regulation of environmental sectors (the general trend in international law is towards integrated approaches to environmental protection), transsectoral problems (toxic products and wastes, radioactivity, global warming), liability for environmental harm, and the concept of environmental security which "would then mean recognizing the right of each person to conservation of his environment." The rights of future generations "also could serve as an alternative conceptual framework for global environmental protection." An International Covenant on the Global Environment could help to frame all existing instruments. **(evolving environmental law)**

417

Transnational Environmental Law and Its Impact on Corporate Behavior. Edited by Eric J. Urbani (NYC), Conrad P. Rubin (Atlanta), and Monica Katzman (Berkeley CA). Irvington-on-Hudson NY: Transnational Juris Publications, Jan 1994/354p/$95.00.

Papers and discussion from an October 1991 symposium sponsored by the Fletcher School of Law and Diplomacy and the Boston law firm of Goodwin, Proctor & Hoar. Topics include the impact of changing international institutions on environmental law and corporate responsibility, environmental law under the nation states, the corporate response to government controls and information gathering, doing business under Mexico's environmental laws and regulations, systems for corporate information disclosure, trends in international environmental enforcement, EC environmental legislation and priorities for the 1990s, enforcement of environmental initiatives by regional institutions, and the impact of environmental treaties and agreements on corporate strategies. Also included is a summary of an August 1991 survey by McKinsey & Company, where more than 400 senior executives reflected on "The Corporate Response to the Environmental Challenge." Throughout all industry sectors and all regions, companies have taken significant steps to respond to increasing environmental demands, but most of the leaders expect a substantial tightening of government regulations and increasing criticism from customers. A large majority of the leaders expressed a readiness to act, and almost half of the respondents saw 50% tighter standards by 2000 as a realistic goal. Responses varied significantly by industry sector and region, with the metals sector most pessimistic about covering increases in environmental costs via higher market prices. Overall environmental costs as a percentage of sales were expected to rise from 2.4% in 1991 to 4.3% in 2000.

(corporations and environmental law)

5. National/Regional Action

*418

Green Plans: Greenprint for Sustainability. Huey D. Johnson (President, Resource Renewal Institute, San Francisco). Our Sustainable Future Series, Vol 7. Lincoln NE: U of Nebraska Press, May 1995/206p/$30.00.

Green plans can solve environmental problems for the world, the nation, businesses, labor, environmentalists, future generations—for everyone. They show what planning can and should be, and rescue the concept of planning from the scrapheap of history. Green plans are comprehensive, integrated, and large-scale—three traits that are key to solving environmental problems. They are government funded and implemented, developed with the input of key players, supported by a broad and diverse public, and flexible enough to change as new information becomes available. Green plans are not a theory but are actually happening. Chapters discuss bringing sustainability from theory to practice, as well as ***1) California's IFP Program as a Green Plan Predecessor***: Johnson was head of California's Resource Agency from 1977 to 1982, when he developed plans for a comprehensive resource strategy called Investing for Prosperity; ***2) The Netherlands' National Environmental Policy Plan***: the "best job of technical environmental planning done by any nation to date," using 25 years (one generation) as a framework—expressed by the phrase "each generation cleans up"; ***3) New Zealand's Resource Management Act***: a "remarkable achievement" that seeks to maintain a high level quality of life, while restructuring resource agencies and laws around the term "sustainable management"; ***4) Canada's Green Plan***: created and adopted by a conservative government, it probably will not be abandoned by any future regime; ***5) Various Countries on the Right Track***: "near-plan" countries that could at any time pull everything together into a comprehensive plan: Norway's Council for Sustainable Development, Sweden's five action programs, Denmark's Action Plan on Environment and Development, Austria's National Environmental Policy Plan, the UK Strategy on Sustainable

Development, various German green policies, the Singapore Green Plan (seeking to become a model green city by 2000), the European Union's action plan; ***6) Various US States***: Washington's Environment 2010, New Jersey's State Planning Act, Florida's Environmental Strategic Plan, Oregon's Benchmarks plan (setting 5, 10, 15, and 20-year goals for people, economy, and quality of life), Minnesota Milestones, Connecticut's Environment 2000 (a few other states are developing, planning, or implementing similar projects). The best of these state groups are similar to the Canadian Round Tables, bringing together various stakeholder groups. "These initial state efforts will catch the attention of national leaders, and the movement will begin to roll." To expand US green planning, a Federal commission on green plans should be set up to help states plan development. On the national level, even a few key green planning measures—a set of environmental priorities and goals, indicators to measure progress, and regular public reporting—would be a great step forward. [**NOTE**: Clearly-written and important. The Resource Renewal Institute (Ft. Mason Center, Bldg A, San Francisco CA 94123; 415/454-3963; rsrnin@well.com), founded by Johnson in 1983, started the Global Green Plan Center as an information clearinghouse and training center. Since 1989, it has published a newsletter, *The International Green Planner*. ALSO SEE: **Within One Generation: The Netherlands Environmental Policy Plan** by Alan AtKisson and Alex Steffen (AtKisson & Associates, 546 N 74th St, Seattle WA; 206/784-5773; Feb 1995/14p/$5.00).]

(green plans worldwide and in US)

419

The History of the Dutch National Environmental Policy Plan, Graham Bennett (Institute for European Environmental Policy, Arnheim, The Netherlands), *Environment*, 33:7, Sept 1991, 6-9 & 31-33.

After a long and difficult birth over the past decade, the National Environmental Policy Plan, adopted by the Netherlands in 1989, is "the first serious attempt by a national government to develop an integrated environmental policy based on explicit control principles and clearly formulated long-term environmental objectives." It contrasts sharply with the traditional incrementalist approach to environmental policy. The goal is to make environmental problems managable within 20-25 years, based on the central precept of "sustainable development." Policy will be based on pollution abatement at the source, the "polluter pays" principle, the prevention of unnecessary pollution, energy conservation, integrated life-cycle management, improvements in production processes, etc. NEPP sets out 50 strategic objectives and 228 specific measures, such as stabilizing CO_2 emissions at 1990 levels by 2000, introduction of environmental auditing, subsidies for energy conservation, reducing pesticide use 50% by 2000, composting 50% of household organic wastes by 2000, and substantial growth in spending for mass transit. But the target of sustainable development for the Netherlands by 2010 will be very difficult to achieve. Implementation of the plan is already behind schedule. Still, NEPP has had influence elsewhere in Europe. The Flemish regional government in Belgium published its **MINA Plan 2000** in 1989. The French government launched its **Plan Nationale pour l'Environnment** in June 1990. The British government published its strategy document, **This Common Inheritance**, in September 1990. And it is likely that the EC will adopt a five-year Environment Action Programme to take effect in 1993. [**NOTE**: The Netherlands and other European nations are also working on a "green" GNP calculation: an alternative system of national accounting to reflect environmental costs (*New York Times*, 9 Dec 1990, E3). ALSO SEE: **Canada's Green Plan** (**FS Annual 1992**, #11117), and ***Venezuela's Organic Law*** (*Environment*, Sept 1991, 16-20), on attempts of an industrializing country to regulate pollution.]

(green plans in Europe)

420

Nations of the Earth Report, Vol II & Vol III. UNCED National Reports Summaries. Geneva: UN Conference on Environment and Development, Oct 1992/353p/$50.00 & Feb 1993/518p/$50.00. (Available from UN Publications, E.92.I.18 & E.92.I.19.)

To prepare for the UNCED Earth Summit in June 1992, governments were asked to submit National Reports on their perspectives and experience, together with information on issues, activities, and policies at the national level. By the end of 1992, a total of 169 reports were received, adding up to more than 19,000 pages. **Volume I** summarizes 47 reports [**FS Annual 1993** #11657]. **Volume II** summarizes 50 reports from such countries or areas as Brazil, Columbia, Germany, Japan, Morocco, Pacific Islands, Pakistan, Poland, and Spain. **Volume III** summarizes 72 reports from such countries or areas as the Arctic, Australia, Cuba, European Community, India, Indonesia, Iran, Iraq, Mexico, Russia, and USA. Each summary presents information on the full title, where available, drafting process, problem areas, capacity building initiatives, recommendations and priorities on environment and development, financial arrangements and funding requirements, environmentally sound technologies, international cooperation, and expectations from UNCED. Each volume provides an overview of the reports. Of the 72 reports in Volume III, for example, 85% referred to problems linked with agriculture (generally soil erosion or degradation), 70% referred to demographic pressure, 60% have waste management problems, and 35% report limited access to freshwater.

(national environmental reports for UNCED)

421

Strategies for National Sustainable Development: A Handbook for their Planning and Implementation. Jeremy Carew-Reid *et al.* London: Earthscan, in association with IUCN and IIED, 1994/203p/free from IUCN (Gland, Switzerland) or IIED (3 Endsleigh St, London).

A distillation of lessons learned from 14 years of comprehensive strategy experience in every continent, which could prove useful for implementing the IUCN Caring for the Earth strategy (#170) and UNCED's Agenda 21 (#164/165). Ten key lessons: 1) strategies seek to maintain and improve the well-being of people and ecosystems; 2) the overall goal of strategies is sustainable development; 3) the choice of strategy objectives should be tactical; 4) the strategy process is adaptive and cyclical; 5) the strategy should be as participatory as possible; 6) communication is the lifeblood of a strategy; 7) strategies are processes of planning and action (main components are assessment, design, and implementation); 8) integrate the strategy into the decision-making systems of society; 9) build the capacity to undertake a strategy at the earliest stage; 10) external agencies that provide financial and technical assistance should be on tap, not on top. Chapters discuss the contribution of strategies to sustainable development (they provide a framework and focus for debate and consensus-building, and can help countries solve problems in an integrated fashion), participation in strategies, the strategy cycle (getting started, planning, implementing, staying on track), and resources to drive the strategy.

(sustainable development: national strategies)

422

Futures by Design: The Practice of Ecological Planning. Edited by Doug Aberley (Heriot-Watt U, Edinburgh). Foreword by David Suzuki. *The New Catalyst* Bioregional Series. Gabriola Island BC and Philadelphia PA: New Society Pubs, June 1994/214p/$39.95;$14.95pb.

An anthology designed to show ecological planning as a technique of applied bioregionalism. Essays include Gary Snyder on a California watershed, Donella Meadows on the world as a village of 1,000 people (of which 584 are Asians, 329 are Christians, 330 are children, 70 own an automobile, 7 are teachers, 3 are refugees, etc.), Helena Norberg-Hodge on the Tibetan model of counter-development, Raymond Dasmann on sound ecological planning, Jane Jacobs on city ecology, Murray Bookchin on ecological politics, Bill Mollison on permaculture, Richard Register on eco-cities, Janice Perlman on sustainable mega-cities, Christopher Canfield on eco-village building, Melanie Taylor on the new suburban village, David McClosky on Cascadia as a bio-region, Reed Noss on land conservation for wildlife, and Frederick Steiner on ecological landscape planning. Concludes with an "Ecological Planning Reading List" of some 250 books, arranged in about 50 categories. **(ecological planning anthology)**

423

Collaborative Planning for Wetlands and Wildlife: Issues and Examples. Edited by Douglas R. Porter (President, Growth Management Institute, Chevy Chase MD) and David A. Salvesen (Kensington MD). Washington: Island Press, April 1995/293p/$29.50pb.

Environmentalists and developers have charged that existing Federal and state regulatory programs do not adequately protect wetlands and endangered species, nor guide urban growth in a rational, consistent manner. Environmentalists have long complained that the case-by-case planning process causes "death by one thousand cuts" of sensitive areas, while developers chafe under a system of delays, inconsistent decisions, redundant and sometimes conflicting review and permitting processes, and overlapping authorities. The concept of collaborative, area-wide planning, like regional planning, was born out of the need to address problems of greater than local significance. Typically, such planning occurs in areas where intense development pressures collide with strong interests in protecting an imperiled natural resource. Several models already exist for this large-scale resource management, e.g.: the Adirondack Park Agency in northern New York [also see #118] and the Pinelands Commission in southern New Jersey. Collaborative plans involve trade-offs and compromises. But, ideally, such plans offer something for everybody: developers gain greater predictability and reduced regulation, environmentalists are assured that valuable natural areas will be protected, and Federal agencies reduce the number of permits they must process.

Chapters discuss the traditional paradigm in land use regulation (one that addresses issues in a fragmented and incomplete manner and promotes conflict), habitat conservation plans, the Balcones Canyonlands Conservation Plan in the Texas hill country, multi-species planning in three Southern California counties, the Bolsa Chica site along the Southern California coast, the Columbia River Estuary Study Taskforce, planning for New Jersey's Hackensack Meadowlands, the Chiwaukee Prairie area of southeast Wisconsin, the Chesapeake Bay Critical Areas Law, the Anchorage Wetlands Management Plan, and the East Everglades planning study. Common factors important to the success of a planning effort include political leadership, participation of all affected interests, funding to initiate and carry out a planning process, continuity of planning and management, anticipation of future conflicts, mediation to resolve conflicts, and integration of area-wide conservation plans with local and/or coastal zone plans. **(area-wide conservation planning)**

424

A Safer Future: Reducing the Impacts of Natural Disasters. National Research Council, US National Committee for the Decade for Natural Disaster Reduction. Washington: National Academy Press, 1991/67p(8x11")/ $12.95pb.

The World Health Organization estimates that natural disasters between 1964 and 1983 killed nearly 2.5 million people and left an additional 750 million injured or homeless. Unless action is taken to reduce the toll, these statistics can be expected to rise as populations increase and concentrate in vulnerable urban and coastal areas. Direct losses from natural disasters in the US, currently averaging $20 billion per year, continue to escalate. The sci/tech advances of the last half century provide many opportunities for mitigating the impacts of natural hazards. The US National Committee, formed in response to the UN International Decade for Natural Disaster Reduction, makes such proposals as: 1) community-wide awareness and education programs as a national priority; 2) a comprehensive long-range plan to reduce natural disaster impacts for every community at significant risk; 3) plans for emergency response, recovery, and reconstruction for every business and local jurisdiction at significant risk (preparedness can reduce immediate losses and minimize long-term social, economic, and environmental damage); 4) systematic collection and sharing of data on physical, biological, social, and health aspects of disasters. [**NOTE**: Obviously of importance in the wake of Hurricane Andrew. Global warming is likely to increase the intensity and perhaps the frequency of such storms (*Newsweek*, 7 Sept 1992, p30).] **(disaster impact mitigation)**

425

Research to Protect, Restore, and Manage the Environment. National Research Council, Committee on Environmental Research (Dale R. Corson of Cornell U, Chair). Washington: National Academy Press, 1993/242p/$28.00.

National and global environmental problems are acute. Science and engineering have the tools to address these problems, but the tools are not being used effectively. The US "system" to address environmental challenges consists of hundreds of programs in more than 20 Federal agencies that invest $5 billion a year in environment-related R&D. The Committee found several weaknesses that need to be addressed: 1) there is no comprehensive national environmental research plan to coordinate the efforts of the 20+ agencies involved; 2) the research establishment is poorly structured to deal with complex, interdisciplinary research on large spatial scales and long-term temporal scales; 3) bridges between science, policy, and management are weak; 4) "the nation's environmental efforts have no clear leadership"; 5) long-term monitoring and assessment of environmental trends and consequences of environmental regulations "are seriously inadequate"; 6) education in US universities is still strongly disciplinary, whereas solving environmental problems requires broadly trained people and multidisciplinary approaches; opportunities for broadly based graduate degrees are few, and faculty are not rewarded for interdisciplinary work; 7) most Federal environmental R&D is narrow, supporting either a regulatory or a management function.

Research must be directed toward protection, restoration, and management of natural, economic, cultural, and human resources that encourages sustainability. Toward this end, the Committee recommends: 1) establishing a *National Environmental Council* in the executive office of the President, to be chaired by the Vice President; the Council would provide national leadership and coordination among the Federal agencies for environmental research; 2) developing a *National Environmental Plan* (overseen by the NEC), to be updated every 2 years and comprehensively reconsidered every 5 years; 3) creating an *Environmental Assessment Center* to analyze large issues that cross agency lines; 4) initiating a *National Environmental Status and Trends Program* to monitor the status of US national resources (the National Biological Survey initiated by the Department of the Interior would be part of this effort); 5) launching a *National Environmental Data and Information System*, to collect and make easily accessible environmental data. Four organizational frameworks are considered to implement these changes, ranging from the current agency structure with enhancements to a National Institute for Environmental Research and a Department of the Environment.

(National Environmental Council needed)

426

Beyond the Horizon: Using Foresight to Protect the Environmental Future. Environmental Futures Committee, EPA Science Advisory Board. Washington: US Environmental Protection Agency (202/260-8414), Jan 1995/60p. (EPA-SAB-EC-95-007)

Traditional responses to environmental problems will not be effective enough, or take effect quickly enough, to protect vital resources. Thinking about the future is needed to initiate thought and analysis well in advance of anticipated change. In July 1993, the EPA formed the Environmental Futures Committee to study environmental foresight under a system of inquiry capable of anticipating possible issues over the next 5-30 years. Committee findings: 1) "large social, economic, technological, and institutional forces will cause future environmental risks that are potentially greater than those currently recognized and managed"; 2) the value of futures research and analysis lies not in making predictions but in organizing and analyzing information that can help shape decisions and actions; 3) "national and international environmental issues are rapidly becoming a matter of strategic national interest"; 4) future environmental issues may emerge in at least five different problem areas: sustainability of terrestrial ecosystems, non-cancer human health effects (loss of fertility, birth defects), total air pollutant loadings, non-traditional environmental stressors (new chemicals, control-resistant microbes and insects, exotic species), health of the deep ocean. Recommends an early-warning system for EPA and coordinating national efforts to anticipate and respond to environmental change. [ALSO SEE the technical annex to this report, **Futures Methods and Issues** (EPA-SAB-EC-95-007A).]

(future environmental issues)

6. Local Action

427

Making Development Sustainable: Redefining Institutions, Policy, and Economics. Edited by Johan Holmberg (IIED, London). Washington: Island Press, Oct 1992/362p/$45.00;$22.00pb. (First published in UK by Earthscan in 1992.)

Since 1987, many political leaders have talked about sustainable development, and much has been published on the subject (some 70 definitions are in circulation). "Sustainable development as a concept has become devalued to the point where, to some, it is now just a cliche." There is no universally agreed definition of sustainable development within IIED (the International Institute for Environment and Development), but there is no felt need for a rigorously consistent definition. Vagueness is no real drawback; the powerful idea underlying the concept is that of intergenerational equity—meeting today's needs without prejudice to future generations. "Primary environmental care" (PEC) is an umbrella term used by IIED and others to describe the process of moving toward sustainability at the grass roots. PEC has three integral elements: meeting basic needs, protecting the environment, and empowering groups and communities. These essays by IIED staff and others largely have PEC as a common denominator. Topics include prerequisites for empowerment, improving economic analysis for sustainable natural resource management, regenerating agriculture through low external input and community-based development, urban sustainability based on individual and community initiatives, restructuring industry for sustainability, sustainable forest strategies, sustainable livelihoods in Africa's drylands, energy efficiency for LDCs, financing sustainable development (through more efficient use of existing aid, reductions of debt, economic incentives, applying the "polluter pays" principle), and a new ethic for reducing profligate consumption in the North. [ALSO SEE: **Green Development: Environment and Sustainability in the Third World** by W. M. Adams of Cambridge U (Routledge, Nov 1992/257p/$19.95pb), which also focuses on empowerment of the poor.]

("primary environmental care" for sustainability)

428

Environmental Stewardship: Studies in Active Earthkeeping. Edited by Sally Lerner (Prof of Environment and Resource Studies, U of Waterloo). Waterloo, Ontario: U of Waterloo, Dept of Geography (Publication #39), Jan 1993/453p/C$30.00;US$25.00pb.

Case studies and surveys of mostly Canadian grassroots groups and coalitions. Such "stewardship groups" are significant because they make a positive difference in hundreds of local environments, and serve as an ethical, emotional, and political training ground for the grassroots environmental vanguard, currently one of the strongest constituencies for environmental interests. These are the people who, in increasing numbers, are demanding comprehensive proactive community, regional, watershed, and provincial/state planning for environmental quality, and who understand how to get what they want. These groups are typically formed because of a common interest in nature or an outdoor activity, or shared concern about an issue. Any forum that brings together concerned individuals or groups can lead to the formation of groups and coalitions (e.g., a single seminar at Brock University led to forming The Preserve Agricultural Lands Society in Ontario). Encouragement and support from academics, politicians and civil servants can provide the impetus for coalition formation and survival (e.g., in Montreal, city planners played a central role in forming the Green Coalition). Groups sometimes form because supporters break away from an organization to protest its goals, tactics, power structure, or inaction (e.g., Sisters for Non-Violent Action, or SNAG, was a splinter group formed to prevent logging of the Carmanah Valley on Vancouver Island). Life stages of the average advocacy group: recognizing a problem, disillusionment with authority, forming a group, getting informed, going public, campaigning (lobbying politicians, working with media, rallies and demonstrations, public hearings and lawsuits, publishing newsletters, raising money). All groups face questions of elitism, cooptation, tokenism, and means and ends. If a group meets its objectives, it winds down or goes on to work for broader goals. [ALSO SEE: **How to Save Your Neighborhood, City, or Town: The Sierra Club Guide to Community Organizing** by Maritza Pick (Sierra Club Books, April 1993/213p/$12.00pb), with tips on forming a group, leadership and teamwork, publishing a newsletter, winning elections, dealing with the media, and fundraising. **Getting Down to Earth: A Call to Environmental Action** by John Heidtke (Paulist Press, Jan 1993/179p/$9.95) focuses on discovering personal values, eco-value clarification, and acting on real values.]

(grassroots stewardship groups)

429

The Politics of Transformation: Local Activism in the Peace and Environmental Movements. Betty H. Zisk (Prof of Political Science, Boston U). Westport CT: Praeger, Oct 1992/256p/$49.95.

Interviews or observations of 156 state and local groups, conducted in 1987 in seven areas of the US. The local level was chosen because most of the "transformational" groups studied (those seeking to make a basic change in the way we conduct our politics, our economics, and our social relations) have chosen to work from the grassroots level up, and the most exciting and cutting-edge work is occurring locally, following the slogan of "Think globally, act locally." Chapters discuss policy issues for local environmentalists and peace activists, organizational structure/goals/activities (a consistent relation was found between the size and bureaucratization of movement groups and their success in regard to membership, media coverage, and goal attainment), tensions between group process and political success (Greens and other consensus-seeking groups are generally less successful than majoritarian groups), lobbying and litigation efforts, movement groups and the electoral process, protest strategies (vigils and civil disobedience are probably most effective as morale builders), movement building, education and outreach (one clear problem is "thinness" of movement resources), coalitions among peace and environmental groups (the gradual convergence of interests is a move toward a broad vision), group formation and dissolution, movement building and adaptation to change, and interconnected structural factors (economic, political, social, and ideological) "that all but paralyze transformational politics in the US." Concludes that "most transformational goals, and many of the more modest hopes of the incremental groups as well, remain unrealized and unlikely to be realized in the foreseeable future." Problems also abound within the movements: "lack of skill in fund raising, communications, coalition building, and, perhaps most important, framing a clear vision and the strategies to implement the vision." [**NOTE**: A scholarly study by "a full-scale activist in Green politics" with a Quaker background. An appendix justifies the benefits of participant observation, and hopes that the two worlds of activism and scholarship can enrich each other.]

(peace and environmental movements in US)

430

The Past, Present, and Future of Citizen Action (20th Anniversary Issue). *The Workbook*, 19:2, Summer 1994/98p/$5.00 (SRIC, Box 4524, Albuquerque NM 87106).

Essays on rural-urban cooperation for sustainable agriculture, economic "growth" vs. economic development in rural America, neighborhoods vs. toxic wastes, barriers to effective citizen action on Native American lands, the fight against redlining by banks, rights for the "psychiatrically labeled," what has been learned about the nuclear waste legacy (no quick solutions, programs motivated by political expediency, the need to say "No" consistently), a Kentucky community's fight for safe water, impacts of the grassroots movement for environmental justice, promoting cycling and undoing motor vehicle dependence, and women community leaders. In the lead essay, **David Walls** (Prof of Sociology, Sonoma State U) asserts that "community organizing must be judged a remarkable if unheralded success over the past 20 years." Builders of grassroots organizations can now draw on well-tested techniques of protest (marches, boycotts), political action (voter registration, lobbying), mutual aid (co-ops, credit unions), organizational development, fund-raising, and media access. Community organizing not only contains the democratic promise of empowering the disenfranchised, but also the vision of the cooperative commonwealth. Community organizing enters the 1990s with no single guru, no vanguard organization, and no hegemonic model; rather, many networks follow many approaches in diverse situations. Grassroots organizers, however, need to work on forging alliances with potential allies, and on building a majority coalition. **(grassroots action in past 20 years)**

431

Enough Is Enough: The Hellraiser's Guide to Community Activism. Diane MacEachern (Vanguard Communications, Washington). NY: Avon Books, 1994/407p/ $10.00pb.

This book had its genesis in the NIMBY movement that energized many environmental struggles over the last two decades; it seeks to empower people to protect themselves from undesirable developments and to pursue their own desirable objectives. Chapters discuss the four keys to successful organizing (a good issue, a good strategy, the right tactics, a solid organization), the importance of prioritizing issues, keys to an effective strategy (a long-range goal, a short-term goal, maximum use of resources, campaign targets, a time frame), refining your message with focus groups ("the way you position your issue will determine whether many people will join you"), basic tactics (letter-writing, boycotts, lawsuits, public hearings, negotiating, petitions, non-violent action, organizing special events), building organizations and coalitions, fundraising, communications, action at the polls (getting out the vote, referenda, initiatives, recalls, helping candidates), and lobbying. **(community activism guide)**

432

Where We Live: A Citizen's Guide to Conducting a Community Environmental Inventory. Donald F. Harker (President, Mountain Assn for Community Economic Development, Berea KY) and Elizabeth Ungar Natter (Director, Kentucky Local Governance Project). Washington: Island Press, Jan 1995/319p(8x11")/$18.95pb.

Too often, environmental decisions are made without the benefit of citizen involvement, to the detriment of communities. When citizens are at the decision-making table, the decisions are better for the environment and the health of the local people. This guide seeks to help citizens deal with government officials and company representatives, and was field-tested by three citizen groups. Chapters consider: 1) the human impact on the environment (surface water, groundwater, land, air); 2) getting started (organizing your group, deciding what to inventory, selecting the study area); 3) mapping your information to pro-

vide a focus for discussing potential impacts; 4) the environmental concerns inventory (who is polluting, what's being discharged, what pollutants are legal); 5) natural resources inventory (rare plants and animals, natural areas, wetlands, etc); 6) examining an individual facility or site (auditing an industry, the public water supply, or the sewage treatment plant); 7) analysis of pollutants; 8) expanding what agencies know (in many instances, citizens know more than government agencies); 9) environmental decision making in your area (evaluating sites, the permitting process). Appendices list addresses of relevant Federal and state agencies, a sample freedom of information request, summaries of key Federal laws (Clean Water Act, Clean Air Act, Emergency Planning and Community Right-to-Know Act, etc), and lists of regulated chemicals and chemicals of concern.
(local environmental inventory)

433
Community Energy Workbook: A Guide to Building a Sustainable Economy. Alice Hubbard and Clay Fong (RMI). Cosponsored by the US Dept of Energy, Urban Consortium Energy Task Force. Snowmass CO: Rocky Mountain Institute (970/927-3851), Spring 1995/264p (8x11")/$16.95pb.

Numerous communities are showing that environmental problems need not be addressed at the expense of the economy. This workbook challenges communities and their residents to ask new questions about how the community meets its energy needs. Using an "end-use/least-cost" approach opens up a wealth of win-win opportunities that can "cost-effectively reduce energy use by typically at least a third, and up to 90% in some cases, with the same or improved quality of service." In a typical town, 70 to 80 cents of every dollar spent on energy leaves the local economy; finding ways to reduce these costs and plug these leaks initiates a series of local economic benefits, because money saved tends to circulate in the local economy. Communities that are leading the way in improving their economy through better energy practices have several assets in common: widespread awareness of energy alternatives, strong local leadership, and an effective organization to sustain the effort. Planning steps involve a steering committee representing local interests, an Energy Town Meeting, energy task forces to address community energy sectors (residential, commercial/industrial, transportation), workshops to examine local resources and to brainstorm project ideas, creating and implementing an Energy Action Plan, and anchoring the plan in local organizations. **(local energy planning)**

434
Natural Connections: Perspectives in Community-based Conservation. Edited by David Western (Director, Wildlife Conservation Society), R. Michael Wright (President, African Wildlife Foundation), and Shirley C. Strum (Prof of Anthropology, UC-San Diego). Washington: Island Press, Dec 1994/581p/$49.00;$28.00pb.

Proceedings of a workshop held in Airlie, Virginia, in October 1993. "Community-based conservation" (CBC) covers both new and traditional conservation methods, as well as conservation efforts that benefit the community, whether they originate within or outside a community. In CBC, the emphasis has moved from the top to the bottom, from the center to the periphery, from the elite to the poor, and from the urban to the rural. Case studies describe CBC approaches in Kenya, India, Indonesia, Peru, Australia, Zimbabwe, New Guinea, Costa Rica, Niger, Nepal, and the UK's North York Moors. Lessons learned include the tension between insiders and outsiders (reflecting problems of trust and communication), rights and responsibilities, the disjunction between short-term and long-term costs and benefits, and the diversity of linkages that are necessary and possible. Some recommendations: 1) externally initiated conservation projects should be compatible with local concepts of culture; 2) broad-based local participation is inherent in the CBC concept; 3) CBC requires clear and unequivocal property rights that create a vested interest in managing resources sustainably (the greatest single obstacle to conservation is lack of secure tenure to land, wildlife, and other resources); 4) centralized national policies must be reversed if local conservation efforts are to flourish; 5) paternalism is antithetical to CBC; the existing donor-recipient relationship must evolve into one of partnership. [**NOTE**: Very timely, in view of growing interest in democratization and what works.] **(community conservation)**

435
Rural Environmental Planning for Sustainable Communities. Frederic O. Sargent (U of Vermont), Paul Lusk (U of New Mexico), José A. Rivera (UNM) and Maria Varela (UNM). Washington: Island Press, Nov 1991/254p/ $39.95;$24.95pb.

First published in 1976 by Sargent, this revised and enlarged edition focuses on the REP process, which seeks to balance economic development and environmental protection in accord with the carrying capacity of the land. Chapters discuss creating a rural environmental plan, types of public goals, inventories of the resource base, protecting natural areas, keeping land in agriculture, lake and river basin planning, planning for rural quality and recreation, social impact analysis, economic development models, elements of sustainable economic development (human resources, local control internal investment, and increasing opportunity), and the legal framework of planning. Includes many case studies, most from New England and New Mexico. **(sustainable rural communities)**

436
Preventing Industrial Toxic Hazards: A Guide for Communities. Marian Wise and Lauren Kenworthy. NY: INFORM Inc, June 1993/199p(8x11")/$25.00pb+$3.00p&h.

Revised edition of **A Citizen's Guide to Promoting Toxic Waste Reductions** (INFORM Inc, 1990; **FS Annual 1992** #11141). Governments have responded to the need to reduce toxic chemical wastes at the source with a growing number of laws. But regulations alone cannot protect human health: 1) many chemicals remain unstudied and unregulated; 2) many regulations are complex and can take years or even decades to implement; 3) lean budgets limit enforcement; 4) generation of hazardous waste is increasing (one study predicts a 75% increase from 1988 to 2000); 5) pollution control systems often end up simply moving pollutants from one medium to another. Industry is realizing the liability for cleaning up old pollutants and managing new ones: of the $100+ billion spent on federally mandated pollution control and clean-up in 1990, business paid two-thirds.

Source reduction is the one strategy that protects health and the environment while saving money. Citizens can play a major role in making it happen. This guide provides information on researching a plant, enlisting the support of management and workers, creating source reduction programs (basic techniques include chemical substitution, product changes, process changes, equipment changes, and operational changes), assessing whether a company has created effective policies, and promoting legislation that supports efforts at the local level through incentives and requirements. Appendices provide information on Federal and state agencies and contact people, tables on the health effects of Toxics Release Inventory chemicals, and sample forms used by plant officials for public reporting. [ALSO SEE: **Toxic Nation: The Fight to Save Our Communities from Chemical Contamination** by Pacific News Service editors Fred Setterberg and Lonny Shavelson (John Wiley, June 1993/301p/ $22.95), which provides several case studies.]

(source reduction of toxics)

437

Chemical Alert! A Community Action Handbook. Edited by Marvin S. Legator (Prof of Preventive Medicine, U of Texas-Galveston) and Sabrina F. Strawn (Director, Toxics Assistance Program, UT-G). Austin: U of Texas Press, Feb 1993/238p/$35.00;$14.95pb.

First published in 1985 as **The Health Detective's Handbook**, this revised edition is addressed to community groups that seek to determine whether health risks exist in their area. Industry has the right to operate and seek profits; the community has the right to clean air, water, and soil. In most cases, the scientific tool of traditional epidemiology (how often disease occurs in how many people) is crude and insensitive, because neighborhoods are often too small for useful studies. Moreover, those wielding this tool will often have an institutional bias against finding effects. It is essential for the community to gather as much information as possible and practical about its possible ill health. This manual seeks to focus these efforts, with chapters on nonprofessional identification of health hazards, organizing the community, known health effects of toxic substances, enlisting health professionals in determining adverse health outcomes, experimental design, questionnaire design and tabulation, assembling information (statistics for nonstatisticians), legal implications of acute and chronic effects, and resources. Concludes that, as more and more information accumulates on the role of a toxic agent in causing harm, "such a causal role may progress from mere speculation to distinct possibility, onward to probability, and at times even to virtual certainty."

To better identify toxic harm and to allocate its costs, a special tax is proposed against the toxic waste generator (TWG) to finance the public performance of its common-law duty. If the public took over the TWG's common-law duty to conduct appropriate studies, data worthy of confidence would be created, and the TWG would be freed of a troublesome conflict of interest. [ALSO SEE: **Calculated Risks: The Toxicity and Human Health Risks of Chemicals in Our Environment** by Joseph V. Rodricks (Cambridge UP, 1992/256p).]

(health hazard identification by communities)

438

Confronting Environmental Racism: Voices from the Grassroots. Edited by Robert D. Bullard (Prof of Sociology, U of California-Riverside). Boston MA: South End Press, May 1993/259p/$40.00;$16.00pb.

African Americans, Latinos, Asians, Pacific Islanders, and Native Americans are the groups most at risk from environmental problems. People of color are subjected to a disproportionately large number of health and environmental risks in their neighborhoods (e.g., child lead poisoning, the leading environmental health problem facing children) and on their jobs (e.g., pesticide poisoning of farm workers). Stringent environmental regulations and NIMBY actions have increased the vulnerability of communities of color to the siting of unpopular facilities such as landfills, toxic waste dumps, lead smelters, and incinerators. This book grows out of the work of scholars and activists at the First National People of Color Environmental Leadership Summit, held in Washington in October 1991. Essays discuss the environmental justice movement, toxic wastes and race, coping with industrial exploitation, removing lead from the community, environmental politics in Virginia and Alabama, creating environmentally sustainable alternatives in northern New Mexico and southern Colorado, farmworkers and pesticides, and global threats to people of color. Bullard concludes with a proposal for a Fair Environmental Protection Act that would prohibit environmental discrimination based on race and class. [ALSO SEE: **Dumping in Dixie: Race, Class, and Environmental Quality** by Robert D. Bullard (Westview Press, 1990).]

(environmental racism)

439

Risk and Rationality: Philosophical Foundations for Populist Reforms. Kristin Shrader-Frechette (Distinguished Research Prof, U of South Florida). Berkeley CA: U of California Press, Nov 1991/312p/$39.95;$15.95pb.

Citizen "NIMBY" protests result in part from the perceived failure of government and industry to protect the health and safety of the people. Industry attributes the blocking of oil refineries, nuclear reactors, and toxic waste dumps to public ignorance and paranoia. Kristin Shrader-Frechette, a widely-published environmental ethicist, seeks a middle path between industrial charges of scientific illiteracy and populist charges of technological oppression, arguing for a new paradigm of when acceptance of public hazards is rational. "We also need to recognize that laypersons are often more rational, in their evaluation of societal risks, than either experts or governments appear to have recognized." Chapters discuss science against the people, rejecting reductionist risk evaluation, objectivity and values in risk evaluation, a new framework for rational risk evaluation, "perceived risk" seen by laypersons and "actual risk" calculated by experts, the highly questionable restriction of risk to "probability of fatality," the case for a "maximin" account of risk and rationality, risks of subjecting Third World peoples to transfers of hazardous technology, methodological reforms for risk evaluation, and more democratic procedural reforms in risk management. **(risk assessment reconsidered)**

440

Risk and Responsibility. William Leiss (Research Chair in Environmental Policy, Queens U) and Christina Chociolko (Centre for Policy Research on Sci/Tech, Simon Fraser U). Montreal & Kingston: McGill-Queen's U Press, July 1994/405p/C$55.00;C$22.95pb.

All of us have a direct and vital interest in the proper allocation of responsibility for risky activity. Health and environmental risks, especially those to which people are exposed involuntarily, are controversial because people believe—with good reason—that they cannot control their exposure to the chance of unfair and uncompensated loss. The dominant institutions of society, which are usually proponents of projects leading to involuntary citizen exposure to risk (new chemicals, power installations, hazardous-waste treatment, etc.), are often evasive or silent when it comes to indicating who will be responsible for resultant losses to individuals. The authors examine case studies of risks and responsibilities associated with electromagnetic fields and power line frequencies, Alar (a chemical growth-regulator for food crops), antisapstain chemicals (pesticides used on softwood lumber products), Alachlor (a herbicide used worldwide for many crops), formaldehyde (a widely-used industrial chemical), and tobacco use. They conclude that it is in everyone's interest to move away from disagreement toward consensus. The most promising route to this end is for major participants in risk debates to accept a fuller share of responsibility. Industry and government should look ahead in the zones where risks are likely to arise, and use the best available scientific methods in a disinterested risk assessment. Government should explore ways in which responsible choices can be made in trade-offs between risks and benefits. And public-interest groups should help citizens move away from unreasonably risk-averse standpoints. [**NOTE**: Nicely complements **Risk and Rationality** (above), which also seeks a middle ground between the experts and the public.] **(toward consensus on risks)**

441

Sustaining the Earth: Choosing Consumer Products That Are Safe for You, Your Family, and the Earth. Debra Dadd-Redalia (Forest Knolls CA). NY: Hearst Books (William Morrow), July 1994/352p/$15.00pb.

Sustainability is advocated as the new benchmark for evaluating everyday products. We should ask if we really need the product, is it safe, is it practical and durable, is it made from renewed or recycled materials, is information available about its manufacturing process, is it safely disposed, what kind of packaging does it have, and how far has it been shipped. Chapters discuss product life cycles, reading labels, sustainability attributes, certification organizations, and social responsibility. About a third of the book is devoted to discussing correct and incorrect buys in some 130 product categories such as air conditioners, automobiles, beer, bread, cookware, dishwashers, furniture, insulation, paint, pet products, skin care, water, and yogurt. Concludes with directories of books, greenstores, mail order catalogs, manufacturers, and organizations.
(sustainable consumption)

7. Leadership

442

Inside the Environmental Movement: Meeting the Leadership Challenge. Donald Snow (The Conservation Fund, Arlington VA). Washington: Island Press, Jan 1992/295p/$34.95;$19.95pb.

An estimated 10,000 conservation and environmental organizations protect wildlife habitat, provide recreational facilities, monitor toxic wastes, work for clean air, or lobby Congress. In addition, the nation's 900+ independent land trusts have a combined membership of over 750,000, and new land trusts have been forming on the average of one a week. The rapid increase in the size of the movement has generated a pressing requirement to find, train, and employ people capable of exercising effective leadership. The Conservation Leadership Project was begun in 1989 to find out what needs to be done to increase the overall effectiveness of the movement. The study found organizations with untrained staff people, strife between organizations, and leaders divorced from their memberships. Recommendations include leadership development centers, taking the cause to new constituencies, less emphasis on white male staffing, and more foundation support and academic programs for developing movement leadership. [ALSO SEE: ***Environmental Studies: Is It Academic?*** by Holly Brough (*World Watch*, 5:1, Jan-Feb 1992), estimating that only 10% of US colleges and universities currently offer environmental studies as a major.]
(Conservation Leadership Project)

443

Environmental Leadership: Developing Effective Skills and Styles. Edited by Joyce K. Berry and John C. Gordon (both School of Forestry and Environmental Studies, Yale U). Washington: Island Press, Sept 1993/286p/$45.00;$19.95pb.

A key element in improving the environment is leadership: the ability of an individual or a group to guide positive change toward a vision of an environmentally better future. Environmental problems and opportunities need a special kind of leadership because of the key characteristics of environmental problems: long times to solution, complex systems, an emotion-charged context, a relatively weak and scattered science base, and a need to integrate a wide array of knowledge and concerns. Berry and Gordon assemble a series of personal accounts of leadership by a diverse array of environmental leaders, on ethics for leaders, communications, managing conflict, policy and legislation, state and regional resource management, fundraising, leadership in state agencies, national leadership, international leadership, local voluntary organizations, corporate leadership, academic leadership, and foundation leadership. Concludes with six important insights: 1) everyone in every environmental organization must both lead and follow; 2) leaders must exert extraordinary effort to think about and prepare for change; 3) environmental leaders need great breadth and flexibility; 4) successful communication is paramount; 5) most successful environmental leaders operate from a strong ethical base; 6) leaders must be lifelong learners.
(environmental leadership)

444

Making Things Happen: How To Be An Effective Volunteer. Joan Wolfe. Foreword by Peter A.A. Berle (President, National Audubon Society). Washington: Island Press, Nov 1991/226p/$14.95pb.

Reprint of an introductory guidebook first published in 1981, with chapters on attracting volunteers, successful publicity and programs, simplified rules of order (for most groups, detailed rules can be counterproductive), raising money, elections, officers and directors, leadership, committees, and staff/volunteer relations. Of interest here are the last two chapters on effective citizens in the public realm. A checklist of *how to be effective*: do your homework, work with others to get the job done, try to get insider support, resolve to succeed, persist, understand the importance of the highest-level staff person in any agency you deal with (decision-makers lean heavily on their advice), look for every opportunity to spread your word, keep the media informed, find out what the opposition is saying and answer their objections, share the leadership and the glory, broaden your base of support, keep current and beware of changes. A final chapter on *being persuasive* recommends be honest with the facts, avoid sounding arrogant, address the other person's concerns, speak with courtesy and modesty, and write short letters. [**NOTE**: Very elementary, but many valuable points.]
(effective citizen activism)

445

Leadership for the Common Good: Tackling Public Problems in a Shared-Power World. John M. Bryson (Prof of Planning and Public Affairs, U of Minnesota) and Barbara C. Crosby (Reflective Leadership Center, U of Minnesota Humphrey Institute). San Francisco: Jossey-Bass Publishers, Nov 1992/436p/$32.95.

We live in a world where no one is in charge. In this "shared-power world," organizations must share objectives, activities, resources, and power in order to achieve collective gains or minimize losses. Successfully tackling public problems requires effective leadership. This book, addressed to policy entrepreneurs and change advocates, considers how public leaders can inspire and mobilize others in a shared-power world to take collective action in pursuit of the common good. Chapters discuss leadership tasks, how innovation happens (the policy change cycle), settings for exercising leadership (forums, arenas, courts), and key steps in tackling public problems (agreeing to act, effective problem definition, searching for solutions in adopting and implementing policies, reassessing policies and programs). Concludes with some leadership lessons about shared power: 1) in a shared-power world, advocates of policy change cannot force outcomes; 2) collective public leadership is vital; 3) the policy change cycle can start anywhere, and is rarely linear; 4) policy change processes are prone to disintegration, and are often characterized by disruption and delay (an important leadership task is thus to influence the flow of action so that problems, solutions, and politics are joined); 5) individuals and groups can engage in effective policy change processes, but success is never guaranteed; 6) acts of public leadership are based on understanting the shared-power context in which leadership occurs, team building, the nurture of networks, the design and use of forums and arenas, and putting it all together. **(public leadership and shared power)**

446

Leadership Without Easy Answers. Ronald A. Heifetz (Director, Leadership Education Project, JFK School of Govt, Harvard U). Cambridge MA: The Belknap Press of Harvard U Press, Sept 1994/348p/$24.95.

A psychiatrist observes that we face a crisis of leadership in many areas of public and private life. But we attribute our problems too readily to politicians and executives, as if they were the cause. Pinning the blame on authority provides a simple answer for predicaments, yet our current crises have more to do with the problems of modern political and economic life. To meet our challenges, we need a new idea of leadership and a new social contract to promote our adaptive capacities. Two images of leadership are offered: "influencing the community to follow the leader's vision" vs. "influencing the community to face its problems." The second image—mobilizing people to tackle tough problems—differs from the culturally dominant view of the leader, usually in high office, standing out in front and influencing others. It places emphasis on the act of giving clarity and articulation to a community's guiding values, and on leadership as adaptive work. Above all, the practice of leadership requires a sense of purpose—the capacity to find the values that make risk-taking meaningful. A leader should engage people in facing the challenge, adjusting values, and developing new habits. The adaptive demands of society requires leadership that doesn't wait for "the vision" or the coach's call. One may lead with no more than a question in hand.

(leadership as facing challenges)

447

Overcoming Obstacles in Environmental Policy-making: Creating Partnerships Through Mediation. John K. Gammon (Principal of CONCUR, Collaborative for Environmental Analysis and Conflict Resolution, Santa Cruz CA). Albany NY: State U of New York Press, Dec 1994/250p/$57.50;$18.95pb.

Despite the recent surge of interest in the environmental problems of the developing world, there has been little success in resolving them. Environmental groups in the US and Europe have pressured their governments to force international agencies such as the World Bank and USAID to reform their lending practices. But these efforts have generally not worked. As a result, limited natural resources under severe pressure to feed fast-growing populations and pay off foreign debt continue to be mismanaged. Environmental policies fail because underlying complexities make reform very difficult to achieve, environmental concerns are treated as tangents to the development process, the closed nature of environmental policy-making promotes a distinct anti-environmental bias, and policies grounded in a Eurocentric worldview fail to understand how the developing world thinks and acts. One way for public policy to respond to the environmental crisis in a meaningful way is to apply the theory and practice of negotiation and conflict resolution, which works on the basic premise that all parties affected by an action should have a voice in an open dialogue; from this sharing of interests emerges a collaborative effort to solve what is perceived as a common problem. Gammon discusses such a "negotiated partnership for sustainable development," preconditions of the negotiation process, steps to promote effective reform (e.g. stakeholder analysis, choose a mediator, create an implementation strategy, agree on time frame and post-project evaluation, etc.), the importance of cultural differences, and advice to donor agencies, political leaders, environmental NGOs, and resource users. [**NOTE**: Dense but rewarding study, draws on Gammon's work in the Caribbean but applicable to all developing countries.] **(environmental partnerships in LDCs)**

448

Future Search: An Action Guide to Finding Common Ground in Organizations and Communities. Marvin R. Weisbord and Sandra Janoff (Future Search Associates, Philadelphia PA). San Francisco: Berrett-Koehler Pubs, June 1995/219p/$19.95pb.

A future search is a large group planning meeting that brings a "whole system" into the room to work on a task-focused agenda. Typically, it is conducted over three days—***Day 1***: people in mixed groups focus on the past, the present, and external trends; ***Day 2—Morning***: people in stakeholder groups continue to identify trends important to them and report to the whole; each stakeholder group makes "prouds" and "sorries" about what they are doing; ***Day 2—Afternoon***: people return to mixed groups and prepare ideal scenarios 10-20 years in the future and how they overcame barriers to get there; the whole conference notes common future themes, potential projects to get there, and unresolved differences; ***Day 3***: whole group reviews the lists and makes action plans on common ground. Brief examples are given of the use of future search in union/management joint planning at a 3M plant, implementing a city-wide consensus in Kansas City, organizing statewide voluntary action in Pennsylvania, regional economic development in the Arctic, rethinking work systems at R.R. Donnelley, community development in Hopkinton MA, rethinking business strategy at Digital Equipment, building relationships across urban cultures in Philadelphia, organizing for the future at Alcan Smelters & Chemicals, and nation building in Bangladesh. Chapters discuss tips and traps, developing a future search design, planning a future search, managing the emotional roller-coaster, conditions for success, and a sample invitation letter. **(future search meetings)**

449

Beyond Machiavelli: Tools for Coping with Conflict. Roger Fisher (Prof of Law Emeritus, Harvard Law School; founder, Harvard Negotiation Project), *et al.* Cambridge MA: Harvard U Press, Feb 1994/151p/$16.95.

As our world becomes increasingly interdependent, "conflict is a growth industry." Machiavelli's *Prince* is still a powerful book after almost 500 years because it asks the important question of what advice one would give a leader. If our purpose is to improve the caliber of decisions, we need to move beyond predictions and unconstructive criticism: "we need to look at a particular choice somebody faces and ask how things might be better done." In most cases, the difficulty lies not in a lack of potential options, but in the failure to design, negotiate, and pursue a process that moves us forward to where we would like to be. It is more useful to think about a good process for handling a flow of problems than to think about "solving" a particular problem. "In fluid and turbulent times, it is better to think in terms of coping with conflicts than resolving them" Understanding our task as *conflict management* rather than *conflict resolution* is a paradigm shift.

Chapters discuss: 1) looking forward with a purpose ("it is most unlikely that we will end up where we want to be unless we have both thought about where that is and directed our actions toward getting there"); 2) three perspectives for observing a conflict (my perspective, their perspective, and neutral "fly-on-the-wall"); 3) considering the other side's choice and constructing a Target Future Choice chart that looks ahead to the consequences of saying Yes or No; 4) generating fresh ideas to reframe a problem; 5) how to reach a decision-maker; 6) barriers to resolution (poor design of third-party roles, limited mediation skills, legal and political constraints, and ineffective institutions), and how to overcome them. [**NOTE**: A wise little book with much to say about constructively shaping the future.] **(coping with conflict)**

450

Getting the Word Out in the Fight to Save the Earth. Richard Beamish (Lake Clear NY; editor, *Adirondack Explorer*). Baltimore MD: Johns Hopkins U Press, March 1995/181p/$45.00;$24.95pb.

Practical advice from the former director of communications of the National Audubon Society and public information director for New York's pioneering Adirondack Park Agency. Chapters discuss the need for clear and compelling communications, the importance of a good logo to put a face on your organization, how to recruit members and donors through the mail (appealing to the emotions, preparing the package), communicating with constituents to keep them active and renewing (designing the newsletter, membership renewals), how to publicize your cause (press releases and press conferences, inviting journalists to field trips, op-ed articles and letters), and successful fundraising (establishing a fundable program, identifying prospects at the top of the support pyramid for major gifts, foundation grants). In sum, "good communications are not a luxury but a necessity." [**NOTE**: Simply written with many examples, especially from Adirondack Park organizations. The lessons, however, are applicable to non-US and non-environmental organizations.]

(green organization communications)

170 PERIODICALS:

An Annotated Listing of 170 Environment-Relevant Journals

ADIRONDACK JOURNAL OF ENVIRONMENTAL STUDIES (AJES) (1994). Bi-annual, $10 from Paul Smith's College, Paul Smiths NY 12970 (518/327-6377). Covers the broad range of issues that concern the Adirondack-Champlain Biosphere Reserve in New York State, an important 6 million acre natural area pioneering in public-private cooperation. Reviews. Gary Chilson, David Vinopal, Editors.

AFRICAN JOURNAL OF ECOLOGY (1962). Quarterly, £179.50 in Europe (£198 elsewhere) from Blackwell Science Ltd., Journals Subscriptions, c/o Marston Book Services, P.O. Box 87, Oxford OX2 0DT, UK (01865-206206). Formerly the *East African Wildlife Journal* this renowned publication covers all aspects of African ecology. Book Reviews. F.I.B. Kayanja, Editor.

AFRIKA-NATUURLEWE (AFRICAN WILDLIFE) (1946). Bi-monthly, 88 Rand from Wildlife Society of Southern Africa, P.O. Box 44189, 2104 Linden, South Africa (27-11-4863294). Covers all aspects of the conservation of African wildlife and natural habitats, which also examines the diverse range of issues relating to the environment. Book Reviews. Sandie Anderson, Editor.

AGRICULTURE, ECOSYSTEMS AND ENVIRONMENT (1974). 15 issues/year (in 3 vols., 3 nos./vol.) 1695 Guilders from Elsevier Science Inc., Box 882, Madison Square Station, New York NY 10159 (212/989-5800). Concerned with the interaction of methods for agricultural production, agroecosystems and the environment. Book reviews. T.L.V. Ulbricht, J.W. Sturrock, Editors.

AGRICULTURE AND HUMAN VALUES (1984). Department of Philosophy, 330 Griffin-Floyd Hall, University of Florida, Gainesville FL 32611-8545.

ALTERNATIVES: Perspectives on Society, Technology and Environment (1971). Quarterly, C$23.50/year ($27.50 elsewhere), from University of Waterloo, Faculty of Environmental Studies, Waterloo, Ontario N2L 3G1 Canada (519/885-1211). Provides critical and informed analysis of environmental problems, related social issues and technological developments. Book reviews. Robert Gibson, Editor.

AMICUS JOURNAL (1979). Quarterly, $10 to non-members from National Resources Defense Council, 40 West 20th Street, New York NY 10011 (212/727-2700). Contains news about the environment and NRDC activities. Book reviews. Francesca Lyman, Editor.

AMBIO: A Journal of the Human Environment (1972). Text and summaries in English. 8 issues/year, $61 from Royal Swedish Academy of Sciences, PO Box 50005, S-104 05 Stockholm, Sweden (46-8-673-95-51). US Subscriptions: Allen Press, Inc., Box 1897, Lawrence KS 66044-8897 (913/843-1234). International journal publishing original articles, reports and comments on developments and issues in environmental research, policy and related activities, including environmental management, sustainable technology, and use of natural resources. Book reviews. Elisabeth Kessler, Editor.

ANIMALS' AGENDA: Helping Animals and the Earth (1981). Bi-monthly, $22 from Animals Rights Network, Inc., Box 25881, Baltimore MD 21224 (410/675-4566). Subscriptions: Box 6809 Syracuse NY 13217 (800/825-0061). Provides a broad range of information about animals and environmental issues, and a forum for discussion of related problems and ideas. Book reviews. Kim W. Stallwood, Editor.

ANNALS OF EARTH (1982). $30 from Ocean Arks International and Lindisfarne Association, One Locust Street, Falmouth MA 02540 (508/540-6801). Seeks to disseminate the ideas and practice of ecological sustainability throughout the world. Dedicated to fostering emergence of a new global culture. Nancy Jack Todd, Editor, William Irwin Thompson, Contributing Editor.

AUDUBON (1899). Bi-monthly, $30 from National Audubon Society, 700 Broadway, New York NY 10003 (212/979-3000). Subscriptions: Box 52529, Boulder CO 80322 (800/274-4201). Features outstanding photography and articles on environmental problems, politics, and the natural world. Book reviews. Michael D. Robbins, Editor.

BIOCYCLE: Journal of Composting and Recycling (1969). Monthly, $63/year from JG Press, Inc., 419 State Ave., Emmaus PA 18049 (215/967-4135). For municipal and industrial managers of solid waste and sludge, focusing on composting and recycling techniques—from collection and processing to materials and marketing. Book reviews. Jerome Goldstein, Editor.

BIODIVERSITY AND CONSERVATION (1991). Quarterly, £60 from Chapman & Hall, 2-6 Boundary Row, London, SE1 8HN, UK (0171-865-0066). Devoted to papers on biodiversity in all its aspects, its description, analysis and conservation, and its controlled and rational use. Alan T. Bull, Editor.

BIOLOGICAL CONSERVATION (1969). Monthly, £645 from Elsevier Science Publishers UK, The Boulevard, Langford Lane, Kidlington, Oxford OX5 1GB, UK (44-1865-843000). Publishes papers dealing with the preservation of wildlife and the conservation, and sensible use of biological and allied natural resources. Book Reviews. Eric Duffey, Editor.

BIOSCIENCE (1950). Monthly, $25 to members, from American Institute of Biological Sciences, 730 11th Street NW, Washington DC 20001-4521 (202/628-1509). Articles from all areas of biological sciences, including animals, humans, plants, and the environment. Book reviews. Julie Ann Miller, Editor.

BIOTROPICA (1969). Quarterly, $85 from the Association for Tropical Biology, Julie Denslow, Department of Ecology & Evolution, Tulane University, New Orleans LA 70118 (504/388-8411). Contains original research in the broad area of tropical biology, including the ecology and management of diverse tropical ecosystems. E. Raymond Heithaus, Editor.

BOSTON COLLEGE ENVIRONMENTAL AFFAIRS LAW REVIEW (1971). Quarterly, $22 from Boston College, School of Law, 885 Centre Street, Newton MA 02159 (617/552-4354). Analysis of environmental issues from legal, political, and scientific experts. Book reviews.

BRUNDTLAND BULLETIN (1988). 100 pp. Quarterly, $60/year from The Centre For Our Common Future, 52, rue des Pâquis, CH-1201 Geneva, Switzerland. Reports on how sustainable development is being achieved around the world, with many reports about the UN system, governments, international organizations, NGOs, media, women, youth, academic and research centers, financial institutions, industry, trade unions and professional associations. Anja Halle, Editor.

BUSINESS AND ENVIRONMENT ABSTRACTS (1994). Bi-annual, £75/year from ERP Environment, PO Box 75, Shipley, West Yorkshire BD17 6EZ, UK (+44 (0)274 530408). A reference guide to papers in the world's business and environment journals.

BUSINESS STRATEGY AND THE ENVIRONMENT (1994). Quarterly, £60/year from ERP Environment, PO Box 75, Shipley, West Yorkshire BD17 6EZ, UK (+44 (0)274 530408). Journal on corporate environmental management and strategy.

BUZZWORM: The Environmental Journal (1988-ceased April 1994). Quarterly from Buzzworm Inc., 2305 Canyon Blvd., Suite 205, Boulder CO 80302. Articles and features aim to provoke environmental awareness. Covers international and national issues, green consumerism, children and the environment, and urban ecology.

CAPITALISM, NATURE, SOCIALISM: A Journal of Socialist Ecology (1988). $22 from CNS, Guilford Publications, Journals Department, 72 Spring Street, New York NY 10012 (212/431-9800). An international red green journal of theory and politics which combines the themes of history and nature, society and environment, and promotes the ideals of ecological socialism and feminism. James O'Connor, Editor.

CHOICES: The Human Development Magazine. Available in English, French, Japanese and Spanish. From Division of Public Affairs, United Nations Development Programme, One UN Plaza, New York NY 10017 (212/906-5315). Lois Jensen, Sid Kane, Editors.

CHRISTIAN SCIENCE MONITOR. 5/week, $159 from Christian Science Publishing Society, One Norway Street, Boston MA 02115 (617/450-2000). Outstanding 20 pp. newspaper with good coverage of global, environmental, and technological issues. Feature stories on the environment by Brad Knickerbocker are often leading-edge. Richard J. Cattani, Editor.

CLEARINGHOUSE BULLETIN (1990). 8-12 pp. Monthly, $20 from Carrying Capacity Network, Inc., 1325 G Street NW, Suite 1003, Washington DC 20005 (202/879-3044). A newsletter to keep Clearinghouse participants informed of the latest developments in environmental protection, population stabilization, resource conservation, actions of Clearinghouse participants and others, and environmental legislation. David F. Durham, Editor.

CONNECT: UNESCO-UNEP Environmental Education Newsletter (1976). Text in English. Quarterly, free from United Nations Environment Programme, 7-9 Place de Fontenoy, 75700 Paris France. Essays on education, news of the International Environmental Education Program (IEEP), and notices of publications. Joseph Barry, Editor.

CONSERVATION BIOLOGY (1987). 6 issues/year, $250 in US ($280 elsewhere) from Blackwell Scientific Publications, 238 Main Street, Cambridge MA 02142 (617/876-7022). Provides a forum for debate among environmental professionals, covering a wide range of topics. Reed Noss, Editor.

CORPORATE ENVIRONMENTAL REGISTER. £77 from Environment Press, 26 Brock Street, Bath, BA1 2LN UK (0225 330312). Detailed company profiles of the environmental management practices in 120 firms, and the state of environmental activity in big business.

CULTURAL STUDIES QUARTERLY (1981). 80 pp. Quarterly, $45/year from Cultural Survival, Inc., 53A Church Street, Cambridge MA 02138 (617/495-2562). Comprehensive coverage of indigenous affairs and related research and activism. Marc Miller, Editor.

DELTA (1990). 4 issues/year, free from Royal Society of Canada, Canadian Global Change Program, P.O. Box 9734, Ottawa, Ontario K1G 5J4 (613/991-0175). The CGCP is a multidisciplinary network for coordinating research and communicating results, ideas and recommendations to the policy and research communities.

DOWN TO EARTH (1992). 60 pp. magazine, 26 issues/year, $100/year (outside India) from Society for Environmental Communications, F-6, Kalish Colony, New Delhi 100 048 India (6438109). A magazine on science, technology, environment and development. Covers national and international policy decisions, and grassroots experiences of people from the developing world who are struggling to protect and improve their environment. Seeks to fuel nformed debate on how the world can manage the transition to a sustainable and equitable future. Anil Agarwal, Editor.

E MAGAZINE: The Environmental Magazine (1990). Bi-monthly, $20 from Earth Action Network, Box 5098, Westport CT 06881 (203/854-5559). Subscriptions: Box 6667, Syracuse NY 13217 (800/825-0061). Provides information on current environmental issues, the environmental movement of the 1990s, issues and policies, interviews with noted environmentalists, news of publications, ecologically sound products and services. Book reviews. Doug Moss, Editor.

EARTH FIRST! The Radical Environmental Journal (1980). 8 issues/year (corresponding with European pagan holidays), $25 from Earth First, Box 1415, Eugene OR 97440-1415 (503/741-9191). "A forum for the no-compromise environmental movement. Although Earth First does not accept the authority of the hierarchical state, nothing within the newspaper is intended to run it afoul of its police power." Book reviews.

EARTH ISLAND JOURNAL (1986). 48 pp. Quarterly, $25/year from Earth Island Institute, 300 Broadway, Suite 28, San Francisco CA 94133 (415/788-3666). An environmental news magazine providing coverage of worldwide developments in the global environmental movement. Features include eye-witness accounts and exposes from the environmental front lines. Gar Smith, Editor.

EARTH JOURNAL (1988-ceased April 1994). 6 issues/year from Buzzworm Inc., 2305 Canyon Blvd., Suite 205, Boulder CO 80302.

EARTH TIMES (1991). Semi-monthly 24-page tabloid newspaper, $36/year from Earth Times Foundation, PO Box 3363 Grand Central Station, New York NY 10163 (212/297-0488). Reports on global environmental and development issues, with emphasis on UN agencies. Occasional book reviews. Pranay Gupte, Editor.

EARTHWATCH (1980). 6 issues/year, $25 from Earthwatch Expeditions, Inc., 680 Mount Auburn Street, Box 403N, Watertown MA 02272 (617/926-8200). Includes compendium of scientific research expeditions, feature stories on leading researchers, cultural heritage questions, and environmental and science-related topics. Mark Cherrington, Editor.

ECO: Business & Environment (1993). Quarterly, $40 from Eco, Inc., 83 S. Bedford Rd., Mt. Kisco NY 10549 (914/242-0140). Business magazine focusing on environmental issues. Book reviews. Igor Gordevitch, Editor.

ECODECISION (July 1991). Quarterly, from Environment and Policy Society, Royal Society of Canada, 276 rue Saint-Jacques Ouest, Montreal Quebec H2Y 9Z9 Canada. An international magazine to help private and public sector as well as professional and elected policy decision makers to make the right choices and better manage the transition towards sustainable development.

ECODESIGN. From Ecological Design Association, 20 High Street, Stroud, Glos GL5 1AS UK.

ECOLOGICAL APPLICATIONS / ECOLOGY (1991). $75 from Ecological Society of America, c/o Centre for Environmental Studies, Arizona State University, Tempe AZ 85287-3211. One of the leading applied ecological journals. From 1995 it has been combined with the journal formerly called *Ecology*. Simon Levin, Editor.

ECOLOGICAL ECONOMICS (1989). 9 issues/year (in 3 vols., 3 nos./vol.) Journal of the International Society for Ecological Economics, PO Box 1589, Solomons MD 20688 (410/326-0794). $40 from ISEE, or 870 Guilders from Elsevier Science Inc., Box 882, Madison Square Station, New York NY 10159 (212/989-5800). Includes neoclassical environmental economics and ecological impact studies as subjects, and encourages new ways of thinking about the linkages between ecological and economic systems. Robert Costanza, Editor.

ECOLOGICAL ECONOMICS BULLETIN (1989). Formerly *International Society for Ecological Economics Quarterly Newsletter*, through 1995. Quarterly, from ISEE, PO Box 1589, Solomons MD 20688 (410/326-0794). $30 as part of ISEE membership. Provides information and news to members and non-academic readers who are interested in learning about ecological economics and the innovative approaches it is developing to help preserve and restore the environment and transform growth into development that is ecologically sustainable and socially just. Brief abstracts of books received. Dean Button, Editor.

ECOLOGICAL ENGINEERING: The Journal of Ecotechnology (1991). Text in English. 4 issues/year, 361 Guilders from Elsevier Science Inc., Box 882, Madison Square Station, New York NY 10159 (212/989-5800). On ecotechnology, including synthetic ecology, bioengineering, pollution control, and sustainable agriculture. William J. Mitsch, Editor.

ECOLOGICAL MONOGRAPHS (1931). Quarterly, $50 from Ecological Society of America, c/o Centre for Environmental Studies, Arizona State University, Tempe AZ 85287-3211. Lee N. Miller, Editor.

THE ECOLOGIST (1970). 45 pp. Bi-monthly, $34/year from Ecosystems, Ltd., Agriculture House, Bath Road, Sturminster, Newton DT10 1OV UK. North American Subscriptions: MIT Press, Journals, 55 Hayward St., Cambridge MA 02142 (617/253-2889). Assesses the human impact on the biosphere and the social, economic, and political barriers to regulating this impact. Provides a forum for the social and environmental activists who are seeking to change development policies in both developing and industrialized nations. Nicolas Hildyard, Editor.

ECO-MANAGEMENT AND AUDITING (1994). 3 issues/year, £50 from ERP Environment, PO Box 75, Shipley, West Yorkshire BD17 6EZ, UK (+44 (0)274 530408). A practitioner journal of environmental auditing and environmental management systems.

ECOSOCIALIST REVIEW (1987). From Environmental Commission of the Democratic Socialists of America, 1608 North Milwaukee, Chicago IL 60647.

THE EGG: An Eco-Justice Quarterly (1980). 20 pp. Quarterly, $10/year from The Eco-Justice Project and Network, Anabel Taylor Hall, Cornell University, Ithaca NY 14853-1001 (607/255-4225). Focuses on "eco-justice" issues such as pollution and depletion, poverty and hunger, sufficiency and sustainability, peace and security, and responsible, fulfilling life-styles. Offers an ethical framework, a theoretical and spiritual under-girding, and a long-range global perspective. Dieter Hessel, Editor.

ENDANGERED SPECIES UPDATE (1983). Monthly, $23 US ($28 elsewhere) from School of Natural Resources, University of Michigan, Ann Arbor MI 48109-1115 (313/763-3243). Provides current news and information on endangered species protection, which includes a reprint of the latest Endangered Species Technical Bulletin, the monthly US Fish and Wildlife Service publication. A primary forum for the exchange of ideas and information on endangered species. Book Reviews. John Watson, Editor.

ENERGY POLICY: International Journal of the Political, Economic, Planning and Social Aspects of Energy (1973). 12 issues/year, £285 inside UK and Europe, £305 outside from Butterworth-Heinemann, Linacre House, Jordan Hill, Oxford OX2 8DP UK (0865-310366). Subscriptions: Turpin Transactions Ltd., Distribution Centre, Blackhorse Road, Letchworth, Herts SG6 1HN UK (0462-672555). Addresses issues of energy supply, demand, and utilization that confront managers, consultants, researchers, planners and politicians. Book reviews. Nicky France, Editor.

ENERGY SYSTEMS AND POLICY: An International Interdisciplinary Journal (1973). Quarterly, $75 from Taylor & Francis, 1900 Frost Road, Suite 101, Bristol PA 19007-1598 (215/785-5800). Deals with energy systems problems from resource assessment to final consumption. S. William Gouse, Jr., Editor.

ENVIRONMENT (1950). Monthly (except Jan.-Feb., July-Aug. combined), $33/year from Heldref Publications, 1319 Eighteenth Street, NW, Washington DC 20036-1802 (202/296-6267). Excellent essays on general issues, many by leading environmentalists. Book reviews. Barbara Richman, Editor.

ENVIRONMENT ABSTRACTS (1970). Monthly, $1070 from Congressional Information Service/Reed Reference/Reed Elsevier Group, 4520 East-West Hwy., Bethesda, MD 20814-8380. Abstracts and indexes information from scientific, technical, and business journals; conferences; academic and government reports; newsletters, newspapers, and general magazines. Arranged in 21 categories; 5,000 or more items per year. Indexes by subject, geographic, and industry "key terms", author, and source. Annual includes chronology and review of the year's events, review articles, directory of government agencies and conservation organizations, and conferences. Full text of most abstracted items available on microfiche. Computer database file available, online and on CD-ROM. Anila Banerjee, Editor.

ENVIRONMENT AND DEVELOPMENT ECONOMICS (1996). Quarterly, £42 from Cambridge University Press, Edinburgh Bldg., Shaftesbury Rd., Cambridge CB2 2RU, UK (44-1223-312393). Positioned at the intersection of economics, environment and development. Articles on theoretical and applied aspects of sustainable development, valuation of environmental resources in low income countries, environmental implications of institutional change, and specific issues such as biodiversity loss. Charles Perrings, Editor.

ENVIRONMENT DIGEST. Monthly, £30 (outside Europe) from Environment Digest, 13-17 Sturton St., Cambridge CB1 2SN, UK. Concise reports on key topics such as global warming and climate change, threatened habitats and endangered species, energy, transport, waste and recycling, atmospheric and marine pollution, environmental politics and business.

ENVIRONMENT INDUSTRY YEARBOOK. £77 from Environment Press, 26 Brock Street, Bath, BA1 2LN, UK. Deals with companies and other bodies offering 'environmental protection' services. Includes services such as education and legal advice, lists 500 environmental consultants and all 2650 organizations known by the editors to be active in the 'environmental protection market,' with a column-inch or two of contact details and description.

ENVIRONMENT INTERNATIONAL: A Journal of Science, Technology, Health, Monitoring and Policy (1978). 6 issues/year, £303 from Elsevier Science, 660 White Plains Road, Tarrytown NY 10591-5153 (914/524-9200). Forum of original environmental literature. Includes data, causes of pollution, methods for protection and various topics on environmental protection. Book reviews. Alan Moghissi, Barbara Moghissi, Editors.

ENVIRONMENT AND PLANNING. From Pion Ltd., 207 Brondesbury Park, London NW2 5JN UK (081-459-0066). Published in sections, with separate subscriptions. Section A: *Urban and Regional Planning* (1969). Monthly, $620. Section B: *Planning and Design* (1974). Bi-monthly, $225. Section C: *Government and Policy* (1983). 4 issues/year, $175. Section D: *Society and Space* (1983). 4 issues/year, $225.

ENVIRONMENT AND URBANIZATION (1989). From International Institute for Environment and Development, 3 Endersleigh Street, London UK.

ENVIRONMENTAL ACTION (1970). Quarterly, $25/year from Environmental Action, Inc., 6930 Carroll Ave., Ste. 600, Tacoma Park MD 20912 (301/891-1106). Current news, analysis and resources for action on the key environmental issues facing the US. Theme section in most issues, such as children and the environment, wasteful packaging, religion and the environment. Book reviews. Barbara Ruben, Editor.

ENVIRONMENTAL CONSERVATION: International Journal Devoted to Maintaining Global Vitality Through Exposing and Countering Environmental Deterioration Resulting From Human Population Pressure and Unwise Technology (1974). Text in English. 100 pp., 4 issues/year, 245 Swiss Francs from Foundation for Environmental Conservation. Subscriptions: Elsevier Science Inc., Box 882, Madison Square Station, New York NY 10159 (212/989-5800). Articles on low-impact development, ecologically sound management, viability of global resources, enlightened policy, antipollution measures, education and law. Substantial editorial section. Book reviews. Nicholas Polunin, Editor.

ENVIRONMENTAL DIGEST (1987). c/o *The Ecologist*, from Ecosystems Ltd., Agriculture House, Bath Road, Sturminster, Newton, Dorset DT10 1DU UK (0258-473476). US Subscriptions: MIT Press, Journals, 55 Hayward St., Cambridge MA 02142 (617/253-2889).

ENVIRONMENTAL EDUCATION RESEARCH (1995). 3 issues/year, $52 from Carfax Publishing Co., PO Box 25, Abingdon, Oxfordshire OX14 3UE, UK. US Subscriptions: 875-81 Massachusetts Avenue, Cambridge MA 02139 (800/354-1420). Takes an international, analytical, theory-based approach to environmental issues and education.

ENVIRONMENTAL ETHICS: An Interdisciplinary Journal Dedicated to the Philosophical Aspects of Environmental Problems (1979). Quarterly, $20 from Environmental Philosophy, Inc., Department of Philosophy, University of North Texas, Box 13496, Denton TX 76203-3496 (817/565-2727). Forum for diverse interests and attitudes. Seeks to bring together the nonprofessional environmental philosophy tradition with the newly emerging professional interest in the subject. Book reviews. Eugene C. Hargrove, Editor.

ENVIRONMENTAL FORUM: The Policy Journal of the Environmental Law Institute (1981-86, resumed 1988). Bi-monthly, $75 from Environmental Law Institute, 1616 P Street NW, Suite 200, Washington DC 20036 (202/328-5150). For professionals in environmental law, policy, and management. Stephen Dujack, Editor.

ENVIRONMENTAL IMPACT ASSESSMENT REVIEW (1980-83, resumed 1985). 6 issues/year, $235 to institutions from Elsevier Science Inc., 655 Avenue of the Americas, New York NY 10010 (212/989-5800). Highlights advances in impact assessment, environmental decision making, and resolution of environmental disputes. Lawrence E. Susskind, Teresa Hill, Editors.

ENVIRONMENTAL MANAGEMENT: An International Journal for Decision-makers and Scientists (1977). 6 issues/year, $299 from Springer-Verlag Journals, 175 Fifth Avenue, New York NY 10010 (212/460-1500). Subscriptions: Journal Fulfillment Services, Box 2485, Secaucus NJ 07096-2491 (800/777-4643). On ecological modeling, resources management, energy, hazard response, environmental monitoring, and hazardous substances management. Consists of three sections. Forum: comments, and opinions about environmental matters. Profile: case histories and articles on events, policies, problems, or organizations. Research: scientific studies and their results. Book reviews. David Alexander, Editor.

ENVIRONMENTAL PERIODICALS BIBLIOGRAPHY: A Current Awareness Bibliography Featuring Citations of Scientific and Popular Articles in Serial Publications in the Area of the Environment (1972). Bi-monthly, price varies. International Academy at Santa Barbara, Environmental Studies Institute, 800 Garden Street, Suite D, Santa Barbara CA 93101-1552 (805/965-5010). Covers tables of contents of more than 500 international periodicals on the environment. Available as computerized database, online and CD-ROM. Miriam Flacks, Editor.

ENVIRONMENTAL POLICY AND LAW (1975). Text in English and French. 6 issues/year, 434 Guilders from International Council for Environmental Law, IOS Press, Van Diemenstraat 94, 1013 CN Amsterdam, Netherlands (31-20-6382189). North American subscriptions: Box 10558, Burke VA 22009-0558 (703/323-5554). Multinational environmental law, policy, and administration for pollution, species protection, waste management, land use, and resource conservation. Book reviews. Wolfgang E. Burhenne, Editor.

ENVIRONMENTAL POLITICS (1992). Quarterly, $38 from Frank Cass & Co. Ltd., Gainsborough House, 11 Gainsborough Road, London E11 1RS UK (081-530-4226). Academic study of environmental politics with a focus on the industrialized countries.

ENVIRONMENTAL STRATEGY. Newsletter of the IUCN Commission on Environmental Strategy and Planning. 4 issues/year, free from IUCN - The World Conservation Union, Gland, Switzerland. Commission Chairman's Office: PO Box 189040, Sacramento CA 95818 (916/442-2472). IUCN brings together sovereign states, governmental agencies, and NGO's in a world partnership to provide leadership and a common approach for the world conservation movement. [ALSO SEE: *IUCN BULLETIN*] CESP, with 264 members in 68 countries, is one of six IUCN commissions that draw together an extensive network of professional volunteers. It works to improve the formulation, implementation, and evaluation of policies and strategies for environmental protection and sustainable development, and to advance a world ethic of living sustainability. Ted C. Trzyna, Commission Chairman.

ENVIRONMENTAL TECHNOLOGY ASSESSMENT NEWSLETTER (1994). 2 issues/year from UN Environment Programme, Industry and Environment Programme Activity Centre, 39-43 quai Andre Citroen, 73739 Paris Cedex 15, France (44-37-14-50). On the use of technology assessment as a tool to support the development and application of environmentally sound technologies. John H. Skinner, Editor.

ENVIRONMENTAL VALUES (1992). 4 issues/year, £32 from White Horse Press, 1 Strond, Isle of Harris PA83 3UD, Scotland (085-982204). International journal concerned with the rational basis and justification of environmental policy. Aims to clarify the relationship between practical policy issues and more fundamental underlying principles or assumptions. Alan Holland, Editor.

ENVIRONMENTALIST: The International Journal Concerned with Environmental Awareness (1980). Quarterly, $180 from Institution of Environmental Sciences, Science and Technology Letters, PO Box 81, Northwood, Middlesex HA6 3DN UK. Official journal of the Institution of Environmental Sciences. Serves as a bridge between industry, government, and environmental professionals, providing guidelines and frameworks for management mechanisms, and as a catalyst for environmental education at all levels. Articles, interviews, editorials, news and commentary. John F. Potter, Editor.

EUROPEAN ENVIRONMENT (1994). 6 issues/year, £40 from ERP Environment, PO Box 75, Shipley, West Yorkshire BD17 6EZ, UK (+44 (0)274 530408). A journal of research and development in European environmental policy.

EUROPEAN ENVIRONMENTAL YEARBOOK (1984). Biennial, English edition £60 from Institute of Environmental Studies, Pirola Editore, Foro Bonaparte 12, 20121 Milan, Italy (02-8693994). Distributed by DocTer International UK Ltd., 5 Manfred Rd., London, SW1S 2RS UK. Analyzes the current status of nature conservancy, town and country planning, and environmental conservation in 12 EEC countries. Includes special surveys of the United States, the former USSR, Australia, and Japan. Lists environmental legislation adopted by EEC countries and documents EEC and other multinational environmental protocols and programs. Book reviews.

FOCUS: Carrying Capacity Selections (1991). 30 pp. Quarterly, $20/year from Carrying Capacity Network, Inc., 1325 G Street, NW Suite 1003, Washington DC 20005 (202/879-3044). Current and classic writings on a range of carrying capacity issues (environmental protection, population stabilization, and resources conservation), point-counterpoint discussions and interviews with thought-provoking personalities. Book reviews. Davis F. Durham, Editor.

FOREST ECOLOGY AND MANAGEMENT: An International Journal (1978). Text in English. $27/year (in 9 vols., 3 nos./vol.), 2601 Guilders from Elsevier Science Inc., Box 882, Madison Square Station, New York NY 10159 (212/989-5800). Scientific articles concerned with forest management and conservation, and applications of biological, ecological and social knowledge to the management of man-made and natural forests. Book reviews. R.F. Fisher, E.P. Farrell, Editors.

FORUM FOR APPLIED RESEARCH AND PUBLIC POLICY: A Quarterly Journal for the Discussion of Policy Issues in Energy, Science and Technology, Environment, and Economic Development (1986). 4 issues/year, c150 pgs/issue, $28 from U of Tennessee, Energy, Environment, and Resources Center, 600 Henley Street, Suite 311, Knoxville TN 37996-4134 (615/971-1251). Seeks to further public understanding and aid public policy in above-mentioned fields. Book reviews. Daniel Schaffer, Editor-in-Chief.

FUTURE GENERATIONS JOURNAL (1988). 28 pp. Quarterly, free from Future Generations Network Programme (IEI), Foundation for International Studies, St. Paul Street, Valleta, Malta (00356-234121/2). Facilitates global networking on issues related to humanity's present responsibilities toward future generations with special emphasis on the environment. Hilary Caruana, Editor.

FUTURE SURVEY: A Monthly Abstract of Books, Articles, and Reports Concerning Trends, Forecasts, and Ideas About the Future (1979). Monthly, with *FUTURE SURVEY ANNUAL* cumulation, $79 from World Future Society, 7910 Woodmont Avenue, Suite 450, Bethesda MD 20814 (301/656-8274). 50 abstracts/issue on world affairs, US issues, technology, methods, and environment-related issues. Michael Marien, Editor.

FUTURES: The Journal of Forecasting, Planning and Policy (1968). 10 issues/year, $447 in North, Central, and South America, £300 rest of world from Elsevier Science Ltd., The Boulevard, Langford Lane, Kidlington, Oxford OX5 1GB, UK (+44 (0) 1483 572678). A multidisciplinary journal covering the methods and practice of long-term forecasting and policy making on the future of mankind, culture and society, economics, technology, politics and environment. Book reviews. Colin R. Blackman, Editor.

FUTURES RESEARCH QUARTERLY (1985). Quarterly, $90 from World Future Society, 7910 Woodmont Avenue, Suite 450, Bethesda MD 20814 (301/656-8274). Occasional articles on environmental concerns. Timothy Mack, Audrey Clayton, and Ken Hunter, Editors.

THE FUTURIST: A Journal of Forecasts, Trends, and Ideas About the Future (1967). Bi-monthly, $42 from World Future Society, 7910 Woodmont Avenue, Suite 450, Bethesda MD 20814 (301/656-8274). Occasional articles on environmental concerns. Book reviews. Edward S. Cornish, Editor.

FUTURO SOSTINIBILE [SUSTAINABLE FUTURE] 1992). 50 pp. Weekly, 200,000 lire/year from Center for a Sustainable Future, Viale Giulio Cesare 49, Rome, 00192 Italy. Bulletin about Italian parliament activities on environmental issues. Some issues focus on a single topic of special importance for the global environment. Roberto Giachetti, Editor.

GAIA MAGAZINE (1990). 20 High Street, Stroud, Glos. GL5 1AS, UK.

GARBAGE: The Practical Journal for the Environment (1989). 6 issues/year, $21 from Old-House Journal Corporation, Box 55197, Boulder CO 80322-6519 (800/234-3797). Covers environmental issues such as recycling, composting, garbage disposal and energy, for a general audience. Patricia Moore, Editor.

GEOJOURNAL: International Journal of Physical, Biological, Social and Economic Geography and Applications in Environmental Planning and Ecology (1977). Text in English. 12 issues/year, 1746 Guilders, North American subscriptions: Box 358, Accord Station, Higham MA 02018-0358. Book reviews. Wolf Tietze, Editor.

GLOBAL BIODIVERSITY / LA BIODIVERSITÉ MONDIALE. Quarterly, in English and French. From Canadian Centre for Biodiversity, Canadian Museum of Nature, P.O. Box 3443, Station D, Ottawa, Ontario K1P 6P4, Canada.

GLOBAL ENVIRONMENTAL CHANGE: Human and Policy Dimensions (1990). Quarterly, $336 in North, Central and South America, £225 rest of world from Elsevier Science Ltd., The Boulevard, Langford Lane, Kidlington, Oxford OX5 1GB, UK (1865/843000). Addresses the human ecological and public policy dimensions of the environmental processes which are threatening the sustainability of life on Earth. Topics include deforestation, desertification, soil degradation, species extinction, sea level rise, acid precipitation and new technological hazards. Book reviews. Martin Parry, Andrew Goudie, Michael Williams, Editors.

GREENING UNIVERSITIES (1994). Quarterly newsletter from The Greening of Higher Education Council, 120a Marlborough Road, Oxford OX1 4LS, UK (0865/204244).

ILLAHEE JOURNAL OF THE NORTHWEST ENVIRONMENT (1984). Formerly *Northwest Environmental Journal* (through Jan. 1994). Quarterly, $25 from University of Washington, Institute for Environmental Studies, Engineering Annex FM-12, Seattle WA 98195 (206/543-1812). Science and policy issues covering the Pacific Northwest region of North America. J.R. Karr, E.W. Chu, Editors.

IN CONTEXT: A Quarterly of Humane Sustainable Culture (1980). Quarterly, $24 from Context Institute, Box 11470, Bainbridge Island WA 98110 (206/842-0216). On social and environmental issues, concentrating on creative solutions. Encourages direct participation in development of sustainable culture. Book reviews. Robert Gilman, Editor.

THE NETWORK. 16 pp. Monthly, $30 from The Centre For Our Common Future, 52, rue des Pâquis, CH-1201 Geneva, Switzerland (4122 732 7177). Newsletter on sustainable development issues, reporting on follow-up activities to the 1987 World Commission on Environment and Development and the 1992 UNCED Earth Summit. Joanne Barr, Managing Editor.

INDUSTRY AND THE ENVIRONMENT (c. 1991). Quarterly, from The Global Industrial and Social Progress Research Institute, Mori Building Number 33, 3-8-21 Toranomon, Minatoku, Tokyo 105 Japan.

INTERCIENCIA: Journal of Science and Technology of the Americas (1976). Text and summaries in English, Portuguese and Spanish. $60 outside Latin America from Intersciencia Association, Apdo. 51842, Caracas 1050A Venezuela (582-92-32-24). Publishes scientific and technological research on Latin America and the Caribbean, emphasizing the social context. Many articles on the regional impact of human activities (deforestation, coral reefs). Book reviews. Marcel Roche, Editor.

INTERNATIONAL ENVIRONMENT REPORTER (1978). Monthly, $1583 from The Bureau of National Affairs, Inc. Subscriptions: 9435 Key West Ave., Rockville MD 20850 (800/372-1033). Loose-leaf legal service on international law and policy in industrial countries. Marlon B. Allen, Editor.

INTERNATIONAL ENVIRONMENTAL AFFAIRS: A Journal for Research and Policy (1989). Quarterly, $50 from University Press of New England, 23 South Main Street, Hanover NH 03755-2048 (603/643-7111). Seeks to improve knowledge of environmental policy and management at the international level. Book reviews. K. von Moltke, Editor.

INTERNATIONAL JOURNAL OF CLIMATOLOGY (1981). 10 issues/year, $595 from Royal Meteorological Society. Subscriptions: John Wiley & Sons Ltd., Journals, Baffins Ln., Chichester, Sussex PO19 1UD UK (0243-779777). Global and regional studies of climate, long-range forecasting, changes of past, present, and future climate. Book reviews. B.D. Giles, Editor.

INTERNATIONAL JOURNAL OF ENVIRONMENTAL HEALTH RESEARCH (1991). Quarterly, $98 from Carfax Publishing Company, PO Box 25, Abingdon, Oxfordshire OX14 3UE, UK. US Subscriptions: 875-81 Massachusetts Avenue, Cambridge MA 02139 (800/354-1420). Journal of the International Federation of Environmental Health. Concerns all aspects of the interaction of human health and the environment. Michael H. Jackson, Paul G. Smith, Joan Rose, Editors.

INTERNATIONAL JOURNAL OF ENVIRONMENTAL STUDIES (1970). From Gordon & Breach Science Publishers, 270 Eighth Avenue, New York NY 10011. Contains recent information studies and research papers on the relationship between people and environment, with a focus on applied sciences, e.g. environmental impact assessment. Book Reviews. J. Rose, Editor.

INTERNATIONAL JOURNAL OF GLOBAL ENERGY ISSUES (1989). Text in English. Quarterly, $155 in North America from Interscience Ltd., World Trade Centre Bldg., 110 Avenue Louis Casai, Case Postale 306, CH-1215 Geneva-Aeroport, Switzerland. Subjects include global limits of energy production, sources of energy, policy and planning, conservation and management, and development and global change. Book reviews.

INTERNATIONAL JOURNAL OF WATER RESOURCES DEVELOPMENT (1985). Quarterly, $70 from Carfax Publishing Company, PO Box 25, Abingdon, Oxfordshire OX14 3UE, UK. US Subscriptions: 875-81 Massachusetts Avenue, Cambridge MA 02139 (800/354-1420). Water development and management in both industrialized and Third World countries, practical implementation of policies for water resources development, monitoring and evaluation of technical projects, and water resources research. Asit K. Biswas, Editor.

INTERNATIONAL GREEN PLANNER (1989). Newsletter from The Resource Renewal Institute, Ft. Mason Center, Bldg. A, San Francisco CA 94123 (415/454-3963). RRI started the Global Green Center as an information clearinghouse and training center. Huey D. Johnson, President.

INTERNATIONAL WILDLIFE (1971). Bi-monthly, $16 in US ($25 elsewhere) from National Wildlife Federation, 1400 Sixteenth Street NW, Washington DC 20036-2266 (202/797-6800). Lavishly illustrated publication; gives authoritative coverage of global environment. Jonathan Fisher, Editor.

ISSUES IN SCIENCE AND TECHNOLOGY (1984). Quarterly, $36 from University of Texas at Dallas, Box 830688, Richardson TX 75083-0688 (214/705-6325). Co-sponsors: Institute of Medicine, National Academy of Engineering, National Academy of Sciences. On policy implications of new developments in science, technology and health. Book reviews. Kevin Finneran, Editor.

IUCN BULLETIN (1952). Editions in English and French. Quarterly, $60 from IUCN - The World Conservation Union, rue Mauverney 28, CH-1196, Gland, Switzerland (41 22/999-0001). [ALSO SEE: *ENVIRONMENTAL STRATEGY*.] Book reviews. Nikki Meith, Editor.

JOURNAL OF AGRICULTURAL AND ENVIRONMENTAL ETHICS (1988). Semi-annual, C$35 ($38 elsewhere) from Agricultural Institute of Canada, 907-151 Stater Street, Ottawa Ontario K1P 5H4 Canada (519/824-4120). Ethical issues confronting agriculture, the environment and related disciplines. Frank Hurnick, Hugh Lehman, Editors.

JOURNAL OF CLEANER PRODUCTION (1993). 4 issues/year, £120 outside UK and Europe from Butterworth-Heinemann, Linacre House, Jordan Hill, Oxford OX2 8DP, UK (0865-310366). Subscriptions: Turpin Transactions Ltd. Distribution Centre, Blackhorse Road, Letchworth, Herts SG6 1HN UK (0462-672555).

JOURNAL OF ENVIRONMENTAL EDUCATION (1969). Quarterly, $33 from Heldref Publications, 1319 Eighteenth Street NW, Washington DC 20036-1802 (202/296-6267). Reports and research, viewpoints, and program descriptions intended to advance the field of environmental education. Martha Fogelman, Editor.

JOURNAL OF ENVIRONMENTAL LAW (1989). 2 issues/year, £53 from Oxford University Press. US subscriptions: 2001 Evans Road, Cary NC 27513 (919/677-0977). Environmental law in relation to pollution, waste management, habitat protection, biotechnology, and natural resources. Book reviews. Richard Macrory, Editor.

JOURNAL OF ENVIRONMENTAL MANAGEMENT (1973). Monthly (2 vols./year), £260 from Harcourt Brace & Co., Ltd., Foots Cray High Street, Sidcup, Kent DA14 5BR UK (44-81-300-3322). Papers on all aspects of management and use of the environment, both natural and man-made; generally case studies (local, regional and global). Many have a mathematical or modeling background. J.N.R. Jeffers, Editor.

JOURNAL OF ENVIRONMENTAL PLANNING AND MANAGEMENT (1948). Quarterly, $112 from University of Newcastle-upon-Tyne, Department of Town and Country Planning. Carfax Publishing Co., PO Box 25, Abingdon, Oxfordshire OX14 3UE, UK. US Subscriptions: 875-81 Massachusetts Avenue, Cambridge MA 02139 (800/354-1420). Covers environmental policy, planning, management, regulatory aspects of natural resources, environmental protection, conservation, and human-environment interactions. Book reviews. John Benson, Editor.

JOURNAL OF ENVIRONMENTAL SYSTEMS (1971). Quarterly, $112 from Baywood Publishing Company, Inc., 26 Austin Avenue, Box 337, Amityville NY 11701 (516/691-1270). Interdisciplinary forum for the environmental professions. Links biological and physical systems with policy, economic, engineering, and social systems. Approaches practical environment problems from a theoretical viewpoint. Book reviews. Sheldon Reaven, Editor.

JOURNAL OF SOIL AND WATER CONSERVATION (1946). Bi-monthly from Soil and Water Conservation Society, 7515 Northeast Ankeny Road, Ankeny IA 50021 (515/289-2331). Book reviews. Sue Ballantine, Editor.

KERR CENTER FOR SUSTAINABLE AGRICULTURE NEWSLETTER. Free from Kerr Center for Sustainable Agriculture, Inc., Hwy. 271 S., P.O. Box 588, Poteau OK 74953 (918/647-9123). Strives to provide leadership, education, and technical assistance to the agricultural community in seeking ecologically and economically sound methods of producing food and sustaining farm livelihood.

LAND DEGRADATION AND REHABILITATION: An International Journal Devoted to Land Degradation, Avoidance, Mitigation and Rehabilitation (1989). Quarterly, $165 from John Wiley & Sons, Ltd., Journals, Baffins Lane, Chichester, Sussex PO19 1UK UK (0243-779777). On recognition, monitoring, control, and rehabilitation of degradation. C.J. Barrow, Editor.

LAND USE POLICY (1984). Quarterly, £178 outside UK and Europe from Butterworth-Heinemann, Linacre House, Jordan Hill, Oxford OX2 8DP, UK (0865-310366). Examines geography, agriculture, forestry, irrigation, environmental conservation, housing, urban development and transport in both developed and developing countries. Penny Street, Editor.

LOCAL ENVIRONMENT (1996). 3 issues/year, $38 from Carfax Publishing Co., PO Box 25, Abingdon, Oxfordshire OX14 3UE, UK. US Subscriptions: 875-81 Massachusetts Avenue, Cambridge MA 02139 (800/354-1420). Bob Evans, Julian Agyeman, Editors.

MARINE CONSERVATION NEWS (1976). Quarterly from Centre for Marine Conservation, 1725 DeSales Street NW, Washington DC 20036 (202/429-5609). News and articles on legislation, policy and regulations pertaining to the conservation of all marine species and their habitats. Book Reviews. Rose Bierce, Editor.

MARINE POLICY (1977). Bi-monthly, £220 outside UK and Europe from Butterworth-Heinemann, Linacre House, Jordan Hill, Oxford OX2 8DP, UK (0865-310366). Legal, political, social and economic analysis by international lawyers, political scientists, fishery specialists and marine economists. Book reviews. Jennifer Nicholson, Editor.

MARINE POLLUTION BULLETIN: The International Journal for Marine Environmentalists, Scientists, Engineers, Administrators, Politicians and Lawyers (1970). 24 issues (in 2 vols.), £278/year from Elsevier Science. US Subscriptions: 660 White Plains Rd., Tarrytown NY 10591-5153 (914/524-9200). Concerned with the rational use of maritime and marine resources in estuaries, seas and oceans. Book reviews. C. Sheppard, Editor.

MILLENNIUM MAGAZINE 2000 (1990). 22 pp. Quarterly, $10/year from Peter Sorensen, 1431 Ocean Ave., Suite 415, Santa Monica CA 90401 (213/458-4448). Trends and ideas that will carry humanity and the world from this century and millennium into the next—with a special emphasis on world unity and ecological underpinnings. Seeks to be a sort of "global suggestion box"—a clearinghouse for ideas, proposals and visions. Peter Sorensen, Gregory Wright, Editors.

NATURAL HISTORY (1900). Monthly, $28 from American Museum of Natural History, Publications Department, Central Park West at 79th Street., New York NY 10024 (212/769-5500). Includes articles written by professional scientists on the biological sciences, ecology, anthropology, archaeology, earth science and astronomy. Book Reviews. Alan Ternes, Editor.

NATURAL RESOURCES FORUM (1976). Quarterly, £140 from UN Centre for Natural Resources, Energy and Transport, Subscriptions: Turpin Transactions Ltd. Distribution Centre, Blackhorse Road, Letchworth, Herts SG6 1HN UK (0462-672555). On development of the energy, mineral, and water resources of developing countries. Examines economic, financial, legal, and technical aspects. Raymond R. Knowles, Editor.

NATURE: International Weekly Journal of Science (1869). Weekly, £160 from Macmillan Magazines Ltd., 4 Little Essex St., London WC2R 3LF, UK (071-836-6633). US Subscriptions: Nature, Box 1733, Riverton NJ 08077-7333 (800/524-0384). Original research on all aspects of science. Book reviews. John Maddox, Editor.

NATURE AND RESOURCES: International News About Research on Environment, Resources, and Conservation of Nature (1958). Editions in English, French and Spanish. Quarterly, $45 from Parthenon Publishing Group, Casterton Hall, Carnforth, Lancs. LA6 2LA UK (577-16-10). Founded by UNESCO and a regular source of information on international scientific programs, such as Man and the Biosphere, until 1988. Resumed publication in 1990. Carries brief articles by scientists on sustainable development and the use of scientific information for making environmental decisions. Book reviews.

NEW ECONOMICS FOUNDATION. Briefings (1993). Irregular, free from New Economics Foundation, 88-94 Wentworth Street, London E1 7SA, UK (071-377-5696). Advocates a broader view of economics that includes environmental concerns.

NEW SOLUTIONS: A Journal of Environmental and Occupational Health Policy (1990). 4 issues/year, $60 from Oil, Chemical and Atomic Workers Union, Box 281200, Lakewood CO 80228-8200 (303/987-2229). Charles Levenstein, Editor.

NEW SCIENTIST (1956). Weekly, £67 from IPC Magazines, Oakfield House, 3 South Perrymount Road, Haywards Heath, West Sussex RH16 3DH UK (444-445511). Popular articles on a wide range of science-related fields and topics including environmental studies. Book reviews. Alun Anderson, Editor.

THE NEW YORK TIMES daily, $280 from New York Times Co., 229 West 43rd Street, New York NY 10036. The US newspaper of record. Occasional environmental topics on front page or Tuesday Science section. Max Frankel, Editor.

OCEAN DEVELOPMENT AND INTERNATIONAL LAW: The Journal of Marine Affairs (1973). Bi-monthly, $92 from Taylor & Francis, 1900 Frost Road, Suite 101, Bristol PA 19007-1598 (215/785-5800). On international aspects of ocean regulation and ocean utilization. Book reviews. Jon L. Jacobson, Editor.

ORION: People and Nature (1982). Quarterly, $18 from Myrin Institute, Inc., 136 East 64th Street, New York NY 10021 (212/758-6475). On our responsibilities to the earth and all forms of life, and the ethic of human stewardship. George K. Russell, Editor.

ORYX: Journal of the Fauna and Flora Preservation Society (1903-1913; new series 1921 and 1950). Quarterly, £65.50 in Europe (£94 elsewhere) from Blackwell Science Ltd., Journal Subscriptions, c/o Marston Book Services, P.O. Box 87, Oxford OX2 0DT, UK (01865-240201). Publishes reports and articles dealing with the conservation of species and habitats world-wide, with particular emphasis on those that are endangered. Book Reviews. J. Morris, Editor.

OUR PLANET (1974). Quarterly, free from United Nations Environment Programme, Information Service, P.O. 30552, Nairobi Kenya (254-2-230800). Popular magazine of the UNEP. Features and news of UNEP activities and some editorial comment. Book reviews. Shane Cave, Editor.

PARKS (1990). 3 issues/year, £18 from Nature Conservation Bureau Ltd., 36 Kingfisher Court, Hambridge Road, Newbury, Berks RG14 5SJ, UK (01635-5500380). Newly revised journal, published by the IUCN Commission on Natural Parks and Protected Areas, which seeks "to strengthen international collaboration among protected area professionals and to enhance their role, status and activities." Book Reviews. Paul Goriup, Editor.

PEOPLE & THE PLANET (1992). Quarterly, jointly sponsored by UNFPA, IPPF, IUCN, WWF, $20 from Planet 21, 60 Turiden Road, London NW5 1DN, UK. Brief essays on impacts of population, technology and consumption on the environment of our planet. A few book reviews. John Rawley, Editor.

POLLUTION PREVENTION REVIEW (1990). Quarterly, $175 from Executive Enterprises Publications Co., Inc., 22 West 21st Street, 10th Floor, New York NY 10010-6904 (800/332-8804). Covers both source reduction and recycling with case studies, application of new technologies, processes, substitute materials and recycling programs in use at companies. Includes federal and state developments, and review of the literature. Jean Stephenson, Editor.

POPULATION BULLETIN (1945). 50 pp. Quarterly, $45/year from Population Reference Bureau, Inc., 1875 Connecticut Ave. NW, #520, Washington DC 20009-5728 (202/483-1100). On demographic trends and their implications both in the US and worldwide. Mary Mederios Kent, Editor.

POPULATION AND DEVELOPMENT REVIEW (1975). Quarterly, $32 from Population Council, One Dag Hammarskjold Plaza, New York NY 10017 (212/339-0500). Provides a forum for discussing demographic and socio-economic issues. Book Reviews. Paul Demeny, Editor.

POPULATION AND ENVIRONMENT: A Journal of Interdisciplinary Studies (1978). Bi-monthly, $215 from Human Sciences Press, 233 Spring Street, New York NY 10013-1578 (212/620-8000). Explores relationships between population and societal, cultural, and physical environments. Covers demographic variables linked to lifestyle, law, health, business, economics, and international relations. Book reviews. Virginia Abernethy, Editor.

PROFIT THE EARTH (1990). 12 pp. Monthly, $100/year from California Futures, Inc., 801 Crocker Road, Sacramento CA 95864 (916/482-5200). Tracks leading-edge trends in legislation, consumer values, technologies and programs that enable businesses to profit from environmental programs and policies. Kelly Smith, Editor.

PRO ZUKUNFT [PRO FUTURE] (1987). Text in German. 40 pp. Quarterly, DM 40/year from Internationale Bibliothek für Zukunftsfragen (Robert-Jungk-Stiftung) Imbergstrasse 2, A-5020 Salzburg, Austria (0043-(0) 662 873 206). Abstracts of current German-language books and articles covering futures-related issues of political, scientific, environmental and cultural concern. Walter Spielmann, Editor.

REMINERALIZE THE EARTH (1985). 3 issues/year, $25 from Remineralize the Earth, 152 South Street, Northampton MA 01060-4021 (413/586-4429). Advocates the remineralization of forests, farms, orchards and gardens with glacial gravel and rock dust to regenerate and fertilize soils, restore ecological balance, and ensure the survival of the earth. Book reviews. Joanna Campe, Editor.

RESOURCE AND ENERGY ECONOMICS: A Journal Devoted to Interdisciplinary Studies in the Allocation of Natural Resources (1978). Formerly *Resources and Energy*. Text in English. 4 issues/year, 501 Guilders from Elsevier Science Inc., Box 882, Madison Square Station, New York NY 10159 (212/989-5800). Empirical studies, theoretical models, and application of specific techniques and methods, such as systems analysis, econometrics, growth models and energy and resource analysis. Book reviews. G.S. Tolley, J.S. Sweeney, Editors.

RESOURCES (1959). 20 pp. Quarterly newsletter, free from Resources for the Future, Inc., 1616 P Street NW, Washington DC 20036 (202/328-5025). Brief staff-written articles on environment, and natural resource issues. Richard Getrich, Editor.

RESOURCES POLICY (1974). 80 pp. Quarterly, £145/year from Butterworth Heinemann, Linacre House, Jordan Hill, Oxford OX2 8DP, UK (0865-310366). Focuses on minerals policy and economics issues—including minerals availability, exploration and development, production, consumption, markets and finance. Roderick Eggert, Editor.

RESURGENCE: Journal of the Ecological and Spiritual Culture (1966). Bi-monthly, £16 (£20 overseas surface, £30 airmail) from Ford House, Hartland, Bideford, Devon EX39 6EE, UK (0237/441293). Center for the "small-is-beautiful" thinking of E.F. Schumacher. Satish Kumar, Editor.

ROCKY MOUNTAIN INSTITUTE NEWSLETTER (1985). 3 issues/year, free from Rocky Mountain Institute, 1739 Snowmass Creek Road, Snowmass CO 81654-9199 (970/927-3851). On ideas that harness the problem-solving power of free-market economics to foster the efficient and sustainable use of resources as a path to global security. The Institute focuses its work in seven areas: agriculture, economic renewal, energy, green development, security, transportation, and water. Its E Source subsidiary (1033 Walnut, Boulder CO 80302-5144, 303/440-8500) is a leading supplier of information on advanced techniques for electric efficiency. Dave Reed, Editor; L. Hunter Lovins, and Amory B. Lovins, RMI Directors.

SCIENCE (1880). Weekly (4 vols/year), $87 from American Association for the Advancement of Science, Washington, DC. Subscriptions: Box 2033, Marion OH 43305-2033. News of recent international developments and research in all fields of science. Publishes original research results, reviews and short features. Book reviews. Daniel Koshland, Editor.

THE SCIENCE OF THE TOTAL ENVIRONMENT: An International Journal for Scientific Research into the Environment and its Relationship with Man (1972). 48 issues/year (in 16 vols., 3 nos./vol.), 4416 Guilders from Elsevier Science Inc., Box 882, Madison Square Station, New York NY 10159 (212/989-5800). Man-made changes in the environment, especially disturbances of the natural levels and distribution of chemical elements and compounds. Particular emphasis on application of environmental chemistry to nutrition, pollution control, environmental medicine, planning and policy. Book reviews. E.I. Hamilton, J.O. Nriagu, Editors.

SCIENTIFIC AMERICAN (1845). Monthly, $36 from Scientific American, Inc., 415 Madison Avenue, New York NY 10017. Contains a variety of original and authoritative articles on topics relating to the impact of science on life and society. Some environmental topics, occasionally policy-relevant. Jonathan Piel, Editor.

SIERRA (1893). Bi-monthly, $15 from Sierra Club, 730 Polk Street, San Francisco CA 94109 (415/776-2211). Covers the preservation group's activities and agenda; articles on conservation and recreational use of natural resources, both domestically and internationally. Book reviews. Jonathan F. King, Editor.

SOCIETY AND NATURE: The International Journal of Political Ecology (1992). 3 issues/year, $25 from Aigis Publications, 1449 W. Littleton Blvd., Ste. 200, Littleton CO 80120 (303/730-6232). A forum for dialogue between social ecologists, eco-socialists, feminists, radical greens and activists in the land-based and indigenous movements. Book reviews. Takis Fotopoulos, Editor.

SOLAR TODAY (1987). Bi-monthly from American Solar Energy Society, 2400 Central Ave., Suite G-1, Boulder CO 80301 (303/443-3130). Solar energy for engineers, scientists, architects, educators, practitioners, researchers, and users. Includes case studies and reviews of different technologies. Book reviews. Maureen McIntyre, Editor.

SUSTAINABLE DEVELOPMENT (1994). 3 issues/year, £65 (in Europe) from ERP Environment, PO Box 75, Shipley, West Yorkshire BD17 6EZ, UK (+44 (0)274 530408).

SUSTAINABLE DEVELOPMENT: International Journal of Global Environmental Conservation and Development (c. 1990). Dr. Ashok Khosia, B-17, Ansal Chambers-1, Cama Place, New Delhi 110 066 India.

S.U.R.E.: Sustainable Urban/Rural Enterprise Exchange (1992). Quarterly, $20 from GrassRoots Resource, Inc., P.O. Box 11343, Takoma Park MD 20913 (301/588-7227). For groups engaged in the development of sustainable urban/rural enterprise. Susanna MacKenzie, Editor.

TECHNOLOGY REVIEW (1899). 80 pp., 8 issues/year, $24/year from The Association of Alumni and Alumnae at the Massachusetts Institute of Technology, MIT, W59-200, Cambridge MA 02139 (617/253-8250). On implications of new science and technology developments. One or two environment-related articles per issue. Steven J. Marcus, Editor.

TOMORROW: The Global Environment Magazine. $48 from Tomorrow Magazine, Kungsgatan 27, S-111 56 Stockholm, Sweden.

21.C (1990). 100 pp. Quarterly, Aust. $40/year from Australian Commission for the Future, Ltd., PO Box 115, Carlton, Melbourne, Victoria, 3053 Australia (03/556-3281). On issues that will influence the future in science and technology, cultural issues, urban development, and the environment. Ashley Crawford, Editor.

THE URBAN ECOLOGIST (1975). 24 pp. Quarterly, $25/year from Urban Ecology, PO Box 10144, Berkeley CA 94709 (510/549-1724). Reprinted and original articles on ecocity planning and ecological rebuilding issues, with news reports on ecological rebuilding projects underway in the San Francisco Bay area, the United States, and throughout the world.

THE URBAN ECOLOGIST: The Newsletter of Environmental and Occupational Health Policy (1991). Work Environmental Program, Lowell University, Lowell MA 01854.

UTILITIES POLICY: Strategy, Performance, Regulation (1990). Quarterly, £138 outside UK and Europe, from Butterworth-Heinemann, Linacre House, Jordan Hill, Oxford OX2 8DP, UK (0865-310366). Analysis, interpretation and discussion of developments and challenges facing public-sector utilities. Book reviews. Nicky France, Editor.

WHOLE EARTH REVIEW (1974). Quarterly, $27 from Point Foundation, 27 Gate Five Road, Sausalito CA 94965 (415/332-1716). Formerly *Co-Evolution Quarterly*. Access to tools and ideas for sustainable communities and livelihoods. Book reviews. Howard Rheingold, Editor.

WINDSTAR JOURNAL (1988). From Windstar Foundation, Snowmass CO 81654.

WORLD BUSINESS ACADEMY PERSPECTIVES (1992). Quarterly, $96 from World Business Academy. Berrett-Koehler Publishers, Inc., 155 Montgomery Street, San Francisco CA 94104-4109 (415/288-0260). Discusses how business can be a creative, positive force for resolving the pressing issues facing global society today; offers profiles of visionary business leaders, and practical examples of the transformation process in action. The Academy is an international network of business leaders who seek to create a positive, sustainable future for the planet. Book reviews. Alis Valencia, Editor.

WORLD WATCH (1988). Bi-monthly, $15 from Worldwatch Institute, 1776 Massachusetts Avenue NW, Washington DC 20036 (202/452-1999). On global economic and social issues related to environmental degradation and conservation. Articles by Institute staff members identify trends, practices, and events threatening to the earth and its physical and social environment. Advocates changes in world population, energy, environmental, and food policies. Lester R. Brown, Editor.

WORLDWATCH PAPERS (1975). 4 or more issues/ year, $25 from Worldwatch Institute, 1776 Massachusetts Avenue NW, Washington DC 20036 (202/452-1999). Single-topic issues analyze global problems; topics include all aspects of the environment, poverty, energy, development, population, and inequality.

PUBLISHER/TITLE INDEX

[NOTE: Certain titles are abbreviated here]

Abingdon Press, 201 Eighth Avenue South, Nashville TN 37202-0801 (800/251-3320)
- *Loving Nature: Ecological Integrity and Christian Responsibility* (Nash, 1991) #212

Adamantine Press Ltd., 3 Henrietta Street, Covent Garden, London WC2E 8LU, UK
- *Science and Technology for Eight Billion People: Europe's Responsibility* (Mettler, ed., 1995) #294

Addison-Wesley, 1 Jacob Way, Reading MA 01867 (800/447-2226)
- *Healing the Planet: Strategies for Resolving the Environmental Crisis* (Ehrlich/Ehrlich, eds., 1991) #176

Alberta Environmental Protection Information Centre, 9920 108th Street, Edmonton Alberta T5K 2M4 Canada
- *Ensuring Prosperity: Implementing Sustainable Development* (Future Environmental Directions for Alberta Task Force, 1995) #189

American Association of Engineering Sciences, 1111 19th Street NW, Washington DC 20036 (202/296-2237)
- *The Role of Engineering in Sustainable Development* (Ellis, ed., 1994) #304

The American Enterprise: A National Magazine of Politics, Business, and Culture, American Enterprise Institute, 1150 17th Street., NW, Washington DC 20036
- "Green Eyeshades: Environmentalism Encounters Economic Reality" (Henderson, Jan-Feb 1995) #388

Anchor/Doubleday, 1540 Broadway, New York NY 10036 (800/223-5780)
- *The Gaia Atlas of Cities: New Directions for Sustainable Urban Living* (Girardet, 1993) #331
- *The Gaia Atlas of Green Economics* (Ekins/Hillman/Hutchison, 1992) #231

Annals of Earth, One Locust Street, Falmouth MA 02540
- "Living Machines: Theoretical Foundations and Design Precepts" (Todd/Josephson, 1994) #224

Arkana Books, div of Penguin, 375 Hudson Street, New York NY 10014 (800/631-3577)
- *Living Philosophy: Eco-Philosophy as a Tree of Life* (Skolimowski, 1992) #206

Ashgate, Old Post Road, Brookfield VT 05036-9704 (800/276-3162)
- *Environmental Economics: Policies for Environmental Management and Sustainable Development* (Tisdell, 1993) #229

The Atlantic Monthly, 745 Boylston, Boston MA 02116
- "Reinventing the Wheels" (Lovins/Lovins, Jan 1995) #325

Atlantic Monthly Press, 19 Union Square West, 11th Floor, New York NY 10003 (212/645-4462)
- *All the Trouble in the World* (O'Rourke, 1994) #380

Avon Books, Hearst Book Group, 1350 Avenue of the Americas, New York NY 10019 (800/238-0658; 212/261-6800)
- *Enough Is Enough: The Hellraiser's Guide to Community Activism* (MacEachern, 1994) #431

Bantam Books, 666 Fifth Avenue, New York NY 10019 (800-223-6834)
- *Wisdom of the Elders: Honoring Sacred Native Visions of Nature* (Suzuki/Knudtson, 1992) #200

BasicBooks, 10 East 53rd Street, New York NY 10022 (800/242-7737)
- *Ecocide in the USSR: Health and Nature Under Siege* (Feshbach/Friendly, Jr., 1992) #024
- *No Turning Back: Dismantling the Fantasies of Environmental Thinking* (Kaufman, 1994) #372

Beacon Press, 25 Beacon Street, Boston MA 02108 (617/742-2110; 800/631-8571)
- *Car Trouble* (Nadis/MacKenzie, 1993) #320
- *For the Common Good: Redirecting the Economy Toward Community, the Environment, and a Sustainable Future* (Daly/Cobb, 1994) #234
- *Green Gold: Japan, Germany, the United States, and the Race for Environmental Technology* (Moore/Miller, 1994) #301
- *Spirit and Nature: Why the Environment Is a Religious Issue* (Rockefeller/Elder, eds., 1992) #210
- *State of the Peoples* (Miller, ed., 1993) #107

Berrett-Koehler Publishers, 155 Montgomery Street, San Francisco CA 94104-4105 (415/288-0260)
- *EcoManagement: The Elmwood Guide to Ecological Auditing and Sustainable Business* (Callenbach *et al.*, 1993) #258
- *Future Search: An Action Guide to Finding Common Ground in Organizations and Communities* (Weisbord/Janoff, 1995) #448

BioScience, American Institute of Biological Sciences, 730 11th Street. NW, Washington DC 20001-4521 (202/628-1509)
- "Environmental and Economic Effects of Reducing Pesticide Use" (Pimentel, *et al.*, June 1991) #286
- "Environmental Refugees in a Globally Warmed World" (Myers, Dec 1993) #059

Brookings Institution, 1775 Massachusetts Ave NW, Washington DC 20036 (202/797-6258)
- *The Capacity for Wonder: Preserving National Parks* (Lowry, 1994) #123
- *Cleaning Up the Mess: Implementation Strategies in Superfund* (Church/Nakamura, 1993) #162

California Institute of Public Affairs, PO Box 189040, Sacramento CA 95818 (916/442-2472)
- *World Directory of Environmental Organizations* (Trzyna/Childers, eds., 1992) #390
- *A Sustainable World* (Ted Trzyna, ed., 1995) #191

Cambridge University Press, 40 West 20th Street, New York NY 10011 (800/221-4512)
- *Drought Follows the Plow: Cultivating Marginal Areas* (Glantz, ed., 1994) #277
- *Global Water Resource Issues* (Young/Dooge/Rodda, 1994) #077
- *Industrial Ecology and Global Change* (Socolow, *et al.*, eds., 1994) #265
- *Preservation of Near-Earth Space for Future Generations* (Simpson, ed., 1994) #143
- *World Soil Erosion and Conservation* (Pimentel, ed., 1993) #278

Cato Institute, 1000 Massachusetts Avenue NW, Washington DC 20001 (202/842-0200)
- *Apocalypse Not: Science, Economics, and Environmentalism* (Bolch/Lyons, 1993) #377
- *Sound and Fury: The Science and Politics of Global Warming* (Michaels, 1992) #386

SUBJECT INDEX

AUTHOR/INSTITUTION INDEX